SMALL MILLETS
IN
GLOBAL AGRICULTURE

Proceedings of the First International
Small Millets Workshop
Bangalore, India, October 29-November 2, 1986

Editors:

A. Seetharam
K.W. Riley
G. Harinarayana

OXFORD & IBH PUBLISHING CO. PVT. LTD.
New Delhi Bombay Calcutta

ISBN 81-204-0434-3

Published in India by Mohan Primlani for Oxford & IBH Publishing Co. Pvt. Ltd., 66 Janpath, New Delhi 110 001, typeset by Composers and printed at Pauls Press, Okhla Industrial Area, New Delhi 110 020.

1-S9-11

CONTENTS

EDITORS' PREFACE

The transformation of agriculture to more productive systems has often been accompanied by increased production of a fewer crops species. Concurrently, the area and production of a great diversity of traditional crops have declined. Yet in many parts of the world, these traditional crops play an important role in maintaining stable and sustainable forms of agriculture.

One such traditional group of cereal crops is the small millets. This group includes finger millet (*Eleusine coracana*), foxtail millet (*Setaria italica*), little millet (*Panicum miliare*), proso millet (*Panicum miliaceum*), kodo millet (*Paspalum scrobiculatum*), barnyard millet (*Echinochloa colosna*), fonio (*Digitaria exilis*), and teff (*Eragrostis tef*).

Although precise estimates on area and production of small millets are not available, these crops may occupy between 18 and 20 million hectares, producing 15-18 million tonnes of grain. The regionwise distribution of area is 6.5 m ha in South Asia, 5 m ha in China, 4 m ha in USSR and 3 m ha in Africa. Finger millet is the principal small millet species grown in South Asia, followed by kodo millet, foxtail millet, little millet, proso millet and barnyard millet in that order. Foxtail millet and proso millet are important in China and proso millet is grown extensively in southwestern USSR. In Africa, finger millet, teff and fonio have local importance. Some small millets are grown in United States and Europe on a very limited scale.

Small millets are grown in arid, semi-arid or montane zones as rainfed crops, under marginal and submarginal conditions of soil fertility and moisture. Even so, it should be appreciated that their average grain yield is almost a tonne per ha. Presently, small millets are cultivated in areas where they produce a more dependable harvest compared with any other crop. This has been largely responsible for their continued presence and cultivation in many parts of the world. There is now an increasing realization of this fact, and a greater awareness that these crops merit more research and development.

As a response to this need, the Indian Council of Agricultural Research, New Delhi, India, University of Agricultural Sciences, Bangalore, India, and the International Development Research Centre, Canada, jointly organized the first International Workshop on Small Millets, in October 1986, at Bangalore, India.

The purpose of the workshop was to bring together scientists working on these millets, from countries where these crops are important; to assess the importance, production, and the place of these crops in traditional and improved agricultural systems; to discuss the status of research on these crops and explore ways to collaborate in strengthening millet research. Over 50 scientists from India, Bangladesh, Nepal, Sri Lanka, China, USSR, Ethiopia, Kenya, Zimbabwe, the ICRISAT SADCC Program, Tanzania, Uganda and IDRC attended. Sessions on production trends, genetic resources, breeding, cropping systems and production technology, physiology, food and forage uses were held. There was discussion following each session, and a great deal of information was exchanged, both formally and informally.

This volume contains the proceedings of that workshop, arranged into seven chapters:

- Overview and Taxonomy of Small Millets
- Importance, Germplasm Resources and Varietal Improvement in Asia
- Importance Germplasm Resources and Varietal Improvement in Africa.
- Physiology, Cropping Systems and Production Technology in Asia.
- Physiology, Cropping Systems and Production Technology in Africa.
- Food and Forage Uses.
- Discussion and Recommendations.

Two papers: "Origin Evolution, and Systematics of Minor Cereals" by J.M.J. de Wet, and "Improvement of Finger millet (*Eleusine coracana*) in Ethiopia" by Yilma Kebede and Abebe Menkir, which were written after the workshop, are also included.

During the final day of the workshop, it was recommended that an International Small Millets Network be formed, to help strengthen research on small millets.

This volume, brings together, perhaps for the first time, information from the scientists actively working to improve the small millets in Africa and Asia. It is hoped that this information will be of use not only to the scientists in the newly formed network, but to all those interested in the role of traditional crops in enhancing sustainability, stability as well as the productivity of agricultural systems.

Our appreciation is extended to Dr. M.V. Rao for his inaugural address, Dr. S.V. Patel, Vice Chancellor, University of Agricultural Sciences, and Mr. V.G. Pande, Regional Director, IDRC, for their opening remarks. To Dr K. Krishnamurthy, Director of Research, University of Agricultural Sciences, goes the credit for the excellent organization of the workshop.

A. SEETHARAM
Co-ordinator, All India Co-ordinated
Small Millets Improvement Project
Bangalore

K.W. RILEY
International Development
Research Centre
New Delhi

G. HARINARAYANA
Co-ordinator, All India Co-ordinated
Pearl Millet Improvement Project
Pune

1. Foxtail millet (*Setaria italica*)

2. Barnyard millet (*Echinochloa colona*)

3. Proso millet (*Panicum miliaceum*)

4. Kodo millet (*Paspalum scrobiculatum*)

5. Dehulling of small millets

6. Little millet (*Panicum miliare*)

7. Finger millet (*Eleusine coracana*)

8. Evaluation of finger millet germplasm in Nepal. High straw yields in these tall land races are valued for feeding livestock.

INAUGURAL ADDRESS

THE SMALL MILLETS: THEIR IMPORTANCE, PRESENT STATUS AND OUTLOOK

M.V. Rao
Deputy Director General (Crops)
Indian Council of Agricultural Research
New Delhi

Millets are small grained cereals, the smallest of them include finger, kodo, foxtail, proso, little and barnyard millets. They are the staple food of the millions inhabitating the arid and semiarid tropics of the world. They are distributed in most of the Asian and African countries and parts of Europe. The grains of small millets, being nutritionally superior to rice and wheat, provide cheap proteins, minerals and vitamins to poorest of the poor where the need for such ingredients is the maximum. Practically devoid of grain storage pests, the small millets have indefinite storage life. The untapped grain yield potential coupled with nutritional superiority makes the small millets potential future food crops particularly in the more difficult rainfed areas.

Small millets in India occupy 4.5 per cent of the cultivated area and are confined to vast stretches of drylands and hilly tracks. The area under finger millet has been more or less stable around 2.6 million hectares while the area under other small millets has shown gradual decline from 5.6 million hectares in 1954-55 to 3.6 million hectares during 1983-84. The productivity of finger millet is the highest among the millets, at 1150 kg/ha during 1983-84. However, the productivity of other small millets has remained around 450 kg/ha. The lower productivity of small millets is largely due to poor fertility of soils and non-adoption of improved package of cultivation. Nevertheless, these crops do have large hidden production potential which could be exploited by judicial blending of varietal, production and protection technologies. These crops respond very well even to small doses of inorganic fertilizers and other crop management inputs which do not involve additional expenditure, such as sowing at optimum time, maintenance of adequate plant stand, timely weeding and intercultivation.

Though the crop improvement work on small millets in India under the co-ordinated programme started in 1964, the launching of a separate co-ordinated millet improvement project in 1969 helped in giving greater attention to these crops with a few centres established in different states for specific millets. Five crop specific lead centres were established in 1978-79 with IDRC assistance; at Dholi in Bihar for proso millet, Dindori in Madhya Pradesh for kodo millet, Nandyal in Andhra Pradesh for foxtail millet, Semiliguda in Orissa for little millet, and at Almora in Uttar Pradesh for barnyard millet. The establishment of these centres have greatly assisted in building the necessary infrastructure and manpower for these crops. These efforts have started bearing fruit and many new technologies and varieties are now undergoing critical evaluation in different parts of the country. The major goal during the Seventh Five-Year Plan is to further consolidate the outcome of these research efforts and stabilize the productivity of small millets at a higher level. As an important step in this direction a separate All India-Co-ordinated Small Millets Improvement Project has been launched in the Seventh Plan in 1986 with the following objectives.

1. Diversification of the varietal base by evolving high yielding, disease resistant and widely adaptable genotypes in various small millets.

2. Development of efficient production technologies using low monetary inputs.

3. Identification of ideal crop mixtures and evolving production systems involving pulses and oilseeds as component crops.

4. Intensification of research on plant health and evolution of cheap and efficient plant protection methods.

5. To identify alternate uses for grain in poultry, dairy and in agro-based industries to enhance their economic value.

The above objectives can be achieved only through more research in various disciplines backed up with efforts on transfer of technology. For stabilizing production of small millets the following aspects deserve high priority.

1. Strengthening of both basic and applied research.

2. Identification and utilization of technologies generated from basic and applied research.

3. Quick transfer of technology to the farmer's field for extension.

Small millets received less priority in the agricultural development in the past both at National and International level as evidenced by the fact that none of the existing International Institutes is endeavouring for the improvement of any of the small millets. This is so in spite of the fact that these crops occupy more than 25 million hectares at global level. Small millets not only have been less researched but also have received negligible developmental support. The scientists who have gathered here in the First International Workshop on small millets may kindly take note of these facts in their deliberations and come out with suitable recommendations for removing these imbalances. Some of the topics which need specific attention are:

1. Small millets possess a wealth of genetic diversity—India has assembled more than 9,000 collections of small millets at Bangalore, the headquarters of the Small Millets Improvement Project. Similarly China maintains a rich source of foxtail millet germplasm; USSR, has excellent proso millet collections. Africa has also assembled teff in Ethiopia, finger millet in Kenya and Uganda; The International Crop Research Institute (ICRISAT), Hyderabad, India also has built up one of the largest and diverse collections of small millets. However, there are many areas in India as well as in other countries still unexplored and there is an urgent need to retrieve the genetic diversity under natural conditions. This acquires significance in countries which are faced with chronic drought. The efforts of Ethiopian scientists in this regard are commendable. Besides cultivated land races there is a need to collect related wild species also. Further, an uptodate inventory of all the available germplasm should be prepared and made available to all small millets researchers to facilitate quick and need based exchange of germplasm.

2. Small millets are highly self-fertilized crops and pure line selection has been primarily used to improve the performance of land races. Quite often the farmer himself has selected varieties in his own way in his fields. Hybridization, however, offers immense potential for combining the desirable features. Contact, hot water and gametocide methods have been used in hybridization with certain amount of success in these crops. The smallness of the spikelets and their delicate nature have been hindering hand emasculation. There is an urgent need to standardize hybridization techniques for changing the genetic background of the local cultivars. The discovery of male sterility in foxtail millet in China augures well for the improvement of this crop. Similar mechanisms and also mechanisms like protogyny which promote cross pollination need to be looked for in other small millets.

3. Traditionally small millets are the constituents of dryland farming system. However, they also respond to irrigation. Therefore there is an immediate need to select genotypes for better water use efficiency.

4. Small millets are low input crops and often grown in infertile depleted soils. Obviously they respond remarkably to fertilizer management. This further demands identification of genotypes which have high fertilizer use efficiency particularly nitrogen whether it is native or applied.

5. Small millets are vulnerable to different spectrum of field pests and diseases. Finger millet is more vulnerable to diseases like blast and viruses and barnyard millet to smuts. Little and proso millets are more susceptible to pests like shootfly while borers occur on finger and barnyard millet. The incorporation of genetic resistance offers the best choice in low input crops like small millets. Cultural controls like early planting and appropriate cropping systems

could also reduce pest and disease incidence. These methods in addition to cheap chemical control methods deserve attention.

6. Small millets are the staple food of the poor and the working classes and hence their health depends on the quality of food consumed. Any improvement made in nutritive quality of small millet grain would indirectly help in bettering the general health of the rural people. So, quality breeding to improve the protein content, mineral composition and aminoacid balance should be given due priority. Quality specific genotypes can also be bred in order to widen consumer base to offer a choice of foods and to augment industrial uses of small millets. The manufacture of value added products from small millets will help to upgrade not only the economic status of growers and also their investment resource base.

7. Small millets are generally cooked like rice. They also find their way to the local specific sweets and savouries. Small millets, however, can be used as substitutes to wheat and rice in the food products. They could also be processed into new foods suitable to infants and invalids alike after necessary fortification. Bread could be prepared from finger millet for diabetics and it can also be used in many bakery products. Most small millets could be popped or flaked.

8. As small millets are well protected in glume encasements the processing of the grain to usable form is not only time consuming but also labour intensive. There is therefore a need to develop post-harvest processing technology in order to reduce human drudgery.

WORKSHOP PARTICIPANTS AND AUTHORS

Dr. S. Appa Rao
Millets Scientist
SADCC/ICRISAT
Sorghum-Millet Improvement Programme
P.O. Box 776, Bulawayo
Zimbabwe

Dr. K.V. Bondale
Joint Director (Millets)
Millets Development Directorate
19, Krishnama Road
Madras, India

Dr. L.B. Choudhary
Senior Scientist
Dept. of Plant Breeding & Genetics
Tirhut College of Agriculture
Dholi 830121, India

Dr. J.M.J. deWet
Director, Cereals Programme
ICRISAT
Patencheru P.O.
502 324, (AP) India

Dr. Hugh Doggett
38a Cottinham Road,
Histon, Cambridge,
CB 4, 4 ES, England.

Dr. John-Peter Esele
Plant Pathologist
Sorghum and Millet Unit
Uganda Agriculture and Forest Research Organisation
P.O. Soroti, Uganda

Dr. S.C. Gupta
Principal Millets Breeder
SADCC/ICRISAT
Sorghum-Millet Improvement Programme
P.O. Box 776, Bulawayo
Zimbabwe

Mr. B.T. Shankare Gowda
Geneticist (Millets)
Univ. of Agril. Sciences
Bangalore 560 065, India

Dr. M.A. Hamid
Chief Scientific Officer
Bangladesh Agricultural Research Institute
Joydebpur, Gazipur
Bangladesh

Dr. G. Harinarayana
Project Coordinator (Pearl Millet)
AICMIP, College of Agriculture
Pune 411 005, India

Dr. B.R. Hegde
Chief Scientist (Dry Farming)
Univ. of Agricultural Sciences
Bangalore 560 065, India

Dr. L.R. House
Leader
SADCC/ICRISAT
Sorghum-Millet Improvement Programme
P.O. Box 776, Bulawayo
Zimbabwe

Dr. V.A. Ilyin
Head, Seed Breeding Laboratory
South East Agricultural Res. Inst.
Saratov, USSR

Shri H.C. Joshi
Sr. Scientist (Millets)
Vivekananda Laboratory
Almora 263 601, India

Mr. Chen Jiaju
Professor and Head
Institute of Plant Germplasm Resources
Laboratory of Sorghum & Millets
Chinese Academy of Agricultural Sciences
30 Bai Shi Quao Lu
West Suburbs, Beijing
The People's Republic of China

Dr. Yilma Kebede
Research Officer
Institute of Agricultural Research
Nazareth, Box 2003
Addis Ababa Ethiopia

Dr. Seyfu Ketema
Teff Breeder
Institute of Agricultural Research
Box 2003, Addis Ababa
Ethiopia

Mr. Bill Williams Khizzah
Principal Research Officer
C/o, Sorghum & Millet Unit
Uganda Agriculture & Forestry Research Organisation
P.O. Soroti, Uganda

Dr. K. Krishnamurthy
Director of Research
Univ. of Agril. Sciences
Bangalore 560 065, India

Shri B.K. Linge Gowda
Agronomist (Millets)
Univ. of Agril. Sciences
Bangalore 560 065, India

Mr. Mohamed Abdul Majid
Principal Scientific Officer (Cereals)
Bangladesh Agricultural Research Institute
Joydebpur, Gazipur
Bangladesh

Dr. N.G. Malleshi
Scientist, CFTRI
Mysore 570 013, India

Mrs. Mannujan
Sr. Scientific Officer (Cereals)
Bangladesh Agricultural Res. Inst.
Joydebpur, Gazipur
Bangladesh

Mr. Christopher Mburu
C/o. Dr. A.E.O. Chabeda
Chief Research Officer
Ministry of Agriculture & Livestock Department
Kilimo House, Cathedral Road
P.O. Box 30028
Nairobi, Kenya

Mr. Abebe Menkir
Research Officer
Institute of Agricultural Research
Nazareth, Box 2003
Addis Ababa, Ethiopia

Dr. B. Mishra
Sr. Millet Pathologist
Dept. of Plant Breeding & Genetics
Tirhut College of Agriculture
Dholi 834 121, India.

Shri T.K. Murthi
Entomologist (Millets)
AICMIP, College of Agriculture
Pune 411 005, India

Mr. Fighur Muza
Millet Breeder
Crop Breeding Institute
Dept. of Research & Specialist
Services
Post Box No. 8100, Causeway
Harare, Zimbabwe

Mr. Reuben O.F. Mwambene
Principal Research Officer (Crops)
Uyole Agricultural Centre
P.O. Box 400
Mbeya, Tanzania

Dr. S. Palanisamy
Professor (Millets)
School of Genetics, TNAU,
Coimbatore 641 003, India.

Dr. S.V. Patil
Vice Chancellor
Univ. of Agril. Sciences
Bangalore 560 065, India

Dr. H.B. Patnaik
Breeder
Regional Research Station
P.B. No. 10, Semiliguda
Sunbeda 763 002, India

Mr. S. Ponnuthurai
Research Officer
Regional Research Centre
Department of Agriculture
Kilinochchi, Sri Lanka

Dr. T.G. Prasad
Department of Crop Physiology
University of Agricultural Sciences
Bangalore 560 065, India

Dr. P. Pushpamma
Plot No. 49, Aravindnagar
Domalguda,
Hyderabad 500 029, India

Shri V. Rama Rao
Asst. Millets Specialist
AICMIP, Millet Research Station
P.B. No. 6
Vizianagram 531 201, India

Dr. M.V. Rao
Special Director General, ICAR
New Delhi 110 001, India

Dr. S.M. Razvi
Ministry of Agriculture
and Rural Development
Krishi Bhavan
New Delhi 110 001, India

Dr. A. Sambhashiva Reddy
Agronomist
Regional Agricultural Research Station
Nandyal 518 103, India

Dr. K.W. Riley
Hill Crops Advisor
Hill Crops Improvement Programme
Kavre Agricultural Centre
C/o ARPP, Box 1336
Kathmandu, Nepal

Mr. Deep Man Sakya
Assistant Agronomist
Dept. of Agriculture
G.P.O. Box 1135
Kathmandu, Nepal

Dr. S.R. Sampath
Head, Southern Regional Institute
National Dairy Research Institute
Audugodi,
Bangalore 560 030, India

Dr. T.V. Sampath
Agricultural Commissioner
Ministry of Agriculture and
Rural Development
Krishi Bhavan
New Delhi 110 001, India

Dr. Kishor Sherchan
Asst. Agricultural Botanist
Hill Crops Improvement Program
Dept. of Agriculture
C/o ARPP, P.O. Box 1336
Kathmandu, Nepal

Dr. A. Seetharam
Project Coordinator (Small Millets)
Univ. of Agril. Sciences
Bangalore 560 065
India

Dr. S. Sashidhar
Jr. Physiologist (Millets)
Univ. of Agril. Sciences
Bangalore 560 065, India
India

Dr. R.A. Sheriff
Plant Scientist (Ragi)
Univ. of Agril. Sciences
Bangalore 560 065, India

Dr. D.N. Singh
Director (Millets)
Millets Development Directorate
19, Krishnama Road
Madras, India

Dr. D.P. Thakur
Professor & Head
Dept. of Plant Pathology
Haryana Agricultural University
Hissar 125 004, India

Dr. D.V.S. Tyagi
Millet Breeder
Dept. of Plant Breeding
G.B. Pant University of Agriculture and Technology
Pantnagar 263 145, India

Dr. M. Udaya Kumar
Department of Crop Physiology
University of Agricultural Sciences
GKVK
Bangalore 560 065, India

Dr. S.N.P. Verma
Senior Millet Breeder
College of Agriculture
Rewa 486 001, India

Mr. S. Viswanath
Virologist (Millets)
Univ. of Agril. Sciences
Bangalore 560 065, India

Mr. Vincent Makumbi Zake
C/o Sorghum & Millet Unit
Uganda Agriculture & Forestry Research Organisation
P.O. Soroti, Uganda

Dr. E.N. Zolutukhin
South East Agricultural Research Institute
Saratov, USSR

I

OVERVIEW AND TAXONOMY

1

SMALL MILLETS—A SELECTIVE OVERVIEW

Hugh Doggett

INTRODUCTION

Small millets may be defined as millets cultivated for their small grains which are borne on short, slender grassy plants. Pearl millets (*Pennisetum*) are excluded.

There are many small millets in the world. Some researchers consider that they were developed from the corresponding wild grasses as the result of continued harvesting for food. Others believe that the prior spread of the idea of agriculture was needed in most cases before domestication could begin. Certainly the Alyawara people of central Australia failed to develop domesticated crops, even though some of the wild grasses (*Panicum* spp.) were morphologically and taxonomically similar to those domesticated elsewhere, and seeds were important in their traditional diet (O'Connell *et al.*, 1983).

The millets considered here are those being cultivated in the subtropical and tropical areas of the Old World. The list contains Finger millet (*Eleusine coracana*), Proso millet (*Panicum miliaceum*), Foxtail millet (*Setaria italica*), Little millet (*Panicum sumatrense*), (formerly *P. miliare*), Barnyard (Sawa) millet (*Echinochloa colona*) [formerly *E. frumentacea*], Kodo millet (*Paspalum scrobiculatum*), Teff (*Eragrostis tef*), Fonio millet (*Digitaria exilis* and *D. iburua*).

Small millets may also be called minor millets, but they are not unimportant. Japanese barnyard millet, proso and foxtail millets have all been important in the past, and are still important today, especially in Asia. Finger millet is an old tropical cereal still widely grown in eastern Africa and south Asia. Kodo and little millets continue to be important in Asia in times of famine or difficulty.

The small millets are often grown in difficult conditions, and it is scarcely surprising that they involve high production risks (Joshi and Agnihotri, 1984).

They have always been crops for situations where there is a risk of famine, as well as offering a low but more reliable harvest—relative to other crops—in low rainfall areas. Kodo millet was traditionally stored in the temples, so that seed would be available in times of crisis. The potential of these millets deserves more study. Food and seed reserves in the village are important, and should not be overlooked. They have good potential for livestock feed in the dry zones. The small millets should be developed both for their potential as good grain producers with modest water needs, and also as producers of forage. They can make good use of any irrigation water available after the main crops have been harvested, and so may be fitted in to more productive cropping patterns.

Finger millet

Finger millet was developed in Africa from *E. coracana* subsp. *africana*, probably in the Ethiopian region. It was introduced to India perhaps more than 3,000 years ago. It is a tropical crop, grown from sea-level to 3,000 m asl. This is the most widely grown small millet in India and Africa, and can be very productive. de Wet *et al.* (1984) recognized five races.

Proso millet

Proso millet is also an ancient crop. It was probably domesticated in central and eastern Asia, and was cultivated in Europe in Neolithic times (Purseglove, 1985). It was well known to the Romans, and became the 'common millet'. This is essentially a crop of the temperate regions, but is also grown in the sub-tropics, and on high ground in tropical winters.

Foxtail millet

Foxtail millet is yet another ancient crop, probably domesticated in eastern Asia, and known to the Chinese as early as 2,700 BC (Purseglove, 1985). It is essentially a crop of the sub-tropical and temperate zones. The main production areas are in Japan, China, India and eastern Europe. Cobley (1976) noted that more than 12 rather variable groups of cultivars have been recognized.

Little millet

Little millet is grown to a limited extent in India, up to altitudes of 2,100 m. It occurs wild in northern India and southeastern Asia. It will yield some grain and useful fodder under very poor conditions. Some forms mature in as little as two-and-a-half months (Purseglove, 1985).

Barnyard millet

Japanese barnyard millet *Echinochloa crusgalli* was domesticated in Japan some 4,000 years ago. It belongs essentially to the temperate zone. Barnyard (Sawa) millet, *E. colona,* was domesticated in India, where it remains an important cereal in some areas. It has also been recorded from the Central African

Republic, Tanzania, and Malawi. The two species have different chromosome numbers ($2n$ = 54 and $2n$ = 36 respectively), and hybrids between them are sterile. The wild form is widespread as a tropical weed. de Wet *et al.* (1983a) distinguished four races.

Kodo millet

Kodo millet is grown as a cereal in India only, although the wild grass is a widespread tropical weed. The crop has been grown for at least 3,000 years, yet de Wet *et al.* could not find any clear racial differentiation. Wild, weed, and cultivated types merged in all the characters studied. Kodo millet is said to be poisonous after rain. This could be due to a fungal infection. Winnowed, clean healthy grain seems to pose no health problem (de Wet *et al.*, 1983b).

Teff

Teff is the most important cereal crop in Ethiopia, particularly in the poorly drained, heavy soils that predominate in the Central Plateau. Nevertheless, the crop has not become important outside Ethiopia.

Fonio

Fonio, also known as hungry rice, is grown as a cereal crop throughout the savanna zone of West Africa. In parts of Guinea and Nigeria it is the staple crop. *Digitaria exilis* is considered to be the oldest West African cereal and its cultivation is thought to date back to 5000 B.C. The crop can grow on poor, shallow and rocky soils, but is not grown outside Africa. *D. iburua* is a white seeded form (Purseglove, 1985).

OBSERVATIONS FROM AFRICA

In Africa south of the Sahara, finger millet occupies the largest area under small millets in the eastern part of the continent, but there is a substantial area of teff in Ethiopia. In West Africa, fonio millet occupies a similar ecological niche to finger millet. There is also a small area of *Brachiaria deflexa* millet there.

Area and production figures for these millets separately are not readily available Data collected on the small millets are often combined with pearl millet and even with sorghum figures. Finger millet is grown abundantly in the Lake Victoria region—Uganda, Kenya, Tanzania, Zaire, Rwanda and Burundi The crop is also important in northeast Zambia and the southern highlands of Tanzania. Significant amounts are also grown in Zimbabwe, Malawi, and Mozambique. It is possible to get an approximate picture for the situation of finger millet in Uganda, because very little of any other millet is grown, so the FAO figures for millets refer pretty closely to finger millet. Table 1 shows the millet areas in Uganda from 1961-1984 (as reported by FAO in 1985).

TABLE 1
Area of millet in Uganda (000 hectares)

Year	Area	Year	Area	Year	Area	Year	Area
1961	519	1967	614	1973	580	1979	313
1962	530	1968	558	1974	510	1980	279
1963	528	1969	528	1975	567	1981	300
1964	529	1970	579	1976	508	1982	330
1965	568	1971	580	1977	527	1983	360
1966	573	1972	580	1978	550	1984	360

It will be seen that the millet area was steady from 1961-64, but increased to a higher level between 1965 and 1973, with the exception of 1969. During the period 1979-84, the area was relatively steady, but much reduced relative to 1961-64. Uganda has been through some hard times since 1970. There is good reason to believe that maize has invaded some of the finger-millet areas, so that there may have been a decline in millet area *per capita* which will persist. African finger-millet growers tend to be conservative, a lot of finger millet is fermented, and for such foods other cereals are seldom as suitable. Maize has other advantages, green cobs are popular and profitable to grow, and the ripe grain is easier to gather and handle.

Little serious research has been done in Africa on finger millet improvement and agronomy, with the exception of the work at Serere in Uganda. There, a small improvement programme has been operating at a low level for more than thirty years. Some good lines have been identified, and a male-sterile which was induced some 15 years ago has permitted more intercrossing. The scientists have been working under very difficult conditions for the past 14 years.

Teff is the most important cereal crop in Ethiopia, with the area planted fluctuating between 1.1 million and 0.87 million hectares. Some improved lines of teff have been isolated. Teff has never been adopted as a grain crop outside Ethiopia, and although it will certainly maintain an important place in the Ethiopian highlands between, 1,700 and 2,300 m, the crop is unlikely to spread into other countries. Since 1976, the area of finger millet in Ethiopia has fluctuated between 200,000 and 250,000 ha, falling to 180,000 in 1984. Studies in Ethiopia are now underway to identify finger millet lines with potential in the erratic rainfall areas.

OBSERVATIONS FROM INDIA

Figures are available from India for finger millet area and production by State (Joshi and Agnihotri, 1984). Table 2 shows the percentage of area and production of finger millet for each of the seven major millet producing States for the 1980-81 period. The last three columns of the table show the annual compound growth rates for production, area, and yield between 1970-71 and 1981-82, given as per cent change per year.

TABLE 2

Area and production of finger millet in India, and growth rates for the period 1970-71 to 1980-81

	Percentage share of India total 1980-81		Annual compound growth rates % per annum		
State	area	prodn.	prodn.	area	yield
Andhra Pradesh	9.86	9.85	+1.36	−0.13	+1.49
Bihar	6.87	3.73	+4.33	+1.30	+3.03
Karnataka	42.2	50.53	+5.82	+0.75	+5.07
Maharashtra	8.91	7.98	+3.38	+0.67	+2.71
Orissa	10.99	6.24	+2.46	+6.77	−4.31
Tamil Nadu	9.93	20.33	+3.22	−0.79	+4.01
Uttar Pradesh	6.82	5.33	−5.10	−4.32	−0.78
All-India	100	100	+3.37	+0.64	+2.73

Finger millet production increased at a rate of 3.37 per cent per year during the period covered by the table, which exceeded the rate of population growth. The annual rate of yield increase at 2.73 per cent is gratifying. The breakdown by States shows that Karnataka and Tamil Nadu achieved yield increases at rates of 5.07 and 4.01 per cent per annum respectively, which respresents excellent progress. Production increase rates were maintained well, in spite of the fall in area in Tamil Nadu. Orissa stands out as showing a substantial rate of area increase accompanied by a marked decline in yield, although production increased at a reasonable rate. Presumably the demand for finger millet was well maintained, so Orissa stands out as one needing inputs, including cultivars, that are better adapted. Uttar Pradesh showed negative area, production, and yield trends. One may guess that finger millet was being replaced by more valuable crops, possibly in the wake of increased areas under irrigation. Grain price is clearly important, but a crop such as finger millet is probably most influenced by very local demand. Other statistics presented by Joshi and Agnihotri (1984) showed that finger millet had a higher probability of crop failure than pearl millet, rice, or sorghum, but the difference between *bajra* and *ragi* was small (0.50 to 0.56).

Table 3 shows the ratio of gross returns of finger millet relative to those of rice (Joshi and Agnihotri, 1984) for the years 1970-71 and 1980-81.

Bihar, Karnataka, Maharashtra and Tamil Nadu showed an encouraging increase in the ratio, while Uttar Pradesh had the highest ratio for both years. However, the ratio was clearly in favour of rice apart from Uttar Pradesh, and

TABLE 3
Ratio of gross returns of finger millet to gross returns of rice

	1970-71	1980-81
Andhra Pradesh	0.72	0.55
Bihar	0.61	0.66
Karnataka	0.61	0.62
Maharashtra	0.87	0.95
Orissa	1.25	0.68
Tamil Nadu	0.64	0.74
Uttar Pradesh	1.35	1.05

Orissa in 1970-71. The procurement price for millets and unhusked rice was the same, so yield differences between the millets and rice must have been important, as well as unit production costs.

Figures for the five south Asian small millets being considered, were only available combined together (Joshi and Agnihotri, 1984). They are shown in Table 4.

Again, there was a good increase in yield in Karnataka, together with an increase in production, though there was some reduction in the crop area. This suggests that these small millets are being grown more efficiently in that State. In Maharashtra, there was little change in area, and a modest increase in production accounted for very largely by the increase in yield. Orissa lost yield, but did achieve increased production. The remaining four States in the table all lost production, and they accounted for 62.8 per cent of the whole country's production. The area under small millets decreased, while yields were

TABLE 4
Area and production of 5 small millets in India (excluding finger millet) and growth rates for the period 1970-71 to 1980-81

State	Percentage share of Indian total 1980-81: area	prodn.	Annual compound growth rates, % per annum: prodn.	area	yield
Andhra Pradesh	14.17	16.14	−2.77	−2.15	−0.62
Karnataka	10.23	12.74	+2.86	−1.95	+6.81
Maharashtra	5.15	5.73	+1.61	+0.06	+1.55
Orissa	5.16	5.00	+2.06	+2.74	−0.68
Tamil Nadu	8.99	17.76	−0.31	−0.61	+0.30
Uttar Pradesh	9.17	12.90	−1.59	−1.84	+0.25
Madhya Pradesh	36.70	16.04	−3.80	−0.62	−3.18
All India	100	100	−0.91	−1.20	+0.29

down in both Andhra Pradesh and Madhya Pradesh. Even Tamil Nadu, which produced quite good figures for finger millet showed no signs of improvement as far as the small millets were concerned.

Potential for small millet improvement in India

In the past, the data needed to assess the comparative merits of the small millets have not been available. Comparisons have been essentially qualitative. In 1978, ICAR expanded the research on small millets, and invited IDRC to contribute towards this. Five small millets were chosen, namely foxtail, barnyard, proso, kodo, and little millets. Work was strengthened at appropriate sites, and multilocational testing was available at these sites, as well as at other locations where AICMIP conducted its trials. Direct comparisons between the different millets were not made, but it is of interest to look at the data now available, from this project and to make tentative assessments.

Foxtail millet

This millet is said to be widely grown in India (Anon 1984), but the figures suggest that it is only important in a few areas. In 1977-78, there were 479,000 ha in Andhra Pradesh, 232,000 in Karnataka, and 20,000 ha in Tamil Nadu, and together these three States grew 95 per cent of the total recorded Indian acreage. Within Andhra Pradesh, three of the 13 districts grew 79 per cent of the State total (Anon 1984). It is evidently useful in Andhra Pradesh, and can grow quite well in other areas also, as will be seen from Table 5.

The All-India figures in Table 5 are averages of a series of multilocational trials, and do suggest an upward yield trend for the breeders' materials going in to these trials. The Nandyal data suggest good stability of yield levels over years, with a bad season in 1980, but there is little sign of an upward trend. The figures from elsewhere may indicate very variable yields, though one needs to have the planting dates relative to the onset of the rains. On an experiment station, other activities may have priority, and a small millet variety trial may be sown late. The data at least suggest that yields in Nandyal are more stable

TABLE 5

Mean Foxtail millet yields from coordinated trials in India (q/ha)

Year	All-India	Nandyal	Dholi	Semiliguda	Almora
1979	13.5	18.3	21.4	4.9	7.4
1980	10.5	· 4.7			4.7
1981	12.8	14.9	3.7		
1982	16.0	16.0, 14.8	9.7	2.7, 2.0	
1983	14.5	16.1	14.5	11.0	5.1
1984	17.8	16.8	5.9	8.1	

than those at Dholi, and may point to the reason why so much foxtail millet is grown in one part of Andhra Pradesh. Conditions there are evidently right for this crop. The maturity lengths of lines at Nandyal lie roughly in the range 75-95 days, mostly in the mid-eighties Nandyal has a strong research team, which is making good progress.

Barnyard millet

Work on this crop in India is based at Almora in the hills of Uttar Pradesh. Barnyard millet is quite important in Uttar Pradesh, where it occupies some 230,000 ha, almost equally divided between hills and plains. Within India, it is also cultivated in Madhya Pradesh, Maharashtra, and Tamil Nadu. Table 6 gives coordinated trial results (Anon, 1979-84).

The yield figures shown in Table 6 are reasonably stable for three of the sites, and the combined All-India figure does not fall below 11 q/ha. The length of maturity at Almora in 1984 was around 100 days, some 55-70 in Coimbatore, and 65-85 in Pune. This millet appears to deserve more attention.

Proso millet

Proso millet is grown in Andhra Pradesh, Maharashtra, Tamil Nadu, Uttar Pradesh, and Bihar. In Bihar, it is cultivated throughout the year, as a catch crop before the main *kharif*, or after the *rabi* crop harvest. Two quick crops may be taken during the summer in March-June. There are some 75,000 ha grown in Bihar. Proso-millet is said to have a very low water requirement, and also a wide range of adaptation to climates, soils, and altitudes. The research on proso in India is centered at Dholi. Lengths of maturity are recorded as 55-85 days for germplasm at Dholi, but lines in trials averaged 65-75 days according to season and location. Some figures from the AICMIP trials during the period 1979-84 are shown in Table 7 (Anon, 1979-1984).

TABLE 6
Barnyard millet yields from coordinated trials in India (q/ha)

Year	All-India	Almora	Dholi	Nandyal	Rewa	Pune	Semi-liguda	Coimb-atore
1979	20.6	23.2	19.4	20.6			6.5	
1979	15.2	18.0	19.0	21.3				
1980	11.3	20.8	7.7	14.4	5.4	16.2	7.4	
1980	11.7	19.3	12.0	11.0	9.3			
1981	15.1	16.2	14.9	10.6	10.9	23.3		5.5
1982	17.1	14.0	17.0	13.1	7.8	18.1		8.4
1983	15.6	21.7	10.4	25.9	9.1	17.7	12.4	
1984	12.5	18.3	8.3	8.6	15.5	15.5	16.9	5.8

TABLE 7

Mean Proso millet yields from coordinated trials in India (q/ha)

Year	All-India	Nandyal	Semiliguda	Dindori	Dholi
1979	1.8	5.5	0.6	2.1	
1979	12.4	6.0	2.1		11.1
1980	10.4	16.5			12.9
1981	8.6	2.9			22.8
1982	12.5	13.5	13.8		17.9
1983	7.5	11.9			16.2
1984		11.5 (rabi)	18.6 (kharif)		14.9 (summer)
1984			12.9 (rabi)		10.3 (kharif)
1984					10.3 (rabi)

Fewer figures are available for this millet, but only Dholi has a consistently steady level of yield. Some years show very low figures at Nandyal, Semiliguda, as well as in the All-India trials average itself. This millet may have problems in certain locations or seasons. Proso millet can be damaged severely by shootfly, and birds can be locally troublesome.

Little millet

Separate data for areas under little millet were not available, figures were either combined with those for kodo, or with other small millets. Yields from the trials (Venkateswarlu *et al.*, 1984) are shown in Table 8.

In the AICMIP trials, duration varied from 70 to 118 days, and yields were unexciting. The rather few figures available suggest neither reliable yields, nor good production per growing day. Little millet is grown in Tamil Nadu, Karnataka, Andhra Pradesh, Maharashtra, Orissa, Bihar, Madhya Pradesh and Uttar Pradesh. It is described as a "quick growing, short duration cereal which withstands both drought and waterlogging" (Anon, 1979-1984). Doubtless this is a valuable crop in difficult situations, but other millets seem to have a greater development potential.

TABLE 8

Mean Little millet yields from coordinafed trials in India (q/ha)

Year	All-India	Semiliguda	Dindori	Nandyal	Rewa	Pune	Ranchi
1980		6.4					
1981	9.5	2.2		2.0	6.2		
1982	9.1	3.7		8.3	2.2		
1982		2.7	7.0			8.6	5.2
1983	7.4	4.1	1.8		3.2		
1984		7.0				7.7	7.5

Kodo millet

The last millet included in the AICMIP programme is kodo millet, and mean trial yields from the All-India programme are shown in Table 9.

TABLE 9
Mean Kodo millet yields from coordinated trials in India (q/ha)

Year	All-India	Dindori	Nandyal	Dholi	Rewa	Pune	Coimbatore	Semiliguda
1979	16.1	2.2	15.0	27.1				12.8
1980	16.5	6.5	15.7	13.8	9.3	44.0		
1980	14.7	3.1	6.3	15.3	12.8			
1981	25.8	8.9	15.0	22.0	16.4	43.1	50.6	
1982	19.4	7.2	19.0		9.1	20.2	15.4	29.7
1983	18.4		13.4		17.2	30.5	10.6	22.9
1984	18.4	10.0	18.3		30.2	25.7		34.3

Kodo millet is said to occupy the largest area of any small millet in India. It is grown for grain in Madhya Pradesh, Uttar Pradesh, Tamil Nadu, Karnataka, Gujarat and Maharashtra. It is referred to as a long-duration crop, hardy and drought resistant. The trial data show maturity periods of 81-135 days, suggesting an unusual flexibility.

COMPARISON OF THE SMALL MILLETS

The Small Millets Project under the All-India Co-ordinated Millet Improvement Project (AICMIP) has done a great service in getting together some data on the performance of this neglected group of cereals from a good series of multi-locational trials across the Indian millet areas. In addition, good progress has been made in varietal improvement, agronomy, and in assessing pest and disease susceptibilities and control. We now have some information upon which a policy for these crops can be based.

These observations are based upon variety trials, not upon farmers' trials. The trials have had different treatments from that which a farmer would have given. The level of research staffing has varied from centre to centre. My presentation of the data is open to criticism, the averages of trial yields presented may have come from various trials, though always done in the same year and site as that stated. Nevertheless, helpful tentative conclusions may still be drawn. Proso millet and little millet have been generally poorer than the other three small millets. Foxtail, barnyard, and kodo millets show promise. Table 10 gives the mean yields of trials at locations where there are three or more years of data. The number of years, and locations contributing to each mean are shown in brackets.

TABLE 10
Experimental mean yields over years of the three highest yielding small millets

	Foxtail	Barnyard	Kodo
All-India	14.2 (6)	14.9 (8)	18.5 (7)
Nandyal	14.5 (6)	15.7 (8)	14.7 (7)
Dholi	11.0 (5)	13.6 (8)	19.5 (4)
Semiliguda	6.7 (4)	10.8 (4)	24.9 (4)
Mean of first 4	11.6 (21)	13.8 (28)	19.4 (22)
Almora	6.1 (3)	18.9 (8)	
Dindori			6.3 (6)
Coimbatore		5.4 (3)	25.5 (3)
Rewa		9.7 (6)	15.8 (4)
Pune		18.1 (8)	32.7 (6)
Mean, all trials	11.3 (24)	16.3 (50)	18.6 (42)

Kodo and barnyard millets show up well in Table 10. Foxtail may be more limited in its ecological range, and also in the yield level it can achieve. This may be an unfair conclusion, as it was only tried at five sites. It may not have had an opportunity to give of its best, but there is good evidence on which to support kodo and barnyard millets. Foxtail millet may be left as locally important and deserving of continued improvement work. From Table 6, we see that barnyard millet chalked up 25.9 q/ha at one site, 23.3 at another, and recorded eight trial mean yields of 20 q/ha or better. Kodo millet in Table 9 recorded one yield of 50.6, two over 40, three over 30, and seven of 20 or better in q/ha. These are good values for trial mean yields. Kodo millet thus shows a good yield potential, while barnyard millet may deserve better marks for stability. The lowest barnyard millet yield was 5.5 q/ha. Kodo recorded one of 2.2, and another of 3.1 q/ha. It would be best to back both, but I would choose kodo millet if a choice had to be made.

POSSIBILITIES FOR FURTHER IMPROVEMENT OF SMALL MILLETS

Finger millet

The figures quoted above show clearly that finger millet is an important cereal both in India and in Africa.

Steady progress in finger millet improvement is being made in India. A recent paper by de Wet *et al.* (1984) has set out the germplasm situation very clearly. As with some other African crops that were moved to India—such as niger and sorghum—the wild progenitor was left behind in Africa. This gave rise to two separated populations, one of which continued to develop alongside the wild type, with some introgression between them. The other population continued to develop in India without any such interaction. The two populations have been separated for at least 3,000 years. There must be a great potential in this situation for the improvement of finger millet. A little of it has already been utilized in the development of the 'Indaf' finger millets, but much more remains to be done.

The finger millet crop contains a wide range of variability. There are types selected at Almora which will yield a crop under very low moisture conditions. In southern India, irrigated finger millet can give 40 q/ha on very little irrigation water. Venkateswarlu *et al.* (1984) reported yields of over 40 q/ha on 365 mm of water. Yields in rainfed trials in Uganda may lie between 30 and 55 q/ha (Anon, 1966-69), while entries in the coordinated trials in India may reach 50 q/ha (Anon, 1982-83). This crop has a big unrealized potential in Africa, where irrigated finger millet is scarcely known. Irrigation with transplanting permits better weed control in the early stages of crop growth, most of the land can be cleaned while the nursery is being raised on only one-sixth of the final area. Evidence from India shows that transplanting usually leads to better yields (Rachie and Peters, 1977).

More interaction and cooperation between the research in Africa and in India is needed. More scientists, more research locations, and more support for finger millet research are required in Africa. Serere in Uganda is still doing finger millet research (Makumbi Zake and Esele, 1984): there is increasing interest in Kenya and trials are being done from time to time elsewhere in Africa. It should not be difficult to strengthen work on finger millet in southern Tanzania, and in Zambia or Zimbabwe. Such cooperation will require rather free interchange of germplasm and early generation material. An oilseeds network has been established for identical reasons. There are oilseed projects in both India and eastern Africa, and a network advisor is based in Ethiopia. He keeps the researchers in good touch with each other, and also with programmes and research scientists elsewhere. Something similar would be very useful for the small millets.

The next requirement is for many crosses to be made between African and Indian material so that its potential may be fully exploited. Those two populations separated for such a long period, but under selection as a cultivated crop will have accumulated gene differences that could give very productive combinations when brought together. Hybridizing finger millets is difficult and fiddling. There is a need for male-steriles. I do not know the situation in India, but there is a good male-sterile at Serere. Four populations have been established, and recurrent selection is in progress there. More male-steriles and more population improvement programmes of this kind are needed. A determined effort to obtain male-steriles should be made. Postgraduate students with access to a radiation source and a supply of chemical mutagens should be able to produce useful steriles.

Other small millets

Work on the best of these should be continued in India. Additional effort into proso or little millets could be questioned. Foxtail millet clearly has a place in the agriculture of certain areas where it flourishes, and work on it should certainly be continued, perhaps strengthened. The analysis of the AICMIP results, offered tentatively above, tends to support what botanists might have suspected: that tropical plants perform better overall in the tropics than do temperate or subtropical plants, although there are niches, especially on raised plateaux and in the hills, where the latter do perform well. The two millets deserving serious attention, additional to finger millet, are barnyard (sawa) and kodo millets.

Both millets are likely to have a good potential in Africa, where there is certainly a lot of wild germplasm available. There could be some cultivated barnyard millet there, it was once grown in Egypt. They offer an exciting challenge. Both millets were domesticated in India, and the introduction of exotic germplasm, even of the wild type, followed by hybridization and introgression, could be useful: It would be really good to see a small millets network set up linking the two continents, as mentioned above, with finger millet, kodo millet, and barnyard millet as the crops of the network.

Kodo millet is cleistogamous, but protogynous types have been selected, and crosses made. In the Indian wild types the stigmas protrude from the spikelets. The observation by de Wet *et al.* (1983b), on the lack of racial differentiation after 3,000 years, suggests that it could be a very interesting crop with which to work. Maize, sorghum, finger millet, barnyard (sawa) millet among many other crops have developed racial differentiation.

Barnyard millet is largely self-pollinated, and as with kodo, the wild germplasm is very widespread in the tropics of the Old World. If both millets can perform as they have done in these trials without much improvement from plant breeders as yet, their potential could be great.

Clearly, these two small millets will not be of much interest to Africa initially. If a network can be established for finger millet, then wild African germplasm of all three small millets could be fed back to India, where it would be of value to the breeding programmes. Promising cultivars of kodo and sawa millet could be sent to Africa for observation and trial. Every effort should be made to develop male-steriles of kodo and barnyard millets, as suggested above for finger millet.

Lastly, I cite to two works which give excellent guidelines to those designing improvement programmes. The first is Eberhart *et al.* (1967), which proved very useful to plant breeders in eastern Africa. The second is Jensen (1978), and has influenced breeding approaches in the USA. Jensen's approach could be really useful in breeding small millets, once the male-sterile problem has been solved. A lot can be done using bulk populations, especially where plenty of semi-skilled help is available.

FINAL REMARKS

The point has been made to me that novel crop improvement techniques may not always be appropriate in the developing world. Let me not attempt to discuss what could be a contentious issue, but rather set out a few principles.

The plant breeder is above everything a manager. He must have defined his objectives clearly, he then needs to review his resources. Two of the important ones are his own working time, and the trained capability of his assistants. Do we all make the maximum use of the labour available to us in the developing world? Three instances occur to me: (1) The production of hybrid seed of cotton in India. It is possible to train ladies from the villages to do this reliably and well, they have no background of scientific or experimental work. (2) The use of school girls during the vacation by the Sorghum programme in Ethiopia to make the crosses. (3) The use of six girls from the local village by Dr. Henry Fernando in Kandy, Sri Lanka to do all his tissue culture work in a large and successful hybridization programme.

Let me add my own tribute: I have always received the most careful, accurate, and diligent help from ordinary village people. There is a wealth of talent on which to draw. A lot of people have the capability to be educated far beyond any level they may have achieved, that may be no level at all. They simply have not had the same chance that we had. This training takes some of the breeder's time initially, but it is time well spent. If the breeder starts with one of his permanent assistants to help him, he can very soon hand over that particular aspect of the training. It is obviously important that he be in at the beginning of training in any new area. He has to learn himself how best to do it, and how much to do, relative to the capabilities of his trainees.

The philosophy of the breeding approach should always be clear, and should take into account the most recent knowledge. The implementation of that knowledge must depend on what the plant breeder feels his team is capable of doing in the light of the overwhelming priority to get better material out to the farmer as soon as possible. Thus, Eberhart *et al*. are not recommended so that everybody can start up programmes of reciprocal recurrent selection. They are recommended to emphasize the value of thinking in terms of exploiting heterosis—and of splitting the material into two basic groups, so that one is working towards the efficient use of heterosis one day. Similarly, Jensen's approach is so time-saving, running crosses in mass-selected bulks until the F_5 before exploding into progeny rows. The use of male-steriles to minimize making deliberate crosses. One of the purposes of workshops is to kick ideas around: there may be opportunities to modify one's programme a little as a result.

LITERATURE CITED

Anonymous. 1966, 1967, 1968 and 1969. *Record of Research*, Sorghum and Millet annual reports of EAAFRO, Nairobi, Kenya.

Anonymous. 1982-83. Progress Reports on Finger Millet. AICMIP, ICAR Co-operating Centres, Bangalore.

Anonymous. 1979-1984. Minor Millets Improvement Project, *Ann. Prog. Reps.* ICAR, New Delhi.

Cobley, L.S. 1976. *Introduction to the botany of tropical crops*, 2nd edition, Longman, London.

de Wet, J.M.J., K.E. Prasada Rao, M.H. Mengesha and D.E. Brink. 1983a. Domestication of Sawa millet (*Echinochloa colona*). *Econ. Bot.* **37:** 283.

de Wet, J.M.J., K.E. Prasada Rao, M.H. Mengesha and D.E. Brink. 1983b. Diversity in kodo millet (*Paspalum scrobiculatum*). *Econ. Bot.* **37:** 159.

de Wet, J.M.J., K.E. Prasada Rao, D.E. Brink and M.H. Mengesha. 1984. Systematics and Evolution of *Eleusine coracana*. *Am. J. Bot.* **71:** 550.

Eberhart, S.A., M.N. Harrison and F. Ogada. 1967. A comprehensive breeding system, *Der Zuchter.* **37:** 169.

FAO. 1985. Production Yearbook. Rome.

Jensen, N.F. 1978. Composite Breeding Methods and the DSM System in Cereals. *Crop Sci.* **18:** 622. e.

Joshi, P.K. and A.K. Agnihotri. 1984. Millet Production in India. Problems and prospects. *Agricultural Situation in India.* **39:** 329.

Makumbi Zake V. and J.P. Esele. 1984. Report on Sorghum and Millet Improvement in Uganda. In: Sorghum and Millet Improvement in Eastern Africa. Proc. 3rd Reg. Workshop on Sorg. & Mill. Improvement in E. Africa 5-8 June 1984 p. 147 OAU/STRC/SAFGRAD/ ICRISAT P.O. Box 30786, Nairobi.

O'Connell, J.F., P.K. Latz and P. Barnet. 1983. Traditional and modern plant use among the Alywara of central Australia. *Econ. Bot* **37:** 91.

Purseglove, J.W. 1972 revsd. 1985. Tropical Crops, Monocotyledons. Logman, London.

Rachie, K.O. and LeRoy V. Peters. 1977. *The Eleusines.* ICRISAT P.O. Patancheru, Andhra Pradesh.

Sampath, T.V., S.M. Razvi, D.V. Singh and K.V. Bondale. 1988. Small millets in Indian agriculture, in this volume.

Venkateswarlu, M.S., K.J. Naidu and G.H.S. Reddi. 1984. Production potential of "Kalyani" finger millet under different irrigation frequencies. *Ind. J. Agric. Sci.* **54:** 5.

2

ORIGIN, EVOLUTION AND SYSTEMATICS OF MINOR CEREALS

J.M.J. de Wet

The Poaceae includes an estimated 8,000 species belonging to some 600 genera. Grasses occur on all continents, and in all habitats that support growth of flowering plants. They serve man in many ways, but it is their use as cereals and feed for livestock that make them essential for human survival. The caryopses of most grasses are edible, and at least 300 species were harvested during historical times as wild cereals by nomadic hunters and herders, and by farmers during times of scarcity. Thirty-five species belonging to 20 genera are known to have been domesticated. Their cultivated races rely on man for seed dispersal (sowing), and for a suitable habitat to reproduce successfully (cultivated field).

Cereals are globally planted on an estimated 730 million hectares, and yield an estimated 1,800 million metric tons of grain annually. Wheat, maize and rice account for approximately 80 per cent of grain produced in the world. These cereals are followed in importance by barley, sorghum, oats, rye and pearl millet which together represent another 19 per cent of the world's cereal production. The remaining cereals account for about 1 per cent of the foodgrain produced in the world today. These minor cereals are not important in terms of world food production, but essential as food crops in their respective agro-ecosystems. They are mostly grown in marginal areas, or under agricultural conditions where major cereals fail to consistently produce an acceptable harvest.

MINOR CEREALS OF THE AMERICAS

Wild cereals played an important role in the diets of native Americans until recent historical times (Palmer, 1871; Ball, 1884). Fifty species were extensively harvested, but only six species were domesticated as cereals. Maize (*Zea mays* L.) is the only New World cereal of commercial significance. *Setaria geniculata* (Lam.) P. Beauv. (brittle grass) from arid Mexico (Callen, 1965), and *Phalaris caroliniana* Walt. (may grass) of the southeastern United States (Chomko and Crawford, 1978) are known as cultivated cereals only in an archaeological context. Brittle grass was grown for at least a millennium (Callen, 1967), but was replaced by maize as a cereal some 4000 years ago (Mangelsdorph, MacNeish and Gallinat, 1967). Two minor cereals, mango (*Bromus mango* Desv.) in Chile and sauwi (*Panicum sonorum* Beal) in Mexico were important crops until recent historical times. American wild rice (*Zizania acquatica* L.) is a recent domesticate in north central Canada and adjacent regions of the USA (de Wet and Oelke, 1979).

Molina (1782) recorded that the Araucano Indians in Central Chile grew a kind of rye that was called el Mango, and a kind of barley that was called la Tuca. Tuca probably refers to *Bromus unioloides* HBK., a wild grass that was extensively harvested as a cereal on the highlands of South America (Ball, 1884). El Mango (*Bromus mango*) was cultivated. Gay (1865) recorded that it was a biennial crop, grown with peppers and beans. Florets were roasted to facilitate removal of the lemma and palea, and the grains were ground into flour to make bread or a fermented drink called chicha. Cruz (1972) cited an unpublished manuscript by Arturo Fontecilla Larrain, a professor of agronomy at the Catholic University of Santiago in Chile during the early twentieth century, who recorded that plants of el Mango produced 40-50 culms, each bearing an inflorescence with 70-100 grains. It required eighteen months to mature, however, and el Mango was replaced during the last half of the twentieth century by wheat for making flour and by apples to make cider (Brucher, 1979).

Sauwi (*Panicum sonorum*) is native to arid western North America. It formed an important part of the Sonoran desert agriculture of northwestern Mexico (Nabhan and de Wet, 1984). It was widely grown well into the twentieth century by Indian tribes who lived along the Colorado river delta (Gifford, 1931). Today it is grown only in southeastern Sonora and adjacent Chihuahua. Although little known outside this area, sauwi is a promising cereal for the semi-arid tropics of Africa and Asia. It is drought tolerant, and has acceptable yield potential under adverse conditions. Plants produce several tillers, each of which produce an inflorescence with as many as 2,500 fertile florets.

The only New World minor cereal of present-day economic importance is American wild rice. This is not a true rice. It belongs to the genus *Zizania* rather than *Oryza*. American rice is commercially harvested as a wild cereal (Dore, 1969), and since the late 1960's planted on a commercial scale. *Zizania acquatica* is the only grass species successfully domesticated as a cereal in historical times (de Wet and Oelke, 1979). It is grown in paddies as is rice. Paddies are flooded and seeded in late fall. Germination is rapid in spring and the water level is maintained until August when the crop matures. Fields are then drained and harvested with a modified rice combine. Since the cultigen retains some degree of natural seed dispersal, no subsequent sowing is needed. Spikelets and straw are worked into the wet soil after harvest, where the caryopses lie dormant until the soil thaws in the next spring Natural populations rarely yield more than 100 kg ha^{-1}, while yields of 1000 kg ha^{-1} are obtained in planted paddies. Yields of at least 3000 kg ha^{-1} are possible from fully non-shattering cultivars.

MINOR CEREALS OF AFRICA

At least 60 grass species were extensively harvested in Africa as wild cereals until recent historical times (Busson, 1965; Jardin, 1967). The widely distributed *Brachiaria deflexa* (Schumach.) C.E. Hubbard, *Oryza barthii* A. Chev., and *Paspalum scrobiculatum* L. were extensively collected in West Africa. *Stipagrostis pungens* (Desf.) de Winter, and *Cenchrus biflorus* Roxb. were harvested by nomadic tribes in the Sahara. These species are often still encouraged as weeds in cultivated fields where they are harvested as wild cereals.

In West Africa, *Oryza barthii* Chev. gave rise under domestication to the cultivated *O. glaberrima* Steudel (Porteres, 1976). Teff, *Eragrostis tef* (Succ.) Trotter, is an important cereal in the Ethiopian highlands and was probably derived from *E. pilosa* (L.) P. Beauvois (de Wet, 1977; Costanza, de Wet and Harlan, 1979). The weedy *Brachiaria deflexa* (animal fonio) is cultivated on the Fouta-Djalon highlands of the west African savanna (Chevalier, 1933; Proteres, 1951). Animal fonio differs from wild *B. deflexa* only in having spikelets that disarticulate tardily at maturity.

Two other minor cereals are important crops in the west African savanna. Black fonio, *Digitaria iburua* Stapf is grown by the Hausa tribe of Nigeria, and occurs sporadically across most of semi-arid west Africa (Porteres, 1955; Clayton, 1972). It is often planted between rows of sorghum or pearl millet, and commonly as a mixture with *Digitaria exilis* (Kippist) Stapf (true fonio). Black fonio (*D. iburua*) is drought tolerant and often yields a harvest when the major cereal it accompanies fails to survive. True fonio (*D. exilis*) is widely grown across the west African savanna. It differs from *D. iburua* which has both glumes conspicuously shorter than the spikelet, in having the upper glume at least as long as the spikelet. Fonios are sown in west Africa during May or June, and harvested in September. Harvested inflorescences need to be protected from moisture since the grains become agglutinated to the lemma and palea when they get wet. Threshed grains are parched or dried in the sun before the chaff is removed by pounding in a wooden mortar. Fonio is used in stews, or the boiled grains are eaten as rice with butter or palm oil.

The most important minor cereal in Africa is finger millet, *Eleusine coracana* (L.) Gaertner. It was domesticated in Africa, but is also widely grown in south

Asia, particularly India (Hilu and de Wet, 1976a). The closest wild relative of finger millet is *E. coracana* subsp. *africana* (Kennedy-O'Byrne) Hilu and de Wet. Spontaneous finger millet is widely distributed along the eastern and southern highlands of Africa (Phillips, 1972). Derivatives of hybrids between cultivated and wild taxa are companion weeds of finger millet across most of its distribution in Africa.

Finger millet first occurs in the archaeological record of early African agriculture dating back some 3,000 years (Hilu and de Wet, 1976b), and was introduced into India at least 3,000 years ago (Vishnu-Mittre, 1968). Cultivated finger millet is extensively variable, and this variation is recognized as five races by de Wet, Prasada Rao, Brink and Mengesha (1984). Race coracana is widely distributed across the range of finger millet cultivation in Africa and Asia. It resembles wild finger millet in having five to nineteen slender inflorescence branches that are 6-11 cm long, digitately arranged, and with the tips often becoming slightly incurved or reflexed at time of maturity. Some genotypes differ phenotypically from subsp. *africana* primarily in being unable to disperse their spikelets without the help of man. Race coracana is particularly well adapted to agriculture in the eastern highlands of Africa and the Ghats of India. Some cultivars are drought tolerant and compete aggressively with weeds under conditions of traditional agriculture. It is often sown as a secondary crop with sorghum or pearl millet.

Races vulgaris, elongata, plana and compacta probably evolved from race coracana under cultivation. They probably evolved in Africa and were introduced into India. Little racial evolution took place in this secondary centre of cultivation. Race elongata is characterized by slender inflorescence branches that are 10-24 cm long. It is grown in eastern Africa and the eastern Ghats of India. Indian and African cultivars cannot consistently be separated on the basis of inflorescence morphology. Race plana is primarily African in distribution. It is grown in Ethiopia and Uganda, and to some extent in the eastern and western Ghats of India. Race plana is characterized by large, 8-15 mm long spikelets that are arranged in two more or less regular rows along the rachis, giving the inflorescence branches a flat ribbon-like appearance. In some genotypes the fertile florets are so numerous that they almost surround the rachis at maturity. These genotypes somewhat resemble race compacta, except that the inflorescence branches are not incurved. Members of race compacta are known as cockscomb finger millets in both Africa and India. Spikelets are up to ten flowered, with the inflorescence branches divided at the base and strongly incurved to form a fist-like inflorescence. Indian cultivars commonly have an inflorescence branch located some distance below the terminal cluster on each primary inflorescence axis. African cultivars usually lack this lower inflorescence branch. Race compacta is grown in northeastern India, Ethiopia and Uganda.

Race vulgaris is the most common finger millet of Africa and Asia. It is grown in Africa from Uganda to Ethiopia and to south Africa, and in Asia from India to Indonesia. Inflorescence branches are twisted or incurved. Some genotypes are drought tolerant, others are well adapted to areas of high rainfall, and still others are sown in nurseries and transplanted to fields with the first rains of the season. In rice-growing areas race vulgaris often follows irrigated rice as a rabi crop. Grains are cooked as rice, or ground into flour to make porridge or unleavened bread.

MINOR CEREALS OF EURASIA

Ten minor millets are grown in Asia. Their survival as cereals in competition with rice and wheat attests to the significance of these cereals in the agro-ecosystems of Asia. Two of these, *Panicum miliaceum* L. (broomcorn or proso millet) and *Setaria italica* (L.) P. Beauv. (foxtail millet) are grown across temperate Eurasia, with foxtail millet extending into the semi-arid tropics of Asia. Other minor cereals are important in specialized agricultural niches in Asia.

Crabgrass or manna, *Digitaria sanguinalis* (L.) Scop. is a common weed in all temperate parts of the world. The species is morphologically variable and variously classified into subgenera and varieties (Gould, 1963). It is an annual grass with prostrate or decumbent stems, and flowering culms that can reach well over one metre in height. The species was harvested as a spontaneous semi-domesticate is southern Europe until the first quarter of this century (Werth, 1937). It is cultivated as a cereal in the Caucasus and Kashmir (Henrard, 1950; Bor, 1955). Crabgrass has been a crop for at least 2,000 years. Plinius who lived in the first century A.D. referred to the species as ischaemon and suggested that it was of Slavic origins (Kornicke and Werner, 1885). Matthiolus (1565) recorded that this cereal was grown in Bohemia, Slavonia and the Ukraine. Kornicke and Werner (1885) reported that in Hungary one hectare yielded 420-520 kg of grain and 780-1200 kg of hay. During the late nineteenth century the species was an important cereal in southeastern Europe (Ascherson and Graebner, 1890). Wild manna was harvested with a sickle before the plants were fully matured (Becker-Dillingen, 1927). It never lost the ability of natural seed dispersal. Little is known about its present cultivation in the Caucasus and Kashmir.

Another crabgrass, *Digitaria cruciata* (Nees) A. Camus, is cultivated by the Khasi tribes in Assam, where it is known as raishan. Hooker and Stapf (1896) recognized this cereal as a species of *Paspalum* and Bor (1940) mistakenly included it in *Digitaria corymbosa* (Roxb.) Merrill. Bor (1955) correctly transferred raishan to *D. cruciata* and recognized the cultivated kinds as var. *esculenta* Bor. Veldkamp (1973) described var. *pectinata* Veldkamp to include a cultivar with glutinous grains that was grown at Cha-Pa in northern Vietnam.

Raishan is an annual grass with prostrate to decumbent culms that root at the lower nodes, and that produce flowering culms up to 1.3 m tall. Inflorescences consist of two to ten racemes arranged on a 1-4 cm long central axis, with the racemes up to 18 cm long reflexed at maturity. The chartaceous lemma and palea tightly enclose the grain at maturity, but the grain is readily freed from this fruitcase by pounding in a mortar.

Raishan persists as a cereal in Assam probably because it provides excellent and essential feed for livestock during winter when grazing is scarce. It is commonly sown on land from which potatoes or other crops have been harvested. It is planted between early April and late June and harvested in November. In October the culms are tied together into bunches and allowed to mature. Individual plants produce as many as 30 flowering culms with each inflorescence maturing at a slightly different time. Spikelets are collected by hand about a month after the main inflorescence matures (Singh and Arora, 1972). Harvested spikelets are dried in the sun before they are stored, and usually parched over a fire before they are pounded in a mortar to remove the lemma and palea. The cleaned grains are boiled in a mixture with rice. Singh and Arora (1972) reported that raishan yields up to 800 kg ha^{-1}. The species was probably domesticated by hill tribes in Assam and southeast Asia.

The genus *Echinochloa* is widely distributed, and includes some 20 species, several of which are aggressive weeds. The most obnoxious weed is *E. oryzoides* (Ard.) Fritsch. It invades paddy rice, mimics the crop in vegetative morphology, and flowers a few days earlier than the cultivar it accompanies as a weed. Natural seed dispersal before the crop is harvested, and seed dormancy ensure a new population of *Echinochloa* weeds when rice is planted in the same field during the next growing season.

Barnyard millet, *E. crusgalli* (L.) P. Beauv, is a common weed of temperate and warm regions of the Old and New Worlds. The species is cultivated in China, Korea and Japan, where it is commonly known as Japanese millet. Archaeological records indicate that it was grown in Japan during the Yayoi period dating back some 5,000 years (Watanabe, 1970). Cultivated plants are erect, tufted annuals up to 1 m tall. Inflorescences are erect or slightly bent at maturity, with the ascending racemes often incurved at the tip. Spikelets are persistent and typically cuspidate.

The related *Echinochloa colona* (L.) Link was harvested as a wild cereal in predynastic Egypt (Dixon, 1969). Intestinal contents of mummies excavated at Naga ed-Der include, among other plant remains, recognizable grains of *E. colona*. It is a minor wild cereal in Central Africa where the grains are fermented to make beer (Tisserant, 1953). This cereal has as yet not been identified from among plant remains of the numerous farming sites excavated in India (Vishnu-Mittre, 1977). The species is, however, extensively grown in central India where it is commonly known as sawa (de Wet *et al.*, 1983a).

Sawa differs from Japanese millet primarily in being a more tropical grass, and in lacking the beak to the spikelet that characterizes *E. crusgalli*. Both species have $2n = 54$ chromosomes, but hybrids between them are sterile (Yabuno, 1966). Sawa is an indigenous cereal in India. The species is weedy, spontaneously invades cultivated fields, and is often unintentionally harvested with other minor millets. Sawa is grown in India from Kashmir to Sikkim in the north, and to Tamil Nadu in the south. Cultivated kinds are extensively variable (de Wet *et al.*, 1983a). The strongly branched racemes of some cultivars suggest affinities with Japanese millet, but *E. crusgalli* does not occur in India as a cereal.

Kodo millet, *Paspalum scrobiculatum* L., is another indigenous cultivated cereal of India. The species is widely distributed in damp habitats across the tropics and subtropics of the Old World. It is known to have been grown in southern Rajasthan and Maharashtra for at least 3,000 years (Kajale, 1977). It is grown today from Uttar Pradesh to Bangladesh in the north, and Kerala and Tamil Nadu in the south. This cereal is known as kodo in Hindi and varagu in Tamil. A small-seeded and large-seeded kind are recognized by farmers in Tamil Nadu.

Raceme morphology allows for the recognition of three cultivated complexes. The most common kodo millets are characterized by racemes with the spikelets arranged in two rows on one side of a flattened rachis, as is also typical of wild *P. scrobiculatum*. Two variations on this spikelet pattern often occur in the same field as the more common phenotype. In the one complex, spikelets are arranged in two to four irregular rows along the rachis. In the other complex, the lower part of each raceme has four irregularly arranged rows of spikelets, while spikelet arrangement becomes more regularly two-rowed in the upper part of the raceme (de Wet *et al.*, 1983b). Hybridization between cultivated kinds, and between weedy and cultivated races is common. This explains the absence of clear racial differentiation, even after some 3,000 years of cultivation as a cereal in India.

Farmers believe that kodo millet is poisonous after a rain. It is known to produce unconsciousness, or delirium with violent tremors of the voluntary muscles. Kodo millet is cooked as rice. Bhide and Aimen (1959) suggested that the glumes, lemmas and paleas contain poisonous alkaloids. It is more likely that the poisoning results from a fungus that often invades, and eventually replaces the developing grain. The spore masses are about the same size as mature grains and are not easy to detect at harvest time. Removing the husks and winnowing scatter the spores, and only healthy grains remain to be used as food. Poisoning only occurs when the grains are damp at threshing, and the spores are not winnowed away.

Adlay, *Coix lacryma-jobi* L., is grown under shifting cultivation as a rainfed crop by the hill tribes of tropical Asia from Assam to the Philippines (Arora,

1977). The grains of wild adlay are enclosed in indurated involucres. Wild. *C. lacryma-jobi* is called Job's tears, and the involucres of wild taxa are used to make rosaries and necklaces. Involucres are papery in most cultivars allowing for the ready removal of the grain. In Assam, the grain is ground into flour, and used to make bread, or a sweet dish is prepared by frying the grain and adding sugar. The whole grain is also eaten raw as a snack, or fermented to produce beer.

Two minor millets, *Setaria pumila* (Poir.) Roem. and Schult., and *Brachiaria ramosa* (L.) Stapf, are indigenous as cereals to the hills of central India. As wild species, however, they are widely distributed in tropical Africa and Asia. Cultivated kinds are distinguished from their close wild relatives only in the absence of efficient natural seed dispersal. Complexes with various degrees of spikelet disarticulation commonly occur in the same field. Both species often occur as encouraged weeds in fields of finger or foxtail millets.

Two Eurasian cereal species are of commercial importance. These are *Setaria italica* (L.) P. Beauv. (foxtail millet), and *Panicum miliaceum* L. (broomcorn millet). Both species are extensively grown across temperate Eurasia, with foxtail millet extending into the tropics and sub-tropics of Asia.

The closest wild relative of foxtail millet is the weedy green foxtail, *Setaria italica* subsp. *viridis* (L.) Thellung. Green foxtail is native to temperate Eurasia, but was introduced and became widely established as a weed in temperate parts of the Americas. Wild and cultivated *S. italica* cross naturally to produce fertile hybrids (Li, Pao and Li, 1942; Li, Li and Pao, 1945; de Wet. Oestry-Stidd and Cubero, 1979). Derivatives of such hybrids are obnoxious weeds in the American corn belt (Pohl, 1966).

The antiquity of foxtail millet cultivation is uncertain. The species could have been domesticated anywhere across its natural range extending from Europe to Japan. It has been grown in China for at least 5,000 years (Ho, 1975). Jars filled with husks of foxtail millet were found at Ban-po in Shanxi province dating from the Yang-shao period (Nai, 1963; Chang, 1973). Foxtail millet also occurs in early agricultural sites from Switzerland and Austria dating back some 3,000 years (Werth, 1937). The species became widespread as a cereal in Europe during the Bronze age (van Zeist, 1970). It is absent from known early farming sites in India (Vishnu-Mittre, 1968). This, however, does not necessarily indicate a late introduction of foxtail millet into the tropics and subtropics of South Asia. Its wide distribution and morphological variation suggest a long history of cultivation in tropical Asia.

Foxtail millet is commonly classified into a European complex (race moharia) and a Far Eastern complex (race maxima). Race moharia includes cultivars with relatively small and erect inflorescences (Kornicke and Werner, 1885), while race maxima is characterized by large and pendulous inflorescences (Dekaprelevich and Kasparian, 1928). Two inflorescence types of race maxima are recognized by Gritzenko (1960). Plants with small, essentially erect, and compact inflorescences occur in northwestern China and Mongolia. Plants from eastern China, Japan and Korea typically have large, compact, and pendulous inflorescences. Cultivars from India are morphologically distinct from those of Europe and the Far East, and are recognized as race indica by Prasada Rao *et al.* (1987). Plants are typically robust, with inflorescences bearing branches that are loosely arranged along the primary axis. Some collections from northern India resemble race maxima and probably represent introductions from China. Collections from Meghalaya have long, slender inflorescences with small spikelets. Two collections of the International Crops Research Institute for the Semi-Arid Tropics (ICRISAT) from Karnataka have slender, erect, inflorescences with short lateral branches, somewhat resembling *S. pumila* in inflorescence and spikelet structure. These collections, however, have one to three bristles below each spikelet, whereas *S. pumila* has four or more bristles supporting each spikelet.

The progenitor of broomcorn millet (*Panicum miliaceum*) is native to Manchuria. The species was probably introduced into Europe as a cereal at least 3,000 years ago. Spikelets and florets of broomcorn millet occur together with remains of foxtail millet in early farming sites of the European neolithic. Neuweiler (1946) dated these sites to around 1600 B.C. During the Bronze age the species rapidly spread across Europe as a cereal (Hjelmquist, 1955).

Cultivated kinds of *P. miliaceum* are commonly subdivided into five subspecies (Lyssov, 1975). These are here recognized as races without taxonomic validity. Race miliaceum resembles wild *P. miliaceum* in inflorescence morphology. It is characterized by large, open inflorescences with suberect branches that are sparingly subdivided. Race patentissimum with its slender and diffused panicle branches is often difficult to distinguish from race miliaceum. These two races occur across the range of broomcorn millet cultivation from eastern Europe to Japan. Highly evolved cultivars of broomcorn millet have more or less compact inflorescences. These are classified into races contractum, compactum and ovatum. Cultivars included in race contractum have compact, drooping inflorescences. Those belonging to race compactum have cylindrical shaped inflorescences that are essentially erect. Cultivars with compact and slightly curved inflorescences that are ovate in shape are included in race ovatum.

Races have no ecogeographic unity. This probably is due to extensive movement of seed of the crop across Eurasia, particularly since early in this century. Lyssov (1975) illustrated 21 inflorescence types that are commonly grown in Eurasia.

A different *Panicum* species (sama) is grown as a cereal in the eastern Ghats of India (Rangaswami Ayyangar and Achyutha Wariar, 1941). This species, *P. sumatrense* Roth. ex Roem. and Schult., represents the domesticated complex of the weedy *P. psilopodium* Trin. (de Wet, Prasada Rao and Brink, 1984). The commonly cultivated kind differs from wild *P. psilopodium* with which it crosses to produce fertile hybrids, primarily in having lost the ability of natural seed dispersal. This race of sama is highly tolerant to heat and drought stress. In more favourable agricultural habitats of the eastern Ghats a robust race of sama is grown. The inflorescences of this race are strongly branched and compact. Sama is often grown as a mixture with foxtail millet, pearl millet or sorghum.

LITERATURE CITED

Arora, R.K. 1977. Adlay (*Coix*) crop in Meghalaya. *Journal of the Indian Botanical Society* **52:** 95-98.

Ascherson, P. and P. Graebner. 1890. *Synopsis der Mitteleuropaischen Flora.* Toma II, Berlin. 445 pp.

Ball, H. 1884. Contributions to the flora of North Patagonia and adjacent territory. *Linnaean Society Journal of Botany* **21:** 203-240.

Becker-Dillingen, J. 1927. *Handbuch des Getreidebaues.* Berlin.

Bor, N.L. 1940. Gramineae. *Flora of Assam* **5:** 205-207.

Bor N.L. 1955. The genus *Digitaria* Heist. in India. *Webbia* **11:** 301-367.

Bhide, N.V. and R.R. Aimen. 1959. Pharmacology of a tranquilizing principle in *Paspalum scrobiculatum* grains. *Nature* **153:** 1735-1736.

Brucher, H. 1979. Das angebliche "Genzentrum Chiloe", 50 Jahre nach Vavilov. *Zeitschrift fur Pflanzenzuchter* **83:** 133-147.

Busson, F. 1965. *Plantes Alimentaires de l'Ouest Africain.* Leconte, Marsailles.

Callen, E.O. 1965. Food habits of some Pre-Columbian Mexican Indians. *Economic Botany* **19:** 335-343.

Callen, E.O. 1967. The first New World cereal. *American Antiquity* **32:** 535-538.

Chevalier, A. 1933. Resources vegetables du Sahara et de ses confins Nord et Sud. *Revue International Botanique Applique d' Agriculture Tropical* **12:** 669-919.

Chang, K. 1973. Radiocarbon dates from China: Some initial interpretations. *Current Anthropology* **14:** 525-528.

Clayton, W.D. 1972. Gramineae. *In:* J. Hutchinson and J.M. Dalziel (eds.). *Flora of Tropical West Africa* **3:** 349-574.

Chomko, S.A. and G.W. Crawford. 1978. Plant husbandry in prehistoric eastern North America: New evidence for its development. *American Antiquity* **43:** 405-408.

Costanza, S.H., J.M.J. de Wet and J.R. Harlan. 1979. Literature review and numerical taxonomy of *Eragrostis tef* (t'ef). *Economic Botany* **33:** 413-424.

Cruz, A.W. 1972. El *Bromus mango,* planta desparecida. Idesia **2:** 127-131.

Dekaprelevich, L.L. and A.S. Kasparian. 1928. A contribution to the study of foxtail millet (*Setaria italica* (L.) P.B. subsp. *maxima* Alef.) cultivated in Georgia (western Transcaucasia). *Bulletin of Applied Botany and Plant Breeding* **19:** 533-572.

de Wet, J.M.J. 1977. Domestication of African cereals. *African Economic History* 3 (spring) 1977: 15-32.

de Wet, J.M.J. and E.A. Oelke. 1979. Domestication of American wild rice (*Zizania acquatica* L., Gramineae). *Journal d'Agriculture Traditional et Botanique Applique* **25:** 67-84.

de Wet, J.M.J., K.E. Prasada Rao and D.E. Brink. 1984. Systematics and domestication of *Panicum sumatrense* (Gramineae). *Journal d'Agriculture Traditional et Botanique Applique* **30:** 159-168.

de Wet, J.M.J., K.E. Prasada Rao, D.E. Brink and M.H. Mengesha. 1984. Systematics and evolution of *Eleusine coracana* (Gramineae.) *American Journal of Botany* **71:** 550-557.

de Wet, J.M.J., K.E. Prasada Rao, M.H. Mengesha and D.E. Brink. 1983a. Domestication of sawa Millet (*Echinochloa colona*). *Economic Botany* **37:** 283-291.

de Wet, J.M.J., K.E. Prasada Rao, M.H. Mengesha and D.E. Brink. 1983b. Diversity in kodo Millet, *Paspalum scrobiculatum*. *Economic Botany* **37:** 159-163.

de Wet, J.M.J., L.L. Oestry-Stidd and J.I. Cubero. 1979. Origins and evolution of foxtail millets (*Setaria italica*). *Journal d'Agriculture Traditional et Botanique Applique* **26:** 53-64.

Dixon, D.M. 1969. A note on cereals in ancient Egypt. *In:* J.P. Ucko and F.W. Dimbleby, (eds.) *The Domestication and Exploitation of Plants and Animals*. Aldin, Chicago. p. 131-142.

Dore, .W.G. 1969. *Wild Price*. Canada Department of Agriculture, Research Branch Publication 1393. 84 pp.

Gay, C. 1865. *Historia Fisica y Politica de Chile*. Agricultura Chilena, 2 volumes, Paris.

Gifford, E.W. 1931. *The Cocopa*. University of California, Publication of American Archaeology and Ethnology **31:** 257-334.

Gould, F.W. 1963. Cytotaxonomy of *Digitaria sanguinalis* and *D. abscendens*. *Brittonia* **15:** 241-244.

Gritzenko, R.J. 1960. Chumiza (Italian millet) taxonomy (*Setaria italica* (L.) P.B. subsp. *maxima* Alef.). *Bulletin of Applied Botany Genetics and Plant Breeding* **32:** 145-182.

Henrard, J. Th. 1950. Monograph of the genus *Digitaria*. University of Leiden Press, Leiden. 899 pp.

Hilu, K.W. and J.M.J. de Wet. 1976a. Domestication of *Eleusine coracana* (L.) Gaertner. *Economic Botany* **30:** 199-208.

Hilu, K.W. and J.M.J. de Wet. 1976b. Racial evolution in *Eleusine coracana* subsp. *coracana* (finger millet). *American Journal of Botany* **63:** 1311-1318.

Hjelmquist, H. 1955. *Die altestes Geschichte der Kulturpflanzen in Schweden*. Almquist and Wiksell, Stockholm.

Ho, P. 1975. *The Cradle of the East*. Chicago University Press, Chicago, IL.

Hooker, J.D. and O. Stapf. 1896. Gramineae. *In:* J.D. Hooker (ed.). *Flora of British India*. Volume 7. Reeve and Co., London.

Jardin, C. 1967. *List of Foods Used in Africa*. Food and Agricultural Organization of the United Nations Publication, Rome.

Kajale, M.P. 1977. Ancient grains from excavations at Nevassa, Maharashtra. *Geophytology* **7:** 98-106.

Kornicke, F. and H. Werner. 1885. *Handbuch des Getreidebaues*. Volumes I and II. Verlag Paul Parey, Paris.

Li, H., W.K. Pao and H.W. Li. 1942. Interspecific crosses in *Setaria*. *Journal of Heredity* **33:** 351-355.

Li, H., C.H. Li and W.K. Pao. 1945. Cytological and genetical studies of the interspecific cross of the cultivated foxtail millet, *Setaria* italiea (L.) Beauv., and the green foxtail millet *S. viridis* L. *Journal of the American Society of Agronomy* **37:** 32-54.

Lyssov, B.H. 1975. Proso (*Panicum* L. *In:* A.S. Krotov (ed.). *Flora of Cultivated Plants*. Vol. III. Croat Crops. Kolos. Leningrad.

Mangelsdorph, P.C., R.S. MacNeish and W.C. Gallinat. 1967. Prehistoric wild and cultivated maize. *In:* D.S. Byers (ed.). *The Prehistory of Tehuacan Valley*. University of Texas Press, Austin. pp. 178-200.

Matthiolus, P.A. 1565. *Commentarii in Libros Sex Pedacii Dioscoridis*. Anazarbei de Medica Materia. Venice.

Molina, G.J. 1782. *Saggio Sulla Storia Naturale del Chili*. Bologna, Italia.

Nabhan, G. and J.M.J. de Wet. 1984. *Panicum sonorum* in the Sonoran desert agriculture. *Economic Botany* **38:** 65-82.

Nai, H. 1963. Archaeology in New China. *Antiquity* **37:** 176-185.

Neuweiler, E. 1946. Nachtrage urgeschichtlicher Pflanzen. Vierteljahrschriften der Nutureforschung Geselshaft, Zurich **91:** 122-236.

Palmer, E. 1871. Food products of the North American Indians. United States Commerce and Agriculture Reports **1870:** 404-428.

Porteres, R. 1951. Une cereale mineure cultivee dans l'Ouest-Africain (*Brachiaria deflexa* C.E. Hubbard, var. *sativa* nov. var.). *L'Agronomique Tropicale* **6:** 39-42.

Porteres, R. 1955. Les cereales mineures du genre *Digitaria* en Afrique et Europe. *Journal d'Agriculture Tropicale et Botanique* **2:** 349-675.

Porteres, R. 1976. African cereals: Eleusine, Fonio, Black Fonio, Teff, *Brachiaria, Paspalum, Pennisetum* and African Rice. *In:* J.R. Harlan, J.M.J. de Wet and A.B.L. Stemler (eds.). *Origins of African Plant Domestication*. Mouton Press, The Hague. pp. 409-452.

Phillips, S.M. 1972. A survey of the genus *Eleusine* in Africa. *Kew Bulletin* **27:** 251-270.

Pohl, R.W. 1966. The grasses of Iowa. *Iowa State Journal of Science* **40:** 341-373.

Prasada Rao, K.E., J.M.J. de Wet, D.E. Brink and M.H. Mengesha. 1987. Intraspecific variation and systematics of cultivated *Setaria italica* (Gramineae)..*Economic Botany* **41:** 108-116.

Rangaswami Ayyangar, G.N. and U.B. Achyutha Wariar 1941. Samas the little millet (*Panicum miliare* Lamk.). *Madras Agricultural Journal* **29:** 451-470.

Singh, H.B. and R.K. Arora. 1972. Raishan (*Digitaria* sp.) a minor millet of the Khasi Hills, India. *Economic Botany* **26:** 376-380.

Tisserant, R.P.Ch. 1953. L'agriculture dans les savanes del'Oubanqui. *Bulletin D'Instute Etude Central Afrique* **5:** 209-274.

van Zeist, W. 1970. Prehistoric and early historic food plants in the Netherlands. *Paleohistoria* **14:** 41-173.

Veldkamp, J.F. 1973. A revision of *Digitaria* Haller (Gramineae) in Malesia. *Blumea* **21:** 1-80.

Vishnu-Mittre. 1968. Prehistoric records of agriculture in India. Transactions of the Bose Research Institute **31:** 87-106.

Vishnu-Mittre. 1977. Changing economy in ancient India. *In:* C.A. Reed (ed.). *Origins of Agriculture*. Mouton Press. The Hague. pp. 569-588.

Watanabe, N. 1970. A spodographic analysis of millet from prehistoric Japan. Faculty of Science, University of Tokyo **5:** 357-379.

Werth, E. 1937. Zur Geographie und Geschichte der Hirsen. *Angewandte Botanie* **19:** 42-88.

Yabuno, T. 1966. Biosystematic study of the genus *Echinochloa* (Gramineae). *Japanese Journal of Botany* **19:** 277-323.

II

IMPORTANCE, GERMPLASM AND VARIETAL IMPROVEMENT IN ASIA

3

SMALL MILLETS IN INDIAN AGRICULTURE

T.V. Sampath, S.M. Razvi, D.N. Singh and K.V. Bondale

INTRODUCTION

The term 'small millets' refers to a group of small-seeded cereal crops. The important small millets grown in India are finger millet (*Eleusine coracana*), foxtail millet (*Setaria italica*), kodo millet (*Paspalum scrobiculatum*), common or proso millet (*Panicum miliaceum*), little millet (*Panicum sumatrense*) and barnyard millet (*Echinochloa colona*).

Like grain sorghum and pearl millet, these crops not only form staple food for the farming community but also provide substantial quantities of palatable fodder for cattle. Small millet grains are nutritionally rich. Dietary surveys carried out by the National Institute of Nutrition, Hyderabad indicate that these grains are particularly low in phytic acid and rich in iron and calcium.

Small millets have a wide adaptation. They can withstand a certain degree of soil acidity and alkalinity, stress due to moisture and temperature, and variations in soils from heavy to sandy infertile soils. Small millets are grown from the extreme southern tip of India at sea level to the temperate north Himalayan areas up to an altitude of 3000 metres with consequent variation in photoperiod from short to long days.

Out of the total area of 126.67 million hectares in 1984-85 under foodgrain, the small millets area in India was just 5.78 million hectares or 4.56 per cent. In production, their contribution was 3.85 million tonnes or 2.63 per cent of the total of 146.22 million tonnes of foodgrain in the country. However, the contribution of small millets to total coarse cereal is 14.76 per cent in area and 12.36 per cent in production.

TRENDS IN AREA, PRODUCTION AND PRODUCTIVITY

The area under small millets has fluctuated between 8.00 and 5.76 million hectares during the period from 1949-50 to 1984-85. During the same period, the production has fluctuated between 2.88 and 5.35 million tonnes. Yields have varied from 397 to 727 kg/ha. The five-year moving averages for area, production and productivity of small millets from 1951 to 1985 are given in Table 1.

TABLE 1

Moving five year averages for area, production and yield of all small millets

Period	Area (million ha)	Production (million tonnes)	Productivity (kg/ha)
1951-55	7.56	3.98	526
1956-60	7.48	3.83	512
1961-65	7.23	3.79	524
1966-70	7.01	3.45	492
1971-75	6.98	3.82	546
1976-80	6.99	4.37	635
1981-84	6.16	4.08	662

Source: Ministry of Agriculture, Government of India.

The crop-wise data on area, production and yield for individual small millets are not available, except for finger millet. Therefore, the statistical data are given separately for finger millet and other small millets.

FINGER MILLET

The area under finger millet has fluctuated from 2.04 to 3.2 million hectares in different years during 1950-85 and the production has fluctuated from 1.31 to 3.20 million tonnes. The increase in production is mainly due to the raise in productivity from 704 kg/ha during 1950-55 to 1056 kg/ha during 1981-84 (Table 2). It is interesting to note that during the last decade, from 1975 to

TABLE 2

Moving five year averages for area, production and yield of finger millet

Period	Area ('000 ha)	Production ('000 t)	Yield (kg/ha)
1951-55	2274	1605	704
1956-60	2454	1873	764
1961-65	2555	1888	743
1966-70	2282	1721	754
1971-75	2523	2343	927
1976-80	2590	2650	1021
1981-84	2492	2636	1056

Source: Ministry of Agriculture, Government of India.

1985, there was negligible growth rate in area under coarse cereals. The annual growth rate in area under finger millet was 0.23 per cent as against 0.71 per cent under total foodgrains. Similarly, the annual growth rate for production was 1.78 per cent for finger millet as against 1.46 per cent for other coarse cereals and 2.62 per cent for total foodgrains. The annual rate of improvement in yield was 1.55 per cent for finger millet, 1.32 per cent for coarse cereals and 1.67 per cent for total foodgrains.

In India, finger millet is cultivated in many States—Karnataka, Tamil Nadu, Andhra Pradesh, Orissa, Maharashtra, Uttar Pradesh, Bihar and Gujarat. These eight States, together account for more than 95 per cent of the total area under this crop. Karnataka has the largest area of more than 40 per cent followed by Orissa, Tamil Nadu and Andhra Pradesh with 12.46, 10.3 and 9.2 per cent respectively. The above mentioned eight States account for more than 98.13 per cent of the total finger millet production in the country. Among them, Karnataka and Tamil Nadu are the major contributors accounting for 56.17 per cent of the total production. The relative contribution of Orissa, Maharashtra, Uttar Pradesh, Bihar and Gujarat to the total production is less than the area occupied by these States. The yields in Tamil Nadu, Maharashtra, Karnataka are higher than the national average. Tamil Nadu has the highest yield followed by Karnataka and Maharashtra. The area, production and yield of finger millet in each state during the years 1980-84 are given in Table 3.

Other small millets

The other small millets—foxtail, kodo, common, little and barnyard are mainly grown in Madhya Pradesh, Andhra Pradesh, Tamil Nadu, Karnataka, Gujarat, Uttar Pradesh, Maharashtra and Orissa. They are also grown in small patches in the hilly regions of Himachal Pradesh, Uttar Pradesh, Jammu & Kashmir and the northeastern States. These millets are grown during Kharif (rainy season) and sown with the onset of the southwest monsoon. The area, production and yield of other small millets during the period from 1975-85 are given in Table 4. During the last 35 years, the area, production and yield of small millets have registered negative growth rates as compared to other coarse cereals and total foodgrain. As a result, the area under these crops is declining. The average area during the five years 1951-56 was 5.29 million ha which gradually declined to 3.66 million ha during 1980-85. This works out to 27 per cent fall in area during this period. Similarly, the average production during this period, came down from 2.18 million tonnes to 1.49 million tonnes. The yield of these crops has remained more or less stagnant.

The major area under small millets is in the States of Andhra Pradesh, Bihar, Gujarat, Karnataka, Madhya Pradesh, Maharashtra, Orissa, Tamil Nadu and Uttar Pradesh. These states account for 96.28 per cent of the area and 93.38 per cent of the production. The largest area is in Madhya Pradesh, which accounts for 42 per cent of the total area in the country, but contributes only

TABLE 3
State-wise area, production and yield of finger millet during 1980-84

	State		1980-81	1981-82	1982-83	1983-84	1984-85	% age of total (1984-85)
1.	Andhra Pradesh	A	254.1	258.9	242.0	255.3	220.8	9.20
		P	245.3	286.9	233.4	262.6	209.4	8.27
		Y	965	1108	964	1029	948	
2.	Bihar	A	177.4	168.8	138.2	150.9	131.0	5.50
		P	114.2	93.6	74.9	104.4	103.0	4.00
		Y	644	555	542	692	786	
3.	Gujarat	A	50.0	47.4	47.8	45.0	44.6	1.80
		P	35.6	48.7	43.2	49.1	44.2	1.75
		Y	712	1027	904	1091	991	
4.	Himachal Pradesh	A	10.6	9.8	8.6	8.2	7.7	0.30
		P	7.8	5.9	6.1	7.1	6.7	0.26
		Y	736	602	790	866	870	
5.	Karnataka	A	1062.9	1148.3	1030.3	1124.6	988.3	41.40
		P	1094.1	1427.8	944.5	1434.1	1112.6	43.93
		Y	1029	1243	917	1275	1126	
6.	Madhya Pradesh	A	20.4	20.2	20.6	20.5	19.6	0.80
		P	5.5	5.2	5.5	5.9	4.9	0.20
		Y	270	257	267	288	250	
7.	Maharashtra	A	221.4	225.2	225.8	228.1	227.4	9.50
		P	209.6	228.0	215.0	240.9	261.9	10.34
		Y	947	1012	952	1056	1152	
8.	Orissa	A	336.4	288.9	298.5	297.5	297.5	12.46
		P	264.8	238.1	244.0	270.9	270.9	10.70
		Y	787	824	817	911	911	

9	Tamil Nadu	A	192.8	245.5	206.1	232.5	247.1	10.30
		P	251.4	448.5	273.0	250.9	310.0	12.24
		Y	1304	1827	1325	1079	1255	
10.	Uttar Pradesh	A	170.5	163.4	162.9	165.4	170.4	7.15
		P	164.2	145.5	155.6	177.0	174.9	6.90
		Y	963	890	955	1070	1026	
11.	West Bengal	A	16.2	16.8	14.3	13.6	14.6	0.60
		P	9.8	9.7	8.3	9.0	10.1	0.40
		Y	605	577	580	662	691	
	All India	A	2525.0	2610.4	2411.7	2558.2	2387.5	
		P	2419.9	2960.4	2223.1	2831.0	2532.2	
		Y	958	1134	922	1107	1060	

A = Area in 000 ha; P = Production in 000 tonnes; Y = Yield in kg/ha

Source: Ministry of Agriculture, Government of India.

TABLE 4
Moving five year averages for area, production and yield of other small millets

Period	Area ('000 ha)	Production ('000 tonnes)	Yield (kg/ha)
1951-55	5290	2177	410
1956-60	5022	1955	389
1961-65	4677	1889	404
1966-70	4729	1733	366
1971-75	4565	1781	390
1976-80	4328	1743	402
1981-84	3574	1465	408

Source: Ministry of Agriculture, Government of India.

25 per cent to the total production. This is because of low yield of 233 kg/ha as against the national average of 389 kg/ha.

The area, production and yield of small millets in each state during the years 1980-84 are given in Table 5.

FACTORS LIMITING PRODUCTIVITY

Production of small millets is subject to wide fluctuations, and the area is declining, except in the case of finger millet. The major constraints limiting small millets production are:

1) These crops are often grown in uneven marginal lands, poor in fertility, shallow and gravelly, with low moisture retention capacity.

2) These crops are grown under rainfed conditions in low rainfall arid regions.

3) Improved crop management practices are not adopted by the farmers due to socio-economic constraints.

4) Research on crop improvement and agro-techniques was neglected till recently.

5) There is no organized programme for production and supply of seeds of improved varieties.

6) There is no ready market for the disposal of surplus produce at a remunerative price.

7) There is lack of extension and development support.

NEED FOR INCREASING PRODUCTION

These constraints create imbalance in the food economy of poor and tribal farmers as small millets are predominantly grown in marginal and submarginal dry lands. The fluctuations in production not only bring hardship to people and animals, but also create instability in the total coarse cereal production.

TABLE 5

State-wise area, production and yield of small millets during 1980-84

	State		1980-81	1981-82	1982-83	1983-84	1984-85	% age of total (1984-85)
1.	Andhra Pradesh	A	551.1	603.5	474.0	473.9	410.6	12.10
		P	170.7	371.0	145.9	239.9	126.2	9.56
		Y	310	615	308	506	307	
2.	Assam	A	8.4	8.1	8.1	9.7	12.0	0.35
		P	4.1	4.1	4.0	5.0	6.2	0.46
		Y	488	506	494	515	517	
3.	Bihar	A	194.4	136.8	112.9	119.2	107.6	3.17
		P	52.8	54.5	36.8	58.6	50.4	3.82
		Y	353	398	326	492	468	
4.	Gujarat	A	135.3	140.5	127.6	116.3	105.6	3.11
		P	106.7	109.4	54.3	79.9	65.1	4.93
		Y	789	779	426	687	616	
5.	Himachal Pradesh	A	22.8	22.0	22.3	21.1	20.6	0.61
		P	15.0	9.1	7.4	12.1	10.3	0.78
		Y	658	414	332	573	500	
6.	Jammu & Kashmir	A	18.0	15.9	16.0	16.4	14.8	0.44
		P	11.7	9.6	8.4	10.5	8.9	0.67
		Y	650	604	525	640	601	
7.	Karnataka	A	364.3	377.7	321.2	377.1	325.0	9.58
		P	140.4	149.2	113.8	149.5	100.4	7.61
		Y	385	395	354	396	309	
8.	Kerala	A	2.4	2.7	2.4	2.7	2.6	0.08
		P	1.6	1.7	1.5	1.8	1.8	0.14
		Y	667	630	625	667	692	

Table 5 (contd.)

			1980-81	1981-82	1982-83	1983-84	1984-85	% age of total (1984-85)
9	Madhya Pradesh	A	1448.5	1453.8	1438.6	1446.5	1426.6	42.05
		P	314.7	338.4	287.1	394.3	330.1	25.02
		Y	217	233	200	273	231	
10.	Maharashtra	A	199.3	194.4	172.4	200.9	158.9	4.68
		P	78.7	78.5	74.9	110.7	75.6	4.36
		Y	395	404	434	551	476	
11.	Meghalaya	A	2.7	2.8	2.8	2.9	2.5	0.07
		P	2.7	2.7	2.7	2.6	2.4	0.18
		Y	1000	964	964	897	960	
12.	Orissa	A	362.3	154.2	162.3	178.2	178.2	5.25
		P	197.5	71.4	74.0	178.0	178.0	13.49
		Y	5454	463	456	999	999	
13.	Rajasthan	A	46.9	48.6	43.6	45.2	37.8	1.11
		P	3.8	7.7	5.2	22.2	15.1	1.14
		Y	81	158	191	491	399	
14.	Tamil Nadu	A	272.6	267.7	276.2	301.7	271.4	8.00
		P	213.9	206.3	201.3	193.1	145.1	11.00
		Y	773	771	729	640	535	
15.	Uttar Pradesh	A	360.2	328.5	291.7	296.0	283.1	8.34
		P	237.3	201.1	188.3	197.6	179.3	13.59
		Y	659	613	646	668	663	
16.	West Bengal	A	10.3	10.1	8.6	7.7	13.3	0.39
		P	7.1	7.3	5.7	4.0	7.8	0.59
		Y	689	723	663	519	586	
17.	Arunachal Pradesh	A	16.9	18.0	18.0	20.8	21.7	0.64
		P	14.8	15.8	17.3	14.8	16.2	1.22
		Y	878	878	961	712	747	

18.	Dadra & Nagar Haveli	A	0.7	0.6	0.6	0.7	0.6	0.02
		P	0.5	0.5	0.4	0.5	0.4	0.03
		Y	714	833	667	714	667	
19.	Pondicherry	A	0.2	0.2	0.1	0.4	0.1	
		P	0.1	0.1	0.1	0.3	0.1	
		Y	500	500	1000	750	1000	
	All India	A	3976.3	3786.1	3499.4	3637.4	3393.0	
		P	1574.1	1638.4	1229.1	1675.4	1319.4	
		Y	396	433	351	461	389	

A = Area in 000 ha, P = Production in 000 tonnes; Y = Yield in kg/ha

Source: Ministry of Agriculture, Government of India.

Realising the role of these crops to Indian agriculture, during the Seventh Five Year Plan, it is proposed to increase the production of finger millet and other small millets from 3.85 million tonnes in 1984-85 to 5.0 million tonnes by 1989-90.

Strategy

The main thrust or strategy for increasing the production of small millets during the Seventh Five Year Plan would be through stabilizing the productivity at a higher level. Crop-wise production plans have been prepared indicating the requirement of various inputs like seed, fertilizer, pesticides and credit. The broad outlines are as follows:

1) Expansion of area under improved varieties through their popularization.

2) Production and supply of seeds of improved varieties in sufficient quantities to the farmers.

3) Conservation of soil moisture and its utilization by the adoption of dry farming techniques.

4) Adoption of integrated watershed concept to enhance production and productivity.

5) Adoption of recommended package of practices with special reference to non-monetary inputs like optimum plant population, timely weed, pest and disease control through cultural management.

6) Popularization of low-cost technology through full extension support.

7) Remunerative price and marketing facilities.

8) Intensification of location specific research for development of varieties and low cost technologies for stabilizing production at higher levels.

Developmental efforts

New production technologies developed by scientists are being popularized, through the Central Sector Schemes consisting of Minikit Demonstrations and State Level Training.

Minikit demonstrations aim at quickly popularising the newly released and pre-release varieties among the farming community all over the country and to get farmers' reaction towards such new plant materials before they are taken up for large-scale cultivation. The millets minikit demonstrations were started in 1974-75. Under this programme, kits containing small quantities of seeds of improved varieties are supplied free of cost for demonstration purposes to farmers.

The scheme has been widely welcomed, particularly by the small farmers and has resulted in the quick acceptance and spread of some of the latest varieties.

In recent years, the minikit programme has been further enlarged and the number of minikits for finger millet has been increased from 2,700 in 1980-81 to 38,000 in 1985-86. The target for 1986-87 is to distribute 37,200 minikits.

Similarly, in other small millets, the number of minikits distributed has increased from 386 in 1980-81 to 4,708 in 1985-86. For 1986-87, it is planned to distribute 31,430 minikits.

State level training programmes are being regularly organized in recent years to disseminate the latest technology developed by the State Agricultural Universities, and by the Indian Council of Agricultural Research to the field staff and to provide feedback information to the research scientists from the practical field workers. State Level Training Programmes on production technology of finger millet and small millets are organized every year in all states, in collaboration with the Agricultural Universities, Central Research Institutions, other Research Organizations and the State Departments of Agriculture, for the benefit of the extension officers.

CONCLUSIONS

The area under small millets has come down from 5.68 million ha to 3.4 million ha. Further reduction in the area of these crops may not be possible as there are no alternate crops for substitution in these areas of very low rainfall and poor soils. Recently, new high yielding varieties have been developed in various small millets but they have not yet reached the farmers. Intensive efforts should be made to popularize the same with the farmers and to replace the low yielding local varieties. For this purpose, systematic follow-up action is required for the production of seeds at various stages, its processing and distribution. The improved seed either should be supplied free or subsidized by the Government.

The role of non-monetary inputs such as line sowing, optimum row spacing, depth of seeding, optimum plant population per unit area, timely cultural practices for higher productivity should be explained and demonstrated to the farmers right in the field.

Agronomic research should bring out efficient low-cost technology which is within the means of farmers and easy to adopt. Increased use of small millets in various ready-to-eat food products should be encouraged as it enhances their value and market price.

4

GENETIC RESOURCES OF SMALL MILLETS IN INDIA

A. Seetharam

INTRODUCTION

In many parts of India six small millet crops are cultivated for grain and fodder. They are finger millet or ragi (*Eleusine coracana*), Italian or foxtail millet (*Setaria italica*), common or proso millet (*Panicum miliaceum*), kodo millet (*Paspalum scrobiculatum*), little millet (*Panicum miliare*) (renamed *P. sumatrense*) and barnyard millet (*Echinochloa frumentacea*) (renamed *E. colona*). These crops occupy 4.5 per cent of the cultivated area and are mostly confined to semi-arid zones and hilly areas. On an average, around 6 million ha are planted under these crops every year, of which finger millet alone occupies around 2.5 million ha, followed by kodo millet, foxtail millet, little millet, proso millet and barnyard millet, in that order. One or more of these crops are grown in each state under diverse agroclimatic conditions (Hegde and Seetharam, 1985).

Systematic improvement in small millets has not been attempted until very recently in India. Obviously, the varieties cultivated are local land races, which are the result of indirect human and natural selection over a long period of time (Seetharam, 1983a). The success of any crop improvement programme depends on the availability of diverse germplasm.

As the small millets are indispensable to Indian agriculture there is increasing realization of the need to improve the productivity of these crops through modern methods of breeding. As the germplasm is the basic raw material, one has to bank upon a broad genetic base now and in the future. Therefore collection of germplasm is necessary, which then needs to be conserved, evaluated and distributed. There is overwhelming evidence in the literature to indicate that genetic diversity is rapidly eroding in many areas of the world. This is especially relevant with small millets, where very little of the available vast natural variations have been explored, studied and utilized. Further, there is an immediate need to conserve the genetic resources of small millets, since areas under these crops are gradually depleting in many states. Realizing the gravity of the situation. The All India Coordinated Small Millets Improvement Project under the aegis of the Indian Council of Agricultural Research has established in Bangalore, a germplasm unit, for the conservation of small millets genetic resources in the country. The Unit has been making efforts since 1980 to pool all the available germplasm in the country and conserve them (Seetharam, 1982). The triple functions of the unit are to:

1) collect the genetic resources of small millets and conserve them,

2) function as a service unit and supply genetic stocks to scientists as and when required, and

3) evaluate, characterize and document the material.

GERMPLASM AVAILABLE

Presently 9,443 collections are maintained. This includes 4,490 in finger millet, 1,951 in foxtail millet, 644 in little millet, 577 in proso millet, 816 in barnyard millet and 965 in kodo millet. Table 1 lists the material assembled by place of origin. In addition to the above 50 collections of brown top millet (*Brachiaria ramosa*) are also maintained.

The collections received initially from the Rockefeller Foundation during the early 1970's formed the base genetic stocks. This was further augmented by pooling stocks that were available at various research centres in the country..Collections made by NBPGR, New Delhi, in their regular collection missions within the country are sent to Bangalore periodically and this has broadened the germplasm base. It is possible that the collections include many duplicates. However, it is rather difficult to eliminate all duplicates. It is also likely that the accessions that appear similar phenotypically might differ in micromorphological and polygenic characters. Hence, elimination of all duplicates by visual observation alone is not advisable.

The germplasm represents reasonably diverse material, collected from various regions within and outside the country. Indigenous finger millet accessions are quite comprehensive and large except for a few deficiencies. For example, representations from Gujarat, Himachal Pradesh, Maharashtra, Madhya Pradesh and northeastern regions of India are inadequate. In other small millets, the collections are much less comprehensive. To fill these gaps priority areas for exploration within the country have been identified. The germplasm Unit at Bangalore, even with its limited resources has been making efforts for collection within the country. At least one collection expedition is undertaken every year and in the last four years Andhra Pradesh, Karnataka, Tamil Nadu, Orissa and parts of Bihar and Uttar Pradesh have been explored (Aradhya *et al.*, 1983 and Patnaik *et al.*, 1983)

TABLE 1
Germplasm collections at small millets germplasm unit, Bangalore 1986

Country/State	Finger millet	Foxtail millet	Little millet	Proso millet	Barnyard millet	Kodo millet
I. *Indian collections*						
Andhra Pradesh	191	241	38	135	2	99
Bihar	140	68	59	28	194	39
Gujarat	5	11	1	—	—	14
Himachal Pradesh	3	14	1	20	—	—
Jammu & Kashmir	3	16	—	4	—	—
Karnataka	425	173	71	5	—	30
Kerala	16	—	—	—	—	—
NEHR regions	65	14	—	—	—	—
Orissa	75	19	83	—	4	4
Punjab	2	12	—	—	—	—
Rajasthan	1	3	—	—	—	—
Tamil Nadu	392	96	35	1	1	109
Uttar Pradesh	856	863	15	6	411	147
West Bengal	12	27	8	—	—	7
Indian—statewise source not known	4	7	—	—	—	—
II. *Exotic collections*						
Afghanistan	—	2	—	13	—	—
Bangladesh	—	9	—	1	—	—
Belgium	—	1	—	—	—	—
China	—	49	—	—	—	—
Ethiopia	—	1	—	—	—	—
France	—	1	—	—	—	—
Indonesia	—	1	—	—	—	—
Japan	—	16	—	12	6	—
Kenya	228	—	—	—	—	—
Malawi	183	—	—	—	—	—
Pakistan	—	2	—	—	—	—
Romania	—	—	—	1	—	—
Sri Lanka	12	—	—	—	—	—
Uganda	14	—	—	—	—	—
USA	—	48	—	—	—	10
USSR	—	5	—	—	—	—
Turkey	—	5	—	—	—	—
Taiwan	—	2	—	—	—	—
Other African	7	3	—	—	—	—
III. *Source not known*	1162	42	17	376	28	18
	4490	1951	644	577	816	965

STORAGE AND SUPPLY

The small millet grains have a relatively, long shelf-life. Even under ambient conditions the seeds stored in plastic containers, remain viable up to three years or even more. In Bangalore, the collections are stored in short-term store rooms with temperature maintained at 15° ± 2°C and relative humidity of 40 ± 5 per cent.

During the period 1981-86, a total of 8,654 accessions were supplied to scientists all over the world on request (Table 2). The operational flow chart in Fig. 1 depicts the way accessions are handled and maintained at Bangalore.

TABLE 2
Supply of small millets germplasm within and outside India, 1981-1986

Crop	1981	1982	1983	1984	1985	1986	Total
Finger millet	615	672	666	568	863	783	4167
Foxtail millet	279	143	48	363	123	311	1267
Kodo millet	184	56	38	135	85	100	598
Little millet	177	3	39	286	351	29	885
Proso millet	177	75	49	365	106	32	804
Barnyard millet	175	13	16	160	104	75	543
Total	1607	962	856	1877	1632	1330	8264

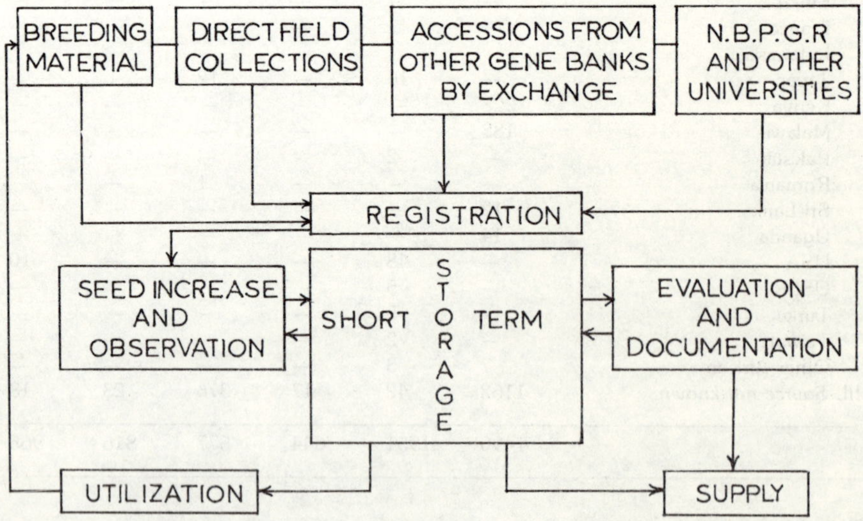

OPERATIONAL FLOW CHART

SMALL MILLETS GERMPLASM UNIT, BANGALORE

SEED INCREASE

At present accessions are grown once in three years for rejuvenation and seed multiplication. Each year about 3,000 accessions are grown. Maintenance of genetic purity has not been a problem in small millets in view of the essentially self-pollinating nature of these crops. It may be possible to rejuvenate accessions once in five years instead of every three years as is done now, once information is obtained on the duration of viability of seeds stored at 15°C with 40 per cent RH.

CHARACTERIZATION OF GERMPLASM

Characterization, classification and cataloguing of the germplasm is slowly gaining momentum. Standard descriptors covering agronomic, morphological, physiological and quality characters have been developed and published for all six millets by the International Bureau of Plant Genetic Resources, FAO, Rome. Using these descriptors, most of the finger millet, and part of the foxtail millet and little millet collections have been preliminarily evaluated and grouped for characters having very high heritability and stability of expression (Tables 3, 7 and 10). However, detailed evaluation of all accessions remains

TABLE 3
Preliminary evaluation and grouping of 4084 accessions of finger millet

I. Ear emergence	Complete	=	3985
	Partial	=	99
II. Nodal tillering	Present	=	4041
	Absent	=	43
III. Synchrony at maturity	Synchronous	=	639
	Non-synchronous	=	3445
IV. Finger branching	Absent	=	3572
	Present	=	512
V. No. of seeds per spikelet	Low (3–5)	=	1753
	Medium (5–6)	=	1881
	High (7)	=	450
VI. Gapiness on ear	Absent	=	4000
	Present	=	84
VII. Ear compactness and shape	Long, dropping	=	53
	open, straight	=	454
	compact tip curved	=	2150
	compact incurved	=	1299
	fist	=	128
VIII. Pigmentation at node	Green	=	2716
	Purple	=	1368
IX. Ear pigmentation	Green	=	2578
	yellow green	=	195
	light purple	=	1234
	dark purple	=	77
X. Glume size	Normal	=	4030
	Long	=	54

to be done for assessing the breeding value of various collections and for identifying sources of useful genes.

A cross-section of 1,941 accessions of ragi germplasm was scored for different forms of blast disease and many stable sources of resistance for blast have been identified (Seetharam, 1983b) (Tables 4 and 5). Finger millet accessions especially from Africa possess genes for blast resistance, robust growth, early vigour, large panicle size, finger number and branching and higher grain density. A few accessions identified already as sources of useful genes are being extensively involved in hybridization programme at many centres. Similarly accessions possessing high protein and desirable physiological attributes, with high carbon dioxide fixation and low leaf area suitable for rainfed conditions have been identified (Sashidhar *et al.*, 1986 and Seetharam *et al.*, 1984). Long glume types with higher test weight will be of special interest for improving seed size (Sashidar *et al.*, 1983) (Table 6).

In foxtail millet, new sources of dwarfing genes controlled by oligogenes have been identified (Byre Gowda *et al.* 1986). The plant type of these accessions are very similar to dwarf wheat or rice and they should form very

TABLE 4

Evaluation of 1941 entries of ragi germplasm for different forms of blast caused by *Pyricularia* sp.

Forms of blast	number	Per cent
A. *Leaf blast*		
resistant	18	0.93
moderately resistant	109	5.62
moderately susceptible	310	15.98
susceptible	671	34.57
highly susceptible	833	42.92
B. *Neck blast*		
up to 1%	281	14.48
1 = 5%	420	21.64
5 = 10%	397	20.46
10 = 20%	431	22.21
more than 20%	412	21.23
C. *Finger blast*		
up to 1%	30	1.55
1 = 5%	372	19.17
5 = 10%	491	25.30
10 = 20%	542	27.93
more than 20%	506	26.07
D. *Neck and finger blast**		
up to 1%	28	1.45
1 = 5%	268	13.81

*Accessions showing less than 5 per cent infection of neck and finger blast only have been considered.

TABLE 5

Identified sources of resistance to all three forms of blast

Accession No.	Leaf blast score (0-9)	Neck blast (%)	Finger blast (%)
GE 281	4	0.00	0.43
GE 568	2	0.00	0.44
GE 669	4	0.00	0.41
GE 705	3	0.00	0.00
GE 1044	4	0.00	0.90
GE 1293	4	0.00	0.00
GE 1409	4	0.00	0.43
GE 1546	4	0.00	0.00
GE 1855	3	0.00	0.00

TABLE 6

Test weight grain density and CO_2 fixation in ears of selected long glume accessions in comparison with normal glume ears

Accession No.	Test weight (mg/100 seeds)	Grain number per cm length of the finger	CO_2 activity fixed	
			Cpm/g	Cpm/organ
1. *Long glumed*				
GE 2970	400	68	16276	27069
GE 3302	445	70	18614	39596
GE 2973	442	71	15098	52958
2. *Normal glumed*				
PES 176	377	70	11131	19394
Indaf 5	386	69	7638	8519

good breeding material. The variability available in foxtail millet for panicle shape, size, arrangement of spikelets, tillering, seed size and colour are very diverse offering great scope for exploitation (Harinarayana and Seetharam, 1981). The variability available for protein content in foxtail millet ranges from 7.16 to 15.73 per cent (Table 8). Similarly seed oil content ranges from 4.0 to 7.1 per cent (Table 9). Thus, identified sources with high protein and high seed oil are available for both direct exploitation and use in breeding (Seetharam *et al.*, 1983).

A cross-section of indigenous little millet germplasm numbering 225 representing different states of India was evaluated, qualitatively and quantitatively grouped (Reddy *et al.*, 1984 and Aradhya *et al.*, 1983).

The preliminary evaluation of other small millets, namely kodo millet, proso millet and barnyard millet is still to be initiated.

TABLE 7
Preliminary evaluation and quantitative grouping of 1,366 accessions of foxtail millet

Character	Group	Frequency
Plant pigmentation (1366)	green	1228
	pigmented	138
Plant height (1366)	tall	688
	medium	627
	short	51
Tillering ability (1366)	heavy	96
	medium	891
	low	379
Leaf size (1366)	broad	331
	medium	981
	narrow	54
Bristles (1366)	long	919
	medium	326
	short	121
Panicle length (1366)	long	28
	medium	1290
	short	48
Ear compactness (1366)	compact	26
	semicompact	1315
	loose	25
Seed colour (1278)	black	64
	yellow	1142
	orange	64
	buff	8
Seed size (1278)	bold	116
	medium	1079
	small	83

DIVERSITY IN DIFFERENT SMALL MILLETS

The germplasm is grown once in every three years for rejuvenation and seed increase. This gives an opportunity to compare the relative diversity existing within and between small millet crops.

Finger millet with a wide range of adaptation for varying temperatures, moistures and soil types exhibited very great diversity. The world collection of about 4,400 accessions includes a wide range of plant types, panicle forms, grain colour, grain quality, disease resistance and maturities. The variability present in African collections is distinctly different from that of the Indian collections supporting both African and Indian origin.

Foxtail millet collections also exhibiting wide variation and extensive diversity are available in 1,900 specimens maintained for ear shape, size, bristling, tillering, pigmentation, grain size, grain colour and maturity duration.

TABLE 8

Range of variation for percent protein and its distribution in 1,309 accessions of foxtail millet

Origin	No. of accessions	Range	Mean	CV (%)
INDIAN				
Andhra Pradesh	28	7.60-13.58	9.98	16
Tamil Nadu	83	8.51-14.06	11.23	14
Karnataka	81	7.97-14.22	11.75	12
Maharashtra	28	9.50-13.63	11.71	8
Gujarat	11	7.27-12.77	10.01	17
Madhya Pradesh	30	7.16-14.98	10.55	21
Bihar	67	9.27-13.31	11.32	8
West Bengal	26	7.22-12.82	9.77	11
Punjab	9	11.21-14.99	12.50	12
Uttar Pradesh (hills)	356	7.24-14.45	11.34	18
Uttar Pradesh (plains)	406	7.47-14.45	10.79	16
Northern hills	37	8.35-12.93	9.84	12
Others	57	7.27-15.73	10.66	16
EXOTIC				
USA	38	7.70-12.82	10.42	14
Others	14	8.24-11.85	10.28	11
Unknown	38	7.75-14.03	11.47	17

TABLE 9

Range of variation for percent oil in seeds and its distribution in 1,309 accessions of foxtail millet

Origin	No. of accessions	Range	Mean	CV (%)
INDIAN				
Andhra Pradesh	28	5.4-7.0	6.33	0.06
Tamil Nadu	83	4.4-5.7	4.93	0.06
Karnataka	81	4.7-6.4	5.64	0.06
Maharashtra	28	4.0-5.7	4.66	0.12
Gujarat	11	5.4-6.1	5.62	0.04
Madhya Pradesh	30	4.8-5.8	5.24	0.05
Bihar	67	4.6-6.5	5.46	0.09
West Bengal	26	5.4-6.8	6.16	0.06
Punjab	9	4.9-5.8	5.48	0.06
Uttar Pradesh (hills)	356	4.3-7.3	5.59	0.10
Uttar Pradesh (plains)	406	4.1-6.9	5.40	0.08
Northern hills	37	5.0-6.2	5.54	0.04
Others	57	4.4-6.7	5.41	0.08
EXOTIC				
USA	38	4.7-6.3	5.54	0.07
Others	14	4.9-5.9	5.34	0.06
Unknown	38	4.4-7.1	5.40	0.11
Overall	1309	4.0-7.3	5.45	0.10

TABLE 10

Preliminary evaluation, qualitative and quantitative grouping of cross-section of 225 accessions of little millet

Characters	Group	Range	Frequency
A. *Quantitative characters*			
Early vigour based on	low	0.10-0.25	205
seedling dry weight (g)	high	0.26-0.50	20
Dry matter on 40th day (g)	low	0.50-1.50	39
	medium	1.51-2.50	156
	high	2.51-3.50	30
productive tillers (no)	low	5.0-15.0	105
	medium	15.1-25.0	114
	high	25.0-above	6
Days to flower	early	30-50	212
	medium	51-70	12
	late	71-90	1
Days to maturity	early	60-70	72
	medium	71-80	138
	late	81-above	15
Plant height (cm)	very dwarf	40.0-55.0	30
	dwarf	55.1-70.0	64
	medium	70.1-85.0	72
	tall	85.1-100.0	50
	very tall	100.1-above	9
Flag leaf length (cm)	short	5.0-15.0	85
	medium	15.1-25.0	128
	long	25.1-above	12
Flag leaf breadth (cm)	narrow	0.10-0.50	21
	medium	0.51-1.00	187
	broad	1.01-1.50	17
Peduncle length (cm)	short	1.00-10.0	2
	medium	20.1-30.0	201
	long	30.1-40.0	22
Main panicle weight (g)	low	0.10-1.50	195
	medium	1.51-3.0	22
	high	3.01-4.5	8
Volume weight (g)	low	13.0-13.85	39
	medium	13.76-14.50	129
	high	14.51-above	57
No. of primary branches/panicle	low	1.0-10.0	151
	medium	10.1-20.0	72
	high	20.1 above	2
No. of secondary branches/	low	1.0-5.0	35
primary branch	medium	5.0-10.0	176
	high	10.1-15.0	14
No. of seeds/secondary branch	very low	1.0-20.0	103
	low	10.1-20.0	98
	medium	20.1-30.0	9
	high	30.1-40.0	10
	very high	40.1-50.0	5

Table 10 (Contd.)

Characters	Group	Range	Frequency
Seed weight/panicle (g)	low	0.10-1.50	199
	medium	1.51-3.0	25
	high	3.01-above	1
Seed yield/plant (g)	low	1.0-5.0	33
	medium	5.1-10.0	158
	high	10.1-above	34
Straw weight/plant (g)	low	1.0-5.0	6
	medium	5.1-10.0	166
	high	10.1-above	33
Harvest Index (%)	low	20.0-30.0	15
	medium	30.1-40.0	143
	high	40.1-50.0	17
panicle length (cm)	short	10.0-20.0	146
	medium	20.1-30.0	72
	long	30.1-40.0	7
B. Qualitative traits			
1. Pigmentation of leaf sheath	green		202
	Purple		23
2. Pigmentation of glume	green		220
	purple		25
3. Pigmentation of stigma	purple		160
	white		65
4. Seed colour	black		43
	light black		42
	grey green		29
	grey brown		38
	yellow		21
	grey yellow		29
	brown		23
5. Plant habit	erect		93
	semi spreading		85
	spreading		47
6. Seed size	small		132
	medium		84
	bold		9

The genetic diversity available in Kodo millet and barnyard millet is moderate whereas in *Panicum* accessions both little millet and proso millet—the diversity is relatively less.

UTILIZATION

The utilization of germplasm in small millets as in most crops, is most important. In the case of small millets, the utilization has been drastically restricted by difficulties in artificial hybridization. Except for ragi and to some extent foxtail millet, hybridization and recombination breeding in small millets has not been attempted in India. Improvement in these crops so far has been through single plant selection, evaluation and release of promising germplasm. During the last five years a number of selected germplasm, have been tested under the All India Coordinated Small Millets Improvement Programme. Large-scale hybridization in ragi involving blast resistant lines has been undertaken in the last five years, and a number of stabilized selections are undergoing large-scale yield tests in different states (Gowda *et al.*, 1986).

Nevertheless, there is still vast scope to utilize the diversity present in these crops through a well planned hybridization programme.

PROPOSED ACTIVITIES

The germplasm unit wishes to strengthen its links with other national and international bodies that are associated with genetic resources to further enrich the collections. There are areas both within and outside the country where large diversity exists and representations from such areas will broaden the existing variability. Some of the priority areas for future explorations within the country are Himachal Pradesh, western Ghats of Maharashtra and Gujarat, northeastern Hill Regions of Assam, Arunachal Pradesh, Meghalaya, Nagaland, Manipur, Tripura, Sikkim and Mizoram besides northern Bihar and western Uttar Pradesh where rich diversity exists for finger millet, foxtail millet, proso millet and barnyard millet. Eastern Madhya Pradesh and Orissa have valuable germplasm of kodo millet and little millet.

African countries—Kenya, Uganda, Tanzania, Malawi, Zimbabwe, Zambia and Zaire, could be potential areas for variability in finger millet. For foxtail millet, China being the centre of origin of this crop could be a potential source. For proso millet USSR accessions may provide the largest diversity.

SUMMARY

The efforts made during the last six years to collect and conserve the small millets genetic resources have resulted in the pooling of more than 9,000 accessions of various small millets. The finger millet and foxtail millet accessions assembled are quite diverse and sizeable. But the diversity collected in other small millets is narrow and not very comprehensive. There are areas both within and outside the country where large diversity might be existing and this needs to be explored before it is lost permanently.

Multilocation evaluation and documentation are equally or more important than conservation, as future utilization of germplasm is depended upon identified sources of useful genes. One of the reasons for lower productivity of small millets has been the lack of directed human selection. So, improvement in yield is the immediate need and easy access to diverse germplasm will be the first step in aiming at a quantum jump in yield.

LITERATURE CITED

Aradhya, K.M., A. Seetharam, D.M. Mahishi and B.T.S. Gowda. 1983. Variability for seed protein in foxtail millet. *MILWAI Newsletter* **2:** 12.

Aradhya, K.M., B.T.S. Gowda, D.M. Mahishi and A. Seetharam. 1983. Minor Millets Collection expedition in Bihar. *MILWAI Newsletter* **2:** 10.

Byre Gowda, M., Y.V. Shetty, S.P. Dinesh Kumar, B.K. Shankare Gowda and A. Seetharam. 1986. Comparison of dwarf and tall genotypes of foxtail millet. *Millets Newsletter* **5:** 35.

Dayakara Reddy, H., A. Seetharam and K.M. Aradhya. 1984. Genetic variability and path analysis in 225 accessions of little millet germplasm. *Indian J. Agric. Sci.* **54:** 365-369.

Gowda, B.T.S., A. Seetharam, S. Viswanath and S. Sannegowda. 1986. Incorporating blast resistance to Indian elite finger millet cultivars from African cv. IE 1012. *Sabrao J.* **18:** 119-120.

Harinarayana, G. and A. Seetharam. 1981. Minor millets genetic resources in India. Paper presented in IBPGR Sorghum and Millets advisory Committee meeting. Dakar, Senegal, October 4-9.

Hegde, B.R. and A. Seetharam. 1985. Ragi and minor millets. In: *Management of Rainfed Crops.* AICRP on Dryland Agriculture, Hyderabad: 79-100.

Patnaik, H.B., D.M. Mahishi and A. Seetharam. 1983 Collection of minor millets in Orissa State. *MILWAI Newsletter* **2:** 10.

Sashidhar, V.R., T.G. Prasad, A. Seetharam and M. Udayakumar. 1983. Ear photosynthesis in long and normal glumed genotypes of finger millet (*Eleusine coracana* Gaertn.). *Indian J. Plant Physiol.* **4:** 359-362.

Sashidhar, V.R., B.R. Gurumurthy, T.G. Prasad, M. Udayakumar, A. Seetharam and K.S.K. Sastry. 1986. Genotypic variation in carbon exchange rate, functional leaf area and productivity in finger millet (*Eleusine coracana*). An approach to identify desirable plant types for higher water use efficiency under rainfed conditions. *Field Crops Res.* **13:** 133-146.

Seetharam, A. 1982. Conservation of minor millets germplasm in India. *MILWAI Newsletter* **1:** 14.

Seetharam, A. 1983a. Minor millets, their importance and some possible approaches for raising their productivity. *Agric. Tech.* **1:** 109-115.

Seetharam, A. 1983b. Identification of sources of resistance to ragi blast and its utilization in breeding. *Proc. of National Seminar on Breeding Crop Plants for Resistance to Pests and Diseases:* 152. Tamil Nadu Agricultural University, Coimbatore, India.

Seetharam, A., K.M. Aradhya and M.R. Laxminarayana. 1983. Variation for seed oil content in a world collection of foxtail millet germplasm. *Sabrao. J.* **15:** 99.

Seetharam, A., K.M. Aradhya, V.R. Sashidhar, D.M. Mahishi and B.T.S. Gowda. 1984. Protein content in white and brown seeded finger millet genotypes. *Sabrao. J.* **16:** 65-67.

5

BREEDING AND VARIETAL IMPROVEMENT OF SMALL MILLETS IN INDIA

G. Harinarayana

INTRODUCTION

Small millets are self-fertilized crops. The degree of selfing varies from near cleistogamy in kodo millet to marginal vicinism in other small millets. The system, however, permits survival of species under natural selection as well as high degree of controlled pollination under directed selection for economic gains. Whatever may be the breeding methodology adopted, it is necessary to ensure a high degree of homozygosity in the end product for realizing maximum production potential. The following paper outlines the objectives, the breeding methodology adopted, varietal gains and the future outlook for small millets improvement in India.

OBJECTIVES

The ultimate goal of breeding small millets remains improvement of grain yield including maximization of biomass and the harvest index.

Most of the small millets particularly little, proso and foxtail millets mature early, and therefore, provide one first harvest for human consumption. Genotypes need to be tailored for maturity-early, mid-late and late, depending on the location specific requirements of soil, rainfall, temperature, humidity, day length and cropping patterns.

Small millets are companions of rainfed farming. Their capability to produce a grain where there is none and storability had earned them an affable adjective of 'famine grains'. Water-use efficiency of these crops deserves breeding efforts.

Being low- or no-input crops, selection needs to be effected for fertilizer use efficiency, particularly nitrogen, both native and applied. Breeding of dwarf varieties is an objective of intensive cultivation.

Small millets have a different spectrum of diseases and pests. Finger millet is more vulnerable to diseases like blast and viruses. Rust frequently occurs in foxtail millet. Smuts are common in barnyard millet. Kodo, little and proso millets are highly vulnerable to shootfly. Borers occur in finger and barnyard millets. Inbuilt resistance to pests and diseases is of vital significance for continued cultivation of small millets.

Being the food of the poorest of the poor segment of the society, quality breeding deserves attention. The evolution of high protein white finger millets *Hamsa* (Mallanna and Rajashekara, 1969) and CO 9 (Hrishi and Ayyamperumal, 1970) deserves mention. Small millets are a rich source of minerals. Processing and food technology also demand varieties with tailor-made characteristics.

Post-harvest technology demands free threshing grain for minimizing human drudgery.

BREEDING METHODS

Intravarietal improvement

Mass selection

Mass selection has been extensively practised in India for the purification of the cultivars (land races) and multiplication of the varieties bred by pure line or pedigree systems. Mass selection has, however, resulted in marginal gain in improving the performance of farmers' finger millet varieties like *Gidda Ragi* and *Hullubele* of Karnataka, *Saluchodi* of Andhra Pradesh, *Udumalpet Ragi* and *Guddapah Rajampet Ragi* of Tamil Nadu, and *Murky* and *Nangkatna* of Sikkim. V 27 and V 306 proso millets adapted to Andhra Pradesh and Koraput Local little millet adapted to Orissa also fall in this category.

Pure line selection

Pure line selection remained the prime breeding method for improving performance, particularly grain yield of small millets. Single plants were selected from land races (germplasm) and 'improved' varieties of farmers', and the progeny was tested. Superior progenies, mostly for earliness, pest and disease resistance, and grain yield were evaluated in multilocation trials, and released as varieties. Pure line selection gained momentum because of the ease with which selection can be practised, because of the variability in the base population, and because of the difficulties in emasculation and hybridization of small millets. The continued success of pure line selection, however, depends on the rate and accumulation of mutations, and the slow release of hidden genetic variability through natural cross pollination, recombination and segregation (Navale *et al.*, 1984) regulated by cytological homoeostasis (Harinarayana and Murty, 1971).

Pure line selection has resulted in the development and release of maximum number of finger millet varieties (41) (Table 1), as well as 13 foxtail millet varieties (Central Seed Committee, 1985). Pure line selection has been the main forte of kodo (13), little (10), proso (12) and barnyard (7) millet breeding in developing and releasing 'new' varieties (Table 2).

Intervarietal improvement

Methods of hybridization

Natural hybridization: Small millets, although self-pollinated, permit a certain degree of cross pollination. Very low degree of natural cross pollination was observed in foxtail millet (Patil, 1952) and little millet (Dwivedi, 1947). This has led to the occurrence of natural hybrids in germplasm. Identification of the natural hybrids and increasing their frequency would considerably enhance the variability in small millets:

1) *Contact method:* The panicles of selected plants are enclosed in a parchment bag before flowering to enhance the chances of natural cross pollination (Ayyangar, 1934). This has resulted in a low frequency (1.5 per cent) of true hybrids in foxtail millet (Mahishi *et al.*, 1982).

2) *Use of marker plant types:* The interplanting of characteristic plant types like foxtail millet dwarfs (Harinarayana *et al.*, 1984) or purple pigmented finger millets or other distinctly distinguishable lines in germplasm and breeding nurseries, collection of seeds and grow out tests reveal natural hybrids.

3) *Protogyny:* Kodo millet is a cleistogamous plant. However, protogyny has been observed in some cultures like IPS 147, IPS 197, IPS 427 etc. Interplanting of protogynous kodo millets in germplasm or in breeding nurseries or in association with chosen lines enhances the prospects of natural hybrids. Controlled pollination could also be effected in protogynous millets. Similar mechanisms that promote cross pollination like protogyny or protandry should be identified in small millets.

4) *Polycrosses:* Planting of selected parents of small millets in greenhouse and forcing pollination through insects at the time of flowering could produce polycross progeny.

5) *Induction and utilization of sterility:* Physical mutagens in low levels induce sterility. Recurrent irradiation and interplanting of selected millets in germplasm and breeding nurseries exposes them to open pollination. Grow out tests would reveal natural hybrids.

6) *Controlled hybridization:* Controlled hybridization aims at sterilization or removal of anthers and controlled pollination to recover hybrids. Besides hand emasculation, hot water was used in emasculating finger millet (Rao and Rao, 1962; Raj *et al.*, 1984) and little millet (Srivastava and Yadav, 1972). Gametocides have also been used to induce male sterility and produce F_1

TABLE 1
Pure line selections of finger millet in India

Duration	Yield (Q/ha)	Before 1950	1950-60	1960-70	1970-80	1980-85
Early	<20	—	—	AKP 4, AKP 5	—	—
	25	—	Aruna	AKP 1, CO 8	AKP 2	Nirmal
	>25	—	—	—	CO 10, Sarada, CO 11	Birsa, Mandua 1
Mid-late	<20	—	VZM 1	AKP 3	VL Mandua 204	PES 176
	25	—	VZM 2	AKP 6	IE 28	—
	>25	—	CO 7	—	—	Simhadri
Late	<20	RO 870, K 1, H 22 ES 11 ES 13	K 2	—	VL Mandua 101	Paiyur 1
	25	CO 2	—	—	Kalyani, Gujarat Nagli	BR 407, PES 110
	>25	CO 1 CO 4	CO 5	Hamsa	Godavari	Nilachal, Ratnagiri, CO 12

TABLE 2
Small millets developed through pure line selection in India

Millet	Duration (days)	Grain Yield (q/ha) Less than 15: Before 1970	Less than 15: 1970-80	Less than 15: After 1980	15-20: Before 1970	15-20: 1970-80	15-20: After 1980	Over 20: Before 1970	Over 20: 1970-80	Over 20: After 1980
Kodo	100	—	—	—	—	JNK 101, JNK 364, K 1	—	—	—	PSC 1
	100-120	—	Niwas 1	CO 3	—	Jawahar Kodo 1	—	—	Jawahar Kodo 1, Gujarat Kodra 1	PSC 10
	120-150	PLR 1, CO 1	CO 2	—	—	—	—	—	—	—
Foxtail	<80	—	CO 4	—	—	—	SIC 3	—	K 2	SIA 326
	80-100	CO 3, Arjuna	—	Chitra	CO 2	—	—	—	—	—
	100-125	CO 1 H 1, H 2	—	—	—	—	—	—	K 221-1, Rs 118	—
Little	<80	—	V15, V17	Dindori 2, PRC 3	—	Dindori 1	—	—	—	—
	80-100	CO 1	Koraput Local, CO 2, K 1	—	—	—	Gujarat Vari 1	—	—	—

Table 2 (contd.)

Millet	Duration (days)	Before 1970	1970-80	After 1980	Before 1970	1970-80	After 1980	Before 1970	1970-80	After 1980
Proso	<70	Ramcheena, Shyam-cheena,	—	K 2	—	K 1	—	—	—	BR 7
	70-80	CO 1, V 27, V 306	—	—	—	—	CO 2, CO 3, CO 4	—	—	—
	80-90	—	—	—	—	Varada	—	—	—	—
Barnyard	<80	—	—	—	—	—	Anurag	—	—	—
	80-90	—	K 1	RAU 3	—	—	—	—	—	Gujarat Banti 1
	90-100	—	—	—	—	V 1 Madira VL 8 CO 1	—	—	—	K 2

hybrids in small millets, particularly finger millet (Reddy *et al.*, 1983). The discovery of male sterility like in foxtail millet (Hu, 1985) is likely to hasten heterosis breeding for commercial exploitation.

The small and delicate spikelets combined with brittleness of the rachis appear to be bottlenecks in hand emasculation. Hot water method requires maintenance of correct temperature and is labour intensive. The contact method depends on the synchronization of female and male florets. The use of gametocides resulted in high degree of infertility. All these methods did not prove a sure success, and moreover resulted in extremely low percentage of hybrids. Considerable need, therefore, exists in developing and standardizing hybridization techniques as in proso millet (Nelson, 1984).

Recombination breeding

Progress through hybridization has been extremely limited, in India in developing small millets except finger millet. This has been chiefly due to difficulties in emasculation and pollination, in identification of true hybrids, limited heterosis in intervarietal crosses (Srivastava and Yadav, 1977), and the availability of unexploited genetic resources.

Pedigree selection: Finger millet occupies pride of place among small millets. The improvement of finger millet has, therefore, received greater attention than other small millets. Until the 1950's, finger millet breeding revolved around improvement of locally adapted material through state located centres. The Project for Implementation of Regional Research on Cotton, Oilseeds and Millets' (PIRRCOM) were established during the late 1950's with some cooperation among state research centres. The institution of All India Coordinated Millets Improvement Project during 1965 provided the necessary fillip for the improvement of millets in India.

Indigenous germplasm of small millets including finger millet was assembled during 1960's. Exotic germplasm, particularly African finger millet germplasm was introduced into India during 1970's. The African germplasm was superior to Indian germplasm in rainy (*kharif*) and post-rainy (*rabi*) seasons, while the performance was comparable during summer (Swaminath, 1979). The African genotypes have also many desirable characters like high initial vigour, large ears, high grain density and thick robust stems with broad dark green leaves (Seetharam, 1982).

Following hybridization among indigenous, exotic and between indigenous and exotic collections, pedigree selection was adopted. Early (less than 100 days), mid-late (100-110 days) and late (more than 110 days) selections were made in Indian × Indian, Indian × exotic or exotic × Indian, and exotic × exotic crosses (Table 3). The best excotic × exotic derivatives (HR 374; CO 6) had the productivity of less than 2000 kg/ha. The best Indian × exotic or vice versa derivatives (Indaf 9; CO 9 and Indaf 5; Indaf 1, Indaf 3, K 5, Indaf 7, Indaf 8, Indaf 6, HR 911 and Indaf 11) had the highest productivity

TABLE 3
Hybrid derivatives of finger millet in India

Hybrid	Duration	1950-60	1960-70	1970-80	1980-85
I × I	Early	Udaya	—	K 7	5-6
	Mid-late	Purna	Annapurna	—	—
	Late	—	Cauvery	Shakti, HPB 7-6	—
E × I	Early	—	—	—	Indaf 9
	Mid-late	—	—	CO 9, Indaf 5	—
	Late	—	—	Indaf 1, Indaf 3, K 5	Indaf 7, Indaf 8, Indaf 6, HR 911, Indaf 11
E × E	Early	—	—	HR 374	—
	Late	CO 6	—	—	—

I × I : Indian × Indian; E × I : Exotic × Indian; E × E : Exotic × Exotic
Early : Less than 100 days; Mid-late : 100-110 days; Late : More than 110 days

of more than 2500 kg/ha. The Indian × Indian derivatives (*Udaya*, K 7 and 5-6; *Purna* and *Annapurna; Cauvery, Shakti* and HPB 7-6) had intermediate productivity.

The Indo-African crosses have provided the real backbone for breaking the grain yield barriers in the improvement of finger millet. They helped in increasing finger millet productivity by more than 50 per cent in Karnataka State (Seetharam, 1982) and by 60 per cent in Tamil Nadu (Nagarajan and Raveendran, 1983).

Compared to finger millet, hybridization efforts were negligible in other small millets. Recognizing the significance of small millets in the hilly and tribal, areas, Indian Council of Agricultural Research established five centres of excellence for the improvement of small millets with the assistance of International Development Research Centre, Canada. Since 1978, these centres at Dindori for kodo millet, Nandyal for foxtail millet, Semiliguda for little millet, Dholi for proso millet and Almora for barnyard millet are devotedly working towards the improvement of small millets. Besides assembling large germplasm collections, varietal improvement through pure line selection and hybridization is receiving priority. Their efforts have already resulted in the release of kodo millets (PSC 1 and PSC 10), foxtail millets (SIA 326 and SIC 3), little millet (PRC 3), proso millet (BR 7) and barnyard millet (RAU 3) varieties through pure line selection. The hybrid derivatives are in advanced stages of testing. One hybrid derivative of foxtail millet, CO 5 (Co 1 × A 113/2) was released in Tamil Nadu (Raveendran *et al.*, 1979).

Mutation breeding

Mutation breeding appears effective and potential in highly self-fertilized small millets to rectify a character, to improve a character, to generate polygenic variation and to induce partial to complete sterility for increasing recombination frequency. Physical, chemical and combination mutagens have been used in small millets.

Rectification of characters

Dwarf, early and bold seeded mutants of small millets are frequently sought after to increase response to nutrient application to fit into cropping systems and to improve the yield of flour. Stabilized dwarf mutants of the finger millet varieties *Annapurna* and *Cauvery* have a better harvest index than the original varieties and outyielded them (Shivshankar and Ranganatha, 1983). Following gamma radiation, early mutants of finger millet, *Hamsa*, with increased finger number and grain-bearing area were obtained (Nayar *et al.*, 1979).

Mutation breeding also led to release of earlier maturing finger millets, *Dibya Sinha* (90 days) originated from *Sarada* (110 days) in Orissa, and CO 3 (110 days) originated from CO 1 (120 days in Tamil Nadu (Table 4). The gamma irradiated mutants 8-2-A-a and 57-A-3 outyielded the standard check B 11 in 1976-82 trials in Maharashtra (Dhonukshe *et al.*, 1983).

Inducing polygenic variation

Raveendran *et al.* (1982) observed increased genetic variance for seven metric traits in M_2 and M_3 generations of finger millet genotypes MS 2698 and *Sarada* following chemical mutagenesis. In addition MS 2698 yielded bold seeded mutants. Gamma irradiation was shown to increase the genetic variance for tiller number, plant height, head length and grain yield in foxtail millet (Rao and Goud, 1973) and in barnyard millet (Mehra *et al.*, 1985).

TABLE 4
Mutant selections of finger millet in India

Duration	Yield (Q/ha)	Before 1950	1970-80
Early	25	—	Dibya Singha
Late	<20	Hagari 1	—
	25	CO 3	—
	>25	—	K 6

Early : Less than 100 days
Late : More than 110 days

Inducing sterility

Fertility (grain set) was reduced by gamma irradiation from 80.2 per cent at 10 kR to 20.5 per cent at 30 kR, whereas EMS had no significant effect on fertility (Sreekantaradhya, 1979).

FUTURE OUTLOOK

Small millets are traditionally, the indispensable components of the dryland farming system. They are ideally adapted to diverse agroclimatic conditions of soil, water, temperature and humidity. They provide the first harvest in the arid and semi-arid tropics. They are adapted to irrigated farming also. Nutritionally superior to wheat and rice, they provide cheap proteins, minerals and essential vitamins to the poorest of the poor.

The area and production of small millets with the exception of finger millet, are coming down. The area under finger millet is constant and fluctuated between 2.25 m ha in 1950-54 and 2.61 m ha in 1975-80. The production and productivity showed continuous increase. The production increased from 1.52 m tonnes in 1950-54 to 2.62 m tonnes in 1980-85, and increase of 72.4 per cent. The productivity crossed the one tonne mark in 1975-80 and is racing towards the one-and-a-half tonne mark, a record for millet crops including sorghum and pearl millet.

The other small millets are showing a reverse trend to finger millet. The area and production are decreasing. The area came down by 39 per cent in 1980-85 (3.65 m ha) over 1950-55 (5.1 m ha). Similarly, the production decreased by 38.7 per cent from 2.09 m tonnes in 1950-55 to 1.49 m tonnes in 1980-85. The productivity showed marginal fluctuations around 400 kg/ha. The reasons are many: The low productivity, poor resources base, lack of input, price and procurement support coupled with no alternate food uses, campaigns for value-added oilseeds and pulses and "urbanization" of food habits are slowly displacing the small millets to more and more marginal, fertilizer-hungry and water-starved abandoned soils.

Millets are, however, a way of life. Millets are grown for food and fodder. They have immense untapped genetic and developmental potential. Some millets have not experienced the sleight of human selection. Finger, kodo and barnyard millets have the capacity to produce more than 2000 kg/ha even under rainfed conditions. They could, therefore, be a stabilizing force in the building up of national foodgrain buffer stocks. Practically devoid of stored grain pests, small millets have a long storage life and keeping quality. Vast potentials do exist for substituting super cereals in industrial and food products. The growing demand for food and a variety of food products also calls for interest and investment in small millets by parliamentarians, plant scientists, people and peasants. Small millets, therefore, hold the key to the future of mankind. They are the potential Food Crops of Tomorrows' World.

LITERATURE CITED

Ayyangar, G.N.R. 1934. Recent work in the genetics of millets in India. *Madras agric. J.* **22:** 16-26.

Central Seed Committee. 1985. Handbook on Cultivars. Department of Agriculture and Cooperation, Ministry of Agriculture and Rural Development, Government of India, New Delhi.

Dhonukshe, B.L., S.P. Birari and N.V. Deorukhakar. 1983. Improvement of finger millet in Konkan. Paper presented at "National Seminar on Finger Millet Genetics and Breeding", UAS, Bangalore. January 12-13, 1983.

Dwivedi, H.P. 1947. Natural cross pollination in *Panicum miliare* Lam. (*Kutki*). *Nagpur agric. Coll. Magazine* **21:** 80-85.

Harinarayana, G. and B.R. Murty. 1971. Cytological regulation of recombination in *Pennisetum* and *Brassica*. *Cytologia,* **36:** 435-48.

Harinarayana, G., P.A. Navale and H.B. Mungse. 1984. Dwarf genotypes of foxtail millet. *J. Maharashtra agric. Univ.* **9**(3): 343-344.

Hrishi, V.K.K. and A. Ayyamperumal. 1970. A high yielding ragi with protein. *Indian Fmg.* **20:** 36-37.

Hu, H.K. 1985. Discovery and utilization of a dominant gene for male sterility in *Setaria italica*. *Hereditas* China **1:** 4.

Mahishi, D.M., K.M. Aradhya, A. Seetharam and B.T.S. Gowda. 1982. Efficacy of contact method for hybridization in foxtail millet. *MILWAI Newsletter* **1:** 16.

Mallanna, K.N. and B.G. Rajashekara. 1969. Hamsa—a protein rich high yielding white ragi. *Mysore J. agric. Sci.* **3:** 1-6.

Mehra, H.S., H.C. Joshi and J. Chikara. 1985. Induced mutations in Japanese millet. *Indian J. agric. Sci.* **55:** 294-295.

Nagarajan, C. and T.S. Raveendran. 1985. Germplasm mobilization and utilization in finger millet in Tamil Nadu. *National Seminar on Finger Millet Genetics and Breeding,* UAS, Bangalore, January 12-13, 1983.

Navale, P.A., G. Harinarayana and H.B. Mungse. 1984. Intrapopulation improvement in a variety of foxtail millet. *J. Maharashtra agric. Univ.* **9:** 354.

Nayar, K.M.D., G. Pushpa and B.G.S. Reddy. 1979. 'Open earhead Hamsa' true breeding early mutants of finger millet. *Curr. Res.* **8:** 212-213.

Nelson, L.A. 1984. Technique for crossing proso-millet. *Crop Sci.* **24:** 205-206.

Patil, J.A. 1952. Natural cross pollination in rala (*Setaria italica* Beauv). *Poona agric. Coll. Magazine* **43:** 151-154.

Raj, S.M., K. Mahudeswaran and A. Shanmuga Sundaram. 1984. Observations on the hot water technique of emasculation of ragi florets (*Eleusine coracana* Gaertn.). *Madras agric. J.* **51:** 71.

Rao, T.M. and J.V. Goud. 1973. Induced polygenic variation in *Setaria*. *Madras agric. J.* **60:** 236-241.

Rao, V.R. and K.V.R. Rao. 1962. Emasculation of ragi (*Eleusine coracana* Gaertn.) flowers by hot water treatment for hybridization work. *Andhra agric. J.* **9:** 290-293.

Raveendran, T.S., K. Meenakshi and R. Appadurai. 1982. Induced polygenic mutation in ragi (*Eleusine coracana* Gaertn.). *Madras agric. J.* **69:** 637-643.

Raveendran, T.S., C. Nagarajan, U.S. Natarajan and R. Appadurai. 1979. Tenai CO 5—a high yielding rust resistant variety suitable for Tamil Nadu. *Madras agric. J.* **66:** 771-773.

Reddy, B.G.S., G. Pushpa, B.A. Satyan and K.S. Prakash. 1983. Preliminary studies on gibberellic acid induced male sterility in finger millet. *National Seminar on Finger Millet Genetics and Breeding,* UAS, Bangalore, January 12-13, 1983.

Seetharam, A. 1982. Finger millet improvement. *Indian Farming* **32**(3): 3-6.

Shivshankar, G. and A.R.G. Ranganatha. 1983. Induced mutations, their role and prospects in finger millet improvement. *National Seminar on Finger Millet Genetics and Breeding,* UAS, Bangalore, January 12-13, 1983.

Sreekantaradhya, R. 1979. Induced mutagenesis in ragi (*Eleusine coracana* Gaertn.) with gamma rays and ethyl methane sulfonate. I. Sensitivity as estimated through M_1 effects. *Mysore J. agric. Sci.* **13**(2): 136-140.
Srivastava, D.P. and A. Yadav. 1972. Emasculation of flowers of *Panicum miliare* Lam. by hot water treatment. *Sci. Cult.* **38:** 450.
Srivastava, D.P. and A. Yadav. 1977. Heterosis in *Panicum miliare* Lam. *Curr. Res.* **6**(4): 66-67.
Swaminath, M.H. 1979. Study on the genetic diversity in the African collection of finger millet (*Eleusine coracana* Gaertn.) and their interaction with environment (Abst.) *Mysore J. agric. Sci.* **13:** 225-226.

6

IMPORTANCE, GENETIC RESOURCES AND BREEDING OF SMALL MILLETS IN BANGLADESH

M.A. Majid, M.A. Hamid and Mannujan

INTRODUCTION

Kaon or foxtail millet (*Setaria italica* Beauv.), Cheena or Proso millet (*Panicum miliaceum* Linn.), shama kaon or bhura kaon or barnyard millet (*Echinochloa colona* Link.), ragi or finger millet (*Eleusine coracana* Gaertn.) and kodo or ditch millet (*Paspalum scrobiculatum* Linn.) are the small-grained minor cereals cultivated in Bangladesh. Ragi and kodo are grown on a very limited area with minimum care and inputs in the districts of Kushtia and Rajshahi, mainly in poor and marginal soils where no other crops can thrive. Kaon, cheena and shama kaon are cultivated all over Bangladesh relatively in larger areas but with little inputs in poor marginal lands including the river beds. These minor cereals are short in duration, drought tolerant, well adapted and less susceptible to diseases and insect pests. In case of failure or damage of major crops due to drought or other natural calamities, some of these minor cereals are grown as a replacement to minimise the loss or to avoid imminent famine. For this reason, they are called famine or disaster crops.

PRODUCTION TRENDS OF CEREAL CROPS

Farmers in Bangladesh grow indigenous, low yielding land races of millets, and consequently, obtain very poor yields. The total acreage and production of millets and other minor cereals have declined considerably in recent years due to increase in acreage and production of high yielding varieties of rice and wheat (Table 1).

TABLE 1
Production trend of cereal crops

Crop	1972-73 Area (lakh ha)	Production (lakh tonnes)	Yield (kg/ha)	1983-84 Area (lakh ha)	Production (lakh tonnes)	Yield (kg/ha)
Rice	96.297	100.889	1047	105.480	145.075	1375
Wheat	1.201	0.910	757	5.260	12.115	2302
Millets	0.679	0.486	716	0.367	0.317	865
Barley	0.259	0.164	631	0.096	0.066	684
Other cereals	0.039	0.032	813	0.024	0.017	718
	98.475	102.481	1040	111.227	157.590	1417

Source: Bangladesh Bureau of Statistics.

The low yield of millets (700-860 kg/ha) can be attributed to poor soil, non-availability of improved varieties, lack of application of fertilizer, irrigation water, pesticide and improved cultural practices.

GENETIC RESOURCES AND RESEARCH ACTIVITIES OF MINOR CEREALS

In the past no research thrust was given for the improvement of these neglected crops. The researchers and the planners were not aware of their importance and role in the cropping pattern and farmers' economy of Bangladesh. In 1981, Bangladesh Agricultural Research Institute initiated a research programme for varietal improvement of two important minor cereals—proso and foxtail millet—and subsequently to develop production technologies for different agro-ecological zones of Bangladesh with the financial assistance of International Development Research Centre (IDRC), Canada. As a first step germplasm was assembled both from home and abroad and so far 222 foxtail millet and 109 proso millet lines were received from ICRISAT and ICAR, India. In addition, 424 foxtail and 163 proso millet lines were collected locally. All these local and exotic lines were grown in the Central Research Station of BARI at Joydebpur for their proper identification, purification and documentation.

A total of 646 lines both exotic and local foxtail millet were tested at two different agroclimatic situations (Joydebpur and Thakurgaon) to evaluate their yield potential, adaptability and other desirable agronomic characters. The variability observed for the plant characters at Thakurgaon Research Station is summarized in Table 2.

The coefficient of variance for grain yield was highest, indicating possible scope for further selection and improvement. Superior lines having high yield potential and other desirable characters were identified and put into a series

TABLE 2

Variability observed for plant characters in foxtail millet germplasm at Thakurgaon in 1984-85

Character	Range	S.D.	Mean ± SE	CV %
No. of effective tillers/plant	1.0 - 4.4	0.50	1.37 ± 0.021	36.20
Plant height (cm)	29.0 - 99.5	13.86	16.11 ± 0.58	20.08
Panicle length (cm)	4.1 - 17.4	2.14	10.55 ± 0.09	20.24
Days to maturity	91.0 - 130.0	30.76	114.69 ± 0.29	26.82
Grain yield/plot (g)	100 - 780	131.41	332.41 ± 5.51	40.79

of field trials at eight different agro-climatic regions—Joydebpur, Ishurdi, Jamalpur, Jessore, Chandina, Thakurgaon, Debiganj and Rangpur (Table 3).

Considering yield and other desirable characters, the local foxtail millet line Bogra-1 performed the best, followed by Lakhmipur-2, Parameshpur and Shibnagar. These four lines are undergoing multilocation trial in a farmer's field. Further, it is proposed to release them through the National Seed Board for commercial cultivation.

Similarly, 272 germplasm of both exotic and local proso millet were grown along with a standard check at two different agroclimatic zones viz., Joydebpur and Jamalpur. The range of variability observed for important agronomic attributes at Joydebpur is presented in Table 4.

The coefficient of variation was again highest for grain yield, indicating possible scope for further selection for yield. Superior lines having high yield potential and other desirable characters were selected and put into a series of yield trials for selection of promising lines (Table 5).

When average yield performance at all the four locations was considered, three locals—Kumardhon (1.339 t/ha), Bagaikandi (1.288 t/ha), Telipara (1.239 t/ha), and one exotic line called white millet (1.209 t/ha) were found high yielding. These four lines are before the National Seed Board for release for commercial cultivation. The white millet variety has already been released and the other three lines are in the process of release.

In the yield trial of 1985-86, some promising lines of proso millet have been identified (Table 6). These lines are BPm-40, BPm-14, BPm-34 and BPm-21.

TABLE 3

Mean grain yield (t/ha) of 10 kaon strains grown at 8 locations (1984-85)

Entries	Joydebpur	Ishurdi	Jamalpur	Jessore	Chandina	Thakurgaon	Debiganj	Rangpur	Mean
Bogra-1	2.005	1.808	2.675	3.250	1.467	1.992	2.898	2.433	2.316
BSi-273	1.328	1.203	1.715	2.667	1.533	1.292	2.546	1.883	1.772
Lakhmipur-2	1.725	1.379	2.302	3.260	2.100	2.142	2.452	2.317	2.208
Parameshpur	1.765	1.600	2.404	2.917	1.800	2.275	2.084	2.583	2.179
Arjuna	1.698	1.783	1.898	3.415	1.867	1.758	2.003	1.917	2.042
Shibnagar	1.665	1.250	1.884	3.167	2.033	2.225	2.531	2.550	2.163
BSi-274	1.760	1.654	2.009	3.083	1.700	1.433	2.383	1.669	1.967
Lohachura	1.660	1.545	2.289	2.333	1.967	1.650	2.397	2.167	1.990
Mogra Kaon	1.917	1.546	2.164	2.000	1.767	2.000	2.775	2.283	2.057
Check (Local)	1.750	0.813	2.275	3.083	1.317	1.667	2.222	2.783	1.989

TABLE 4

Variability observed for plant characters in proso millet germplasm at Joydebpur in 1983-84

Character	Range	S.D.	Mean ± SE	CV %
Days to 50% flowering	63.00 – 104.00	8.998	73.568 ± 1.36	12.231
Days to maturity	86.00-132.00	7.550	119.750 ± 1.14	6.304
Plant height (cm)	41.50-86.40	9.330	62.52 ± 1.41	14.92
Panicle length (cm)	10.80-26.2	2.690	19.39 ± 0.41	13.89
Tiller Nos./plot	2.80-8.20	1.20	4.43 ± 0.18	27.12
Grain yield/plot (g)	115.00-573.00	105.24	362.952 ± 15.80	28.995

TABLE 5

Mean yield (t/ha) of 10 strains of proso millet at four locations in 1983-84

	Locations				
Variety	Hathzzari	Jessore	Ishurdi	Bogra	Mean
Bagaikandi	1.633	1.150	1.360	1.008	1.288
Islampur	1.598	1.133	1.155	0.708	1.149
Kumardhon	1.525	1.400	1.388	1.042	1.339
Lakshmipur	1.308	1.300	1.147	0.692	1.112
Mejo-Sel-Rew-5	1.500	1.200	1.217	0.908	1.206
Telepara	1.400	1.450	1.430	0.675	1.239
White millet	1.725	1.067	1.367	0.675	1.209
BPm-51	1.071	1.350	1.430	0.633	1.121
BPm-52	0.929	1.167	1.305	0.683	1.021
Local (check)	1.058	1.033	0.667	0.633	0.848

TABLE 6

Results of advanced yield trial of proso millet—1985-86

	Grain yield (t/ha)				
Variety	Joydebpur	Jamalpur	Thakur-gaon	Ishurdi	Mean
BPm-2	1.721	1.270	2.233	1.867	1.773
BPm-4	1.878	1.716	1.767	2.317	1.920
BPm-12	1.394	1.587	1.917	2.333	1.808
BPm-14	2.105	1.722	2.083	2.410	2.080
BPm-16	1.919	1.727	2.300	1.833	1.945
BPm-21	1.538	1.973	2.267	2.200	1.995
BPm-34	1.517	1.933	2.167	2.383	2.000
BPm-40	1.908	1.893	1.917	2.404	2.031
BPm-60	1.649	1.781	1.883	2.200	1.878
Bagaikandi	1.759	1.667	1.500	2.217	1.786

BIBLIOGRAPHY

Agricultural Year Book of Bangladesh, 1982. Bangladesh Bureau of Statistics, Dhaka.

Annual Report, Millet Improvement Project (1983-84), Bangladesh Agricultural Research Institute. Joydebpur.

Annual Report, Millet Improvement Project (1984-85), Bangladesh Agricultural Research Institute, Joydebpur.

Annual Report, Millet Improvement Project (1985-86), Bangladesh Agricultural Research Institute, Joydebpur.

Statistical Year Book of Bangladesh, 1984-85. Bangladesh Bureau of Statistics, Dhaka.

7

IMPORTANCE, GENETIC RESOURCES AND BREEDING OF SMALL MILLETS IN SRI LANKA

S. Ponnuthurai

Small millets grown in Sri Lanka are: *Eleusine coracana* (finger millet), *Panicum miliaceum* (common millet) and *Setaria italica* (foxtail millet).

AREA AND IMPORTANCE

Area under finger millet ranges from 16,000 to 44,000 ha and the annual production from 7,000 to 18,000 tonnes. Area under common millet ranges from 900 to 2,500 ha and the production from 500 to 1,100 tonnes. Foxtail millet has the lowest acreage among the millets and ranges from 24 to 425 ha and the production from 24 to 195 tonnes (Table 1). Both area and production of millets are much less compared to rice which is the staple of food and the area under millets showed a further decline over the six year period under consideration. Decline in millet production could be attributed to the liberalization of imports of rice, wheat flour, etc. by the Government and increase in rice production, since 1977.

Finger millet is grown traditionally under the shifting forest fallow system (chena), in the dry zone. The extent cultivated per farming family is about half to one hectare. Millets are grown mixed with the other cereals such as maize, sorghum and also vegetables. Yield of finger millet under this system of cultivation varies from 1000-1200 kg/ha with minimal inputs. Under these conditions the farmer gets a net return of about Rs. 4,000 per ha. In recent times a stabilized system of agriculture has come to stay, due to increased irrigation facilities. Although the commissioning of the river diversion projects such as Mahaweli and others has increased settled agriculture, there is still a sizeable portion of unirrigable highland of about one million hectares in the dry zone (Fernando, 1981) where millets and other highland crops could be cultivated.

TABLE 1
Trends in millet production compared with rice in Sri Lanka

Year	Finger millet		Common millet		Foxtail millet		Rice	
	Area (ha)	Production (tonnes)	Area (ha)	Production (tonnes)	Area (ha)	Production (tonnes)	Area (ha)	Production (tonnes)
1975	43,836	17,873	2,501	489	425	173		
1976	39,526	17,640	1,072	841	284	195		
1977	34,617	16,360	1,103	1,083	173	134	782,618	1,680,861
1982	16,035	13,300	—	—	—	—		
1983	20,595	11,662	934	934	24	24	825,833	2,499,567
1984	16,489*	6,570	—	—	—	—		

*Area grown in the wet season only
Source: Dept. of Agriculture, Sri Lanka

Important areas of millet production are: Anuradhapura and Hambantota districts in the dry zone (where the average annual rainfall is 1270-1900 mm), Matale, Monaragala and Kurunegala districts in the intermediate zone (rainfall of 1,900-3175 mm), and Ratnapura district in the wet zone (rainfall of over 3,175 mm) (Table 2, and Fig. 1). In all these areas finger millet and foxtail millet are grown in the wet season during the northeast monsoon rains that occur mid September to January (Fig. 2B). In Rantapura district of the wet zone finger millet is cultivated in the wet season, but is confined to the relatively drier areas in the southeast. In the Jaffna district where cropping intensity is high, millets in general are grown under irrigation during the dry season in the period April to July (Fig. 2A). Traditional varieties and limited area of improved varieties of finger millet are grown by farmers in this district, with yields ranging from 1200 to 2000 kg/ha.

Common millet constitutes the commonest cereal component of the dry season in the dry zone, and is grown under the shifting cultivation system, with the low (about 200 mm) and uncertain rainfall received. The period of cultivation is from March to July. Average yield obtained ranges from 400 to 700 kg/ha.

Maximum temperatures (monthly mean) ranges from 87-90°F in Ratnapura, 84-88°F in Kurunegala, and 81-90°F in Anuradhapura (Fig. 2B). Minimum temperatures are about the same (70-74°F) in all the three locations. In the dry zone location Anuradhapura, common millet is cultivated during the period March to July, and the maximum temperature is higher, ranging from 91 to 93°F, and the minimum from 73 to 77°F.

Soils vary in the different millet growing areas. In the dry zone reddish-brown earths, predominate. In the Jaffna district which is also a part of the dry zone, soil consists of calcic red yellow latosols and sandy regosols. In the

TABLE 2
Important areas of millet production and their coverage in 1985 (Area in ha)

District	Finger millet	Common millet	Foxtail millet
Anuradhapura	7,600	13	—
Hambantota	1,165	14	23
Ampara	750	—	—
Jaffna	390	91	—
Matale	2,070	8	—
Monaragala	1,965	40	49
Kurunegala	1,440	6	2
Ratnapura	470	57	11
Kandy	400	—	—

Source: Dept. of Agriculture, Sri Lanka.

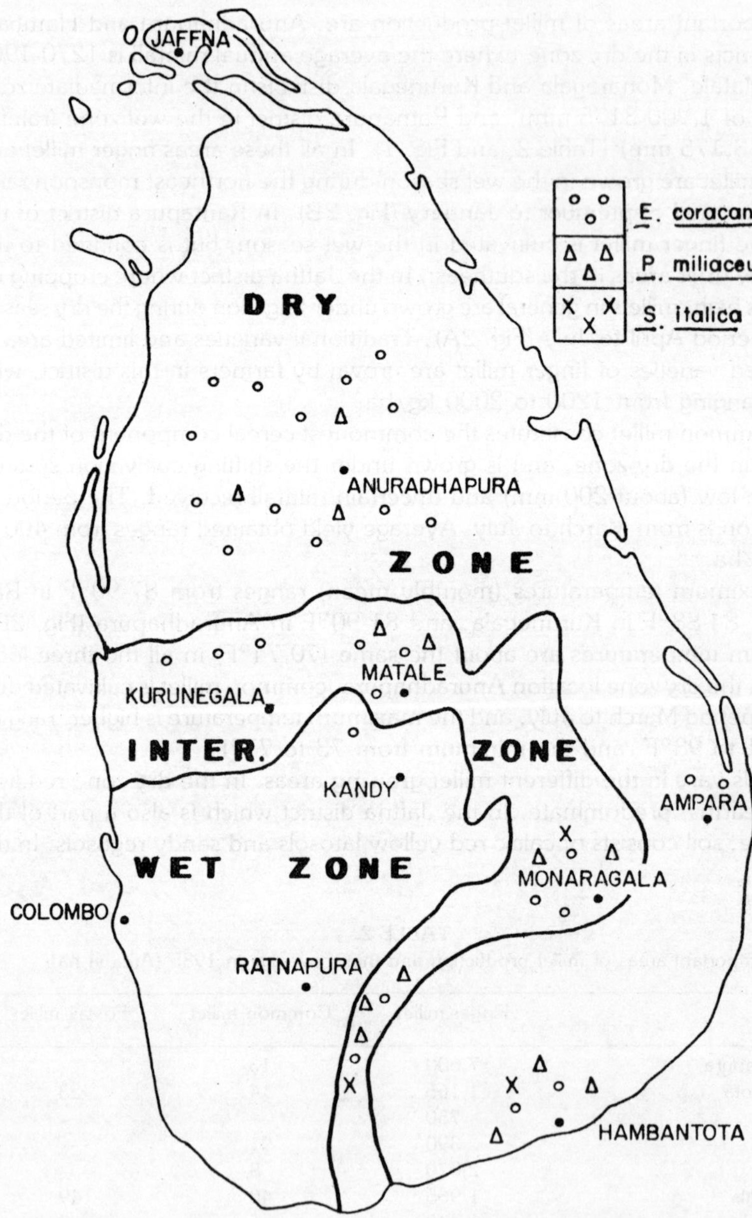

Fig. 1. Important areas of millet production in Sri Lanka

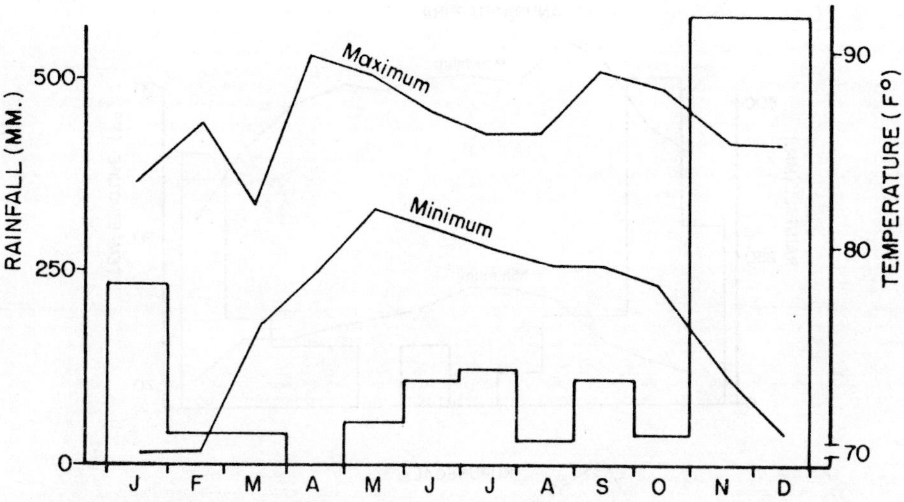

Fig. 2A. Rainfall and temperature in Jaffna (monthly mean)

intermediate zone, soils belong to reddish-brown earths and immature brown loams, reddish brown latosolic soils and immature brown loams, and red yellow podzolic soils with weakly developed laterites.

GERMPLASM RESOURCES

Several indigenous varieties of millets are being cultivated by farmers. Some local germplasm in millets is maintained in the dry zone research stations viz. Maha Illuppallama, Karadian Aru, Kilinochchi and Angunukolapelessa. There appears a tendency for the decline in the area under millets, which could lead to loss of much valued indigenous genetic resources. Concerted efforts are therefore needed to collect and conserve this germplasm in an appropriate manner for use in crop improvement work in the future. Nationally and internationally due attention should be devoted to this important service.

In addition to the local germplasm, millet introductions were made from time to time from other countries, mainly India. MI 302, a finger millet selection from an introduction is recommended for cultivation by the Department of Agriculture since the mid-seventies. The yield of this variety ranged from 800 to 1,000 kg/ha under rainfed and from 1,000 to 1,500 kg/ha under irrigated conditions. Some introductions made subsequently have been found to be better adapted. One such introduced variety is CO-10 which has outyielded MI 302 and the local varieties. Its yield ranged from 2.0 to 2.5 t/ha.

In common millet local varieties are popular and their yield range from 600 to 700 kg/ha. Evaluation of introductions (MS 2420, MS 1491) is in progress at the research stations. Likewise, in foxtail millet, local varieties are

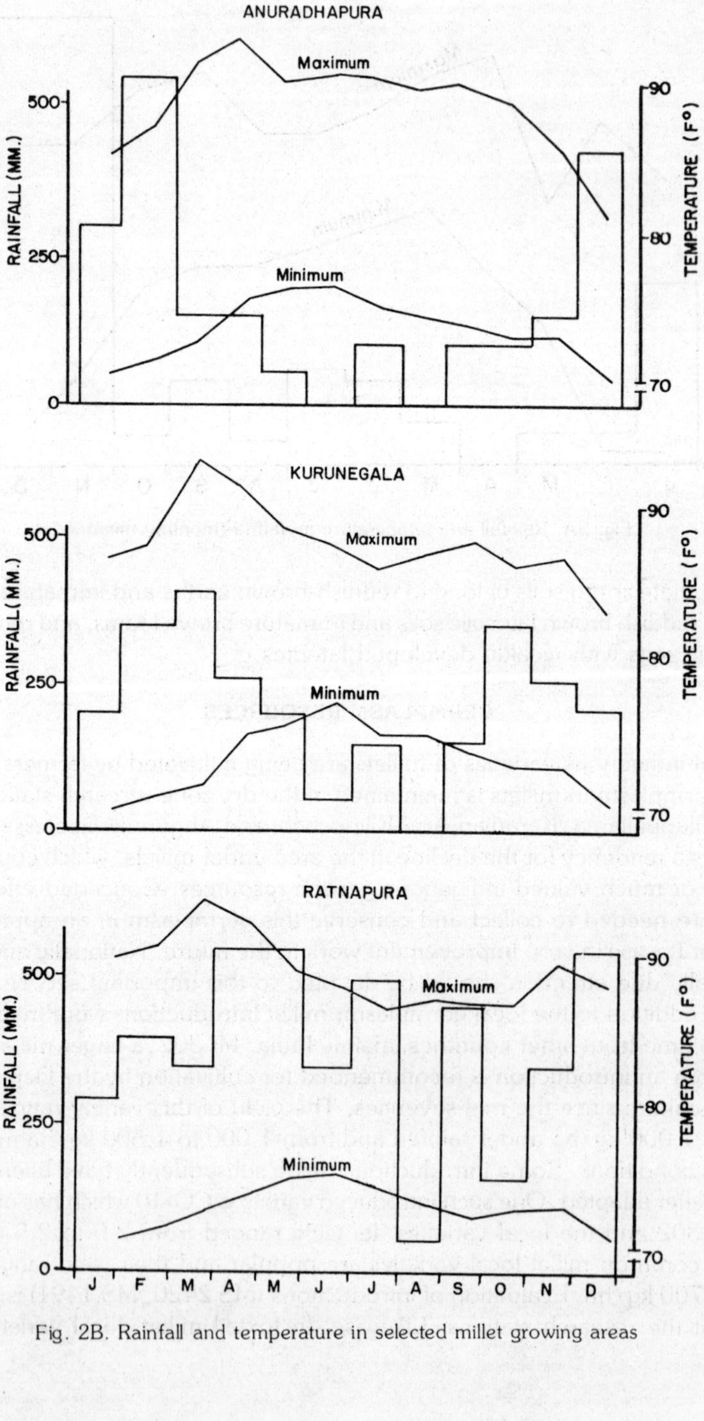

Fig. 2B. Rainfall and temperature in selected millet growing areas

cultivated by the farmers and their yields range from 450 to 700 kg/ha. Introductions such as KHS-1 and ISC 40, evaluated in the research plots show promise.

BREEDING AND VARIETAL IMPROVEMENT

Breeding and varietal improvement work in millets is confined to varietal evaluation for yield in finger millet. Varieties are evaluated in the National Coordinated Varietal Trial and its main objective is to identify high yielding and widely adapted varieties with superior agronomic and morphological characteristics. These trials began in the 1981 wet season. Most varieties evaluated were introductions, with MI 302 as the check variety. Selected varieties and their mean yields over several locations in the different agro-ecological regions are presented in Table 3. In most instances CO-10 has performed the best, followed by JNR-3B-1008 and HPB 83-4. In the dry season, yields tend to be higher compared to the wet season, due to more light, uniform availability of soil moisture through irrigation and less pests and diseases. Lower yields recorded at Makandura (in 1981 and 1984 wet seasons) and Girandurukotte are attributed to erratic rainfall and poor drainage conditions, respectively. Apart from varietal evaluation not much work has been done by way of breeding. Selection of single plants is made in local populations for desirable characteristics like shorter plant height, early maturity (three months), loose panicle structure, uniform maturity, resistance to pests, diseases and lodging. Hybridization following the conventional methods and induced mutation breeding to develop new and widely adaptable varieties of millet in general, is important to intensify research in crop improvement for the future.

BIBLIOGRAPHY

Agricultural Implementation Programme (1984-85) of the Ministry of Agricultural Development and Research. Sri Lanka.

Agricultural Implementation Programme (1985-86) of the Ministry of Agricultural Development and Research, Sri Lanka.

Agricultural Statistical Information Vol. II (1979) Department of Agriculture, Peradeniya, Sri Lanka.

Fernando, G.W.E. (1981). A review of Agricultural Research at Maha Illuppallama, Tropical Agriculturist Vol. 127, p. 13-17. Sri Lanka.

Research Highlights No. 23 (1985), Research Division, Department of Agriculture, Peradeniya, Sri Lanka.

TABLE 3

Yield (t/ha) of finger millet varieties in coordinated varietal trials

	1981 WS			1982 WS			1982 DS	1982 WS	1984 WS	1985 WS
Varieties	*MI	KOC	MK	AP	KOC	MK	TN	KOC	GK	KOC
CO 10	2.02	4.73	0.87	2.36	1.92	0.48	4.61	1.14	0.70	2.10
KM 1	1.82	3.47	0.80	2.29	1.46	0.43	4.16	0.69	0.59	2.62
HPB 83-4	1.59	3.77	0.71	2.63	2.24	0.27	3.71	0.70	0.80	2.54
HR 231	1.73	5.58	0.66	2.31	1.68	0.81	—	1.00	0.90	2.21
PR 1091	1.87	3.65	0.44	2.14	1.18	0.14	5.45	0.70	0.89	2.02
JNR 3B-1008 1008	1.83	5.22	0.62	2.65	2.44	0.40	—	0.95	0.90	2.47
MI 302 (CHECK)	1.89	4.24	0.29	1.22	1.68	0.78	—	0.63	0.72	2.38

* Locations:

MI = Maha Illuppallama
KOC = Kilinochchi
AP = Angunukolapelessa
TN = Thinnaveli
} Dry Zone

MK = Makandura
GK = Girandurukotte
WS = Wet season
DS = Dry season
} Intermediate Zone

8

IMPORTANCE, GENETIC RESOURCES AND VARIETAL IMPROVEMENT OF FINGER MILLET IN NEPAL

Kishor Sherchan

Finger millet (*Eleusine coracana* Gaertn.) is a very important crop in Nepal ranking fourth in area and production after rice, maize and wheat. In spite of its importance, it has received little attention from researchers, and is frequently referred to as one of the neglected crops of Nepal. This is in sharp contrast to the three major crops (rice, maize and wheat) which, fortunately, have received great deal of research support for crop improvement, both nationally and internationally. Also, whereas finger millet is often grown under stressful conditions, the three major cereal crops are commonly grown under more favourable conditions (Lohani, 1980).

Finger millet grain is an important food crop, especially for those living in the hills and mountains of Nepal where approximately 90 per cent of the millet production is concentrated. In Nepal, finger millet is rarely grown solely as a forage crop. However, the straw is virtually always used for forage, and the earlier the grain and straw can be harvested the higher the quality of the fodder. Green straw makes nutritious fodder while yellow-brown straw from fully matured and weathered plants is, of course, much less nutritious.

In Nepal, finger millet is commonly referred to as kodo. This terminology is in contrast to that in India where kodo is *Paspalum scrobiculatum* L. and *Eleusine coracana* Gaertn. is referred to as mandua, or ragi.

MILLET PRODUCTION

According to information from the Agricultural Statistics Division, of the Department of Food and Agricultural Marketing Services, Ministry of Agriculture, the area and production of finger millet in 1984 were estimated at 134,000 ha and 124,000 mt, respectively, giving an average yield of 926 kg/ha. (Anonymous,

1986). This was equivalent to about 5 and 3 per cent of the total cereal area and production, respectively. However, the land and aerial surveys conducted in 1984 by the Land Resources Mapping Project, Topographical Survey Branch, Survey Department, Ministry of Land Reform, projected a much higher area and production level of 235,000 ha and 252,000 mt, respectively, giving an average yield of 1074 kg/ha (Sherchan *et al.*, 1986).

The area, production and average yields of the five most important cereal crops of Nepal are compared in Table 1.

TABLE 1
A comparison of the area, production, and average yields of five important cereal crops based on information from the Agriculture Statistics Division (ASD) and the Land Resources Mapping Project (LRMP); 1984

	Area '000 ha		Production '000 mt		Yield kg/ha	
Crop	ASD	LRMP	ASD	LRMP	ASD	LRMP
Paddy	1377	1417	2709	2177	1968	1536
Maize	579	688	820	763	1417	1108
Wheat	452	504	534	484	1181	961
Millet	134	235	124	252	926	1074
Barley	28	74	24	57	854	775

Geographically, Nepal can be divided into three major zones; the terai and inner terai in the southern part of the country; the hills in the middle; and high mountains in the north. Finger millet is grown as a summer upland crop up to 2500 m above sea level. Its major area of production lies in the mid hills, between 600 and 2000 metres (Table 2), which accommodates more than 50 per cent of the total population. The distribution of finger millet cultivation is shown in Fig. 1.

TABLE 2
Finger millet area, production and yield in three different ecological zones; 1985

	Area		Production		Yield
Ecological Zone	'000 ha	%	'000 mt	%	kg/ha
Terai/Siwalik (up to 600 m)	11.6	7.7	10.5	7.6	906
Mid hills (600-2000 m)	116.9	77.3	106.2	77.0	909
Mountains (2000-2500 m)	22.6	15.0	21.2	15.4	941
Total	151.1	100	137.9	100	913 (Ave.)

Source: Agricultural Statistics Div., Dept. of Food and Agricultural Marketing Services, Ministry of Agriculture.

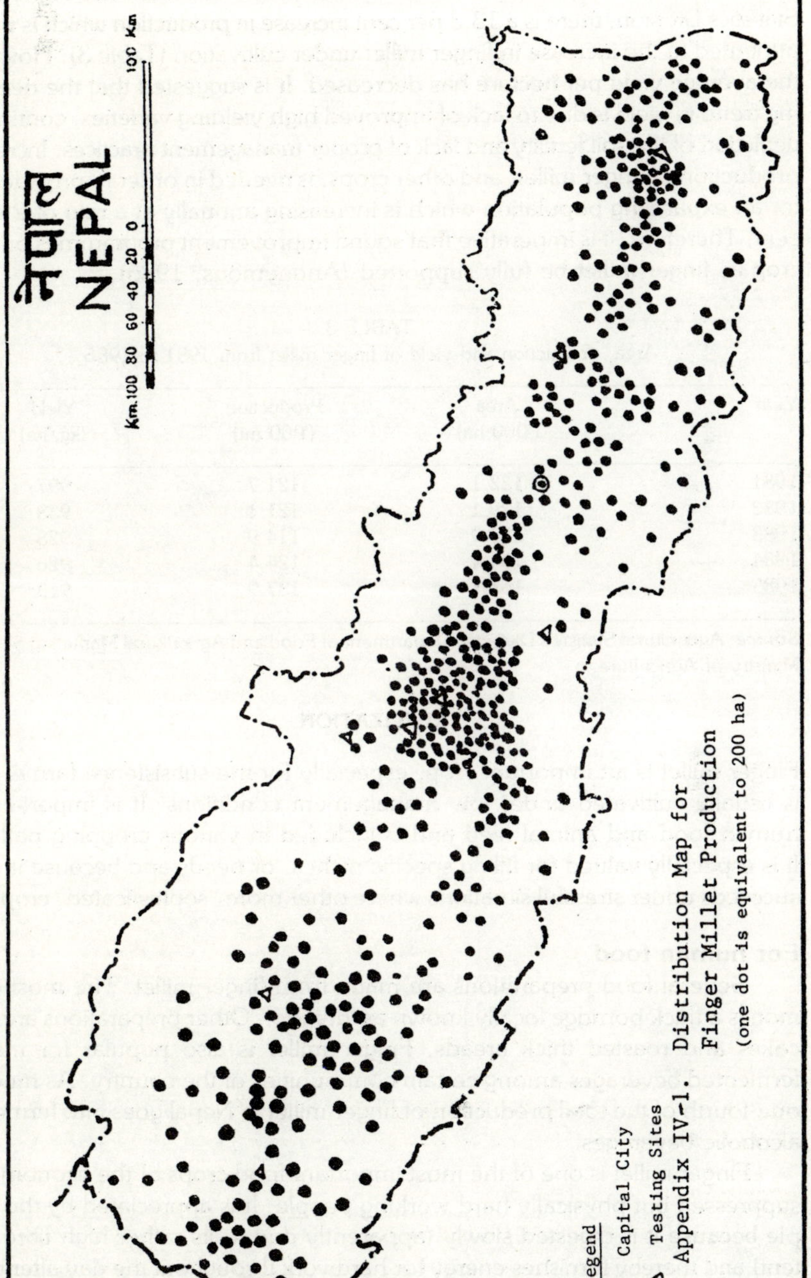
नेपाल
NEPAL
Km.100 80 60 40 20 0 100 Km
Legend
Capital City
Testing Sites
Appendix IV-1. Distribution Map for
Finger Millet Production
(one dot is equivalent to 200 ha)

Fig. 1.

Based on available data for five years (1981-85) from the Agricultural Statistics Division, there is a 13.3 per cent increase in production which is mainly attributed to the increase in finger millet under cultivation (Table 3). However, the average yield per hectare has decreased. It is suggested that the decreasing trend in yield is due to lack of improved high yielding varieties, continuous depletion of the soil fertility and lack of proper management practices. Increased production of finger millet, and other crops, is needed in order to provide food for an expanding population which is increasing annually at a rate of 2.6 per cent. Therefore, it is imperative that sound improvement programmes on such crop as finger millet be fully supported (Anonymous, 1985).

TABLE 3
Area, production and yield of finger millet from 1981 to 1985

Year	Area ('000 ha)	Production ('000 mt)	Yield (kg/ha)
1981	122.1	121.7	997
1982	129.1	121.1	938
1983	123.9	114.9	928
1984	134.4	124.4	926
1985	151.1	137.9	913

Source: Agricultural Statistics Division, Department of Food and Agricultural Marketing Services, Ministry of Agriculture.

UTILIZATION

Finger millet is an important crop, especially for the subsistence farmer, and is usually cultivated under low management conditions. It is important for human food and animal feed and is included in various cropping patterns. It is especially valued for filling specific niches, or needs and because it often succeeds under stressful situations where other more "sophisticated" crops fail.

For human food

Several food preparations are made from finger millet. The most common is a thick porridge locally known as 'dhindo'. Other preparations are pancakes and roasted thick breads. Finger millet is also popular for making fermented beverages among certain communities of the country. As much as one-fourth of the total production of finger millet in Nepal goes into fermented alcoholic beverages.

Finger millet is one of the most important food crops of the economically suppressed but physically hard working people. It is appreciated by the people because it is digested slowly (apparently due to its rather high fibre content) and thereby furnishes energy for hardwork throughout the day after being eaten at a single morning meal.

Finger millet grain is reported to contain 9.2 per cent protein, 1.29 per cent fat, 76.32 per cent carbohydrates, 2.24 per cent minerals, 3.90 per cent ash and 0.33 per cent calcium. Vitamins A and B and phosphorus are also present in smaller quantities (Rachie and Peters, 1977).

For animal feed

The harvest residue of finger millet is extensively used as animal feed. Since finger millet harvest usually takes place at a time when other fodders are in short supply, at the beginning of the dry season and before the availability of rice straw, it is especially appreciated by farmers having livestock. Finger millet grain as such is not used as cattle feed but the spent brewer's grain is an important animal concentrate.

FINGER MILLET IMPROVEMENT

In Nepal, there have been and are separate commodity stations and programmes for the three major cereal crops—the National Rice Improvement Programme (Parwanipur), the National Maize Development Programme (Rampur) and the National Wheat Improvement Programme (Bhairahawa). However, limited attention has been given to the improvement of *Eleusine coracana*.

There has been some screening of kodo germplasm at Rampur and Khumal research stations, since the early 1970's. Approximately 700 exotic lines were obtained from India and 100 local land races were collected in Nepal. From the evaluation of these materials, two improved varieties, Dalle-1 (IE980 from India) and Okhale-1 (selection from local material), were released in 1980. Dalle-1 is adapted to the Inner Terai and mid-hills. Okhale-1 was released for use in the mid-hills and higher elevations. However, it was soon determined that these two varieties are limited in their range of adaptability.

There is need for an extensive collection and testing of both local and exotic sources of germplasm with emphasis on selection for broad adaptability. Some promising new lines have been identified such as NE 1703-34 which has better tolerance to waterlogged conditions; Rampur local for lower elevations; and NE 3801-2 has been found to do well at mid-hill testing locations at Khumal, Kavre Farm and Lumle (Singh and Tamulonis, 1985). The yield performance of some of the promising lines is given in Table 4 (mono-cropped) and Table 5 (relay cropped with maize).

GERMPLASM RESOURCES

People in the hills of Nepal have been growing finger millet since time immemorial. There has not been any authentic record of its introduction. Farmer's have been selecting genotypes in accordance with their own specific needs.

TABLE 4
Yield performance of some promising lines of finger millet at two locations when tested as a sole crop (1985)

Cultivar	Location yield (kg/ha) Khumal	Kavre	Mean
NE 1703-34	2422	1157	1789
NE 6401-26	1753	1341	1547
Dalle-1	1762	1086	1424
NE 1001-1	1501	1329	1415
NE 1104-13	1365	1158	1261
NE 1102-12	1142	1183	1162
NE 1304-1	901	1217	1059
Mean	1549	1210	1380

TABLE 5
Yield performance of finger millet lines when tested as a relay crop with maize (1985)

Lines	Location yield (kg/ha) Khumal	Kavre	Mean
NE 3801-2	1559	1233	1396
Okhale-1	1126	1309	1217
NE 6401-26	873	1528	1200
NE 1304-1	664	1197	930
Mean	1056	1317	1186

The type of finger millet found in Nepal is considered to be of the Afro-Asiatic type with short glumes, lemmas and spikelets (Rachie and Peters, 1977), The panicles may be top curved, incurved or open type. At higher elevations, where rainfall is low, the more dominant type is top curved; whereas in the mid hills, where rainfall is high during the growing season, the open type is more common.

Recently, about 350 local land races from Nepal and 150 exotic introductions, mainly from East Africa, were collected and were tested at Khumal (1360 m.a.s.1.). Considerable genetic variation is noted in terms of maturity, plant type and pigmentation, panicle type and reaction to important diseases such as blast (*Pyricularia* sp.) and blight (*Helminthosporium* sp.). Most of the local land races are early types and short-statured but susceptible to blast and blight. The sources from East Africa, especially from Ethiopia, are free from diseases.

However, most of them are tall and late maturing under Nepal conditions. Recombining local elite lines with those of the Ethiopian highlands should be a good start in generating high yielding, promising lines of wider adaptability for the future improvement programme.

FINGER MILLET IMPROVEMENT STRATEGY

Thus far the finger millet programme has been neglected in the eyes of agricultural scientists and planners. However, recently the Ministry of Agriculture, His Majesty's Government, has paid some attention to the development of this crop, and other minor crops, under the umbrella of the "Hill Crops Improvement Programme (HCIP)". The following improvement objectives are being considered.

1) Collection and conservation of local and exotic germplasm

Although some collections were made within and outside the country during the early seventies, systematic collections could not continue thereafter. Collection and documentation of germplasm of wide genetic base will be extensively carried out. Some 500 finger millet lines were collected in 1985.

2) Selection and immediate release of finger millet varieties

From the local and exotic sources testing, evaluation and selection will be practised and seed of selected lines will be increased. Selection may be carried out on a single plant basis. From these lines, superior high yielding lines with desirable characters such as synchronous maturity, blast and blight resistance and non-lodging types will be selected and released.

3) Long-term improvement programme

The major drawbacks of the Nepal cultivars are low yield and susceptibility to blast and blight diseases. Exotic sources, especially from East Africa, are found to be free from these diseases. Elite lines of local sources will be recombined with those of exotic sources to improve yield and other agronomic characters such as earliness, standability, plant type, threshability and disease resistance, especially to blast and blight.

4) Finger millet genotypes adapted to relay cropping

Relay cropping with maize is the most dominant cropping pattern of finger millet cultivation. In Nepal advanced testings such as Advanced Varietal Trials, Farmers' Field Trials and other outreach testing will be closely integrated with the existing farming/cropping systems. This may help to identify superior genotypes that fit into the relay cropping patterns.

In order to accomplish the above objectives, the Ministry of Agriculture, Nepal, is initiating a phasewise programme with the help of external agencies such as USAID and IDRC.

OTHER MILLETS

The other small millets reported by Regmi (1982) in Nepal are:

1) Cheenu or proso millet (*Panicum miliaceum* L.)
2) Kaguno or Italian millet (*Setaria italica* L. Beauv.)
3) Sawan or Barnyard millet (*Echinochloa colona* L.)

The extent of cultivation and the yield potential of these crops have not yet been studied. However, they are mostly found to have been cultivated in the drier regions of the western and far western hills and mountains of Nepal. They are usually grown in association with other upland crops such as finger millet, amaranth or maize. Among them, cheenu (*Panicum miliaceum*) is found to be widely cultivated and used both as human food and animal feed.

LITERATURE CITED

Anonymous. 1985 Seventh Five Year Plan of Nepal (in Nepali). National Planning Commission, HMG, Kathmandu, Nepal.

Anonymous. 1986. A Handbook of Agriculture Statistics of Nepal. Agriculture Statistics Division, Department of Food and Agriculture Marketing Services, Ministry of Agriculture, His Majesty's Government, Nepal.

Lohani, S.N. 1980. Field Crops of Nepal (in Nepali) Kumultar Research Centre, Kathmandu.

Rachie, K.O. and L.V. Peters. 1977. *The Eleusines*, ICRISAT, Hyderabad, A.P., India.

Regmi, P.P. 1982. *An Introduction to Nepalese Food Plants*. Royal Nepal Academy, Kathmandu, Nepal.

Sherchan, K., T.B. Shrestha and K.O. Rachie. 1986. A Consultancy Report on Hill Crops Improvement Program in Nepal. IDRC/ARPP, DOA, HMG, Nepal.

Singh, K.M. and John Tamulonis. 1985. Reports on Finger Millet, Papers Presented at the 13th Annual Summer Crop Workshop, March 10-13, Janakpur, Nepal.

9

IMPORTANCE AND GENETIC RESOURCES OF SMALL MILLETS WITH EMPHASIS ON FOXTAIL MILLET (*Setaria italica*) IN CHINA

Chen Jiaju

From the numerous crops species grown in China, six crops are listed as millets in the literature.

1) Foxtail millet (*Setaria italica* Beauv.)
2) Proso millet (*Panicum miliaceum* Linn.)
3) Finger millet (*Eleusine coracana* Gaertn.)
4) Japanese millet (*Echinochloa frumentacea* Link)
5) Job's tears (*Coix lachryma-jobi* Linn.), and
6) Pearl millet (*Pennisetum americanum* (L.) Leeke)

Besides, green foxtail (*Setaria viridis*) grows as a wild weed. Little millet (*Panicum sumatrense*) and ditch millet (*Paspalum scrobiculatum*) are not seen and even if they exist in China, they are very rare.

ECONOMIC IMPORTANCE OF MILLET CROPS

Foxtail millet is an important crop grown for food and feed in China. It is one of the main cereal crops in northern China, where the most important crops are wheat and corn. Foxtail millet, sorghum, sweet potato and soybean are of secondary importance. In southern China, foxtail millet is a minor crop.

Proso millet grown in northwest China is an important crop and in certain regions it is the main crop. There is considerable production in northeast China as well but considering the entire country, it is only of minor importance.

Finger millet is a minor cereal in China, mainly scattered in the provinces of the southern and southwestern parts of the country, such as in the southeast of Tibet, Yunnan, Guizhou, Sichuan, Hubei, Jiangxi, Zhejiang, Fujian and Guangdong provinces.

Japanese millet is cultivated in scattered patches in the lowlands and semi-arid regions in the north, distributed in Heilongjiang, Jilin, Liaoning, Hebei, Shandong, Jiangsu provinces.

Job's tears is a medicinal crop produced mainly in the wide area south of 33°N latitude in China, such as Hebei, Shanxi, Henan and Hubei. It is also grown in Shandong, Anhui, Sichuan, Yunnan, Guizhou, Hunan, Jiangxi, Jiangsu, Zhejiang, Fujian, Guangdong, Guangxi and Taiwan provinces. It can grow on uncultivated land where drought and waterlogging often take place.

FOXTAIL MILLET

Production and distribution

China ranks first in the production of foxtail millet in the world. Foxtail millet is grown across the entire country, but the principal growing region is within latitude 32°N to 48°N, and longitude 108°E to 130°E. Twelve provinces (regions) are included in this region, holding more than 95 per cent of the total growing area in the country. The distribution map of area under foxtail millet is shown in Fig. 1.

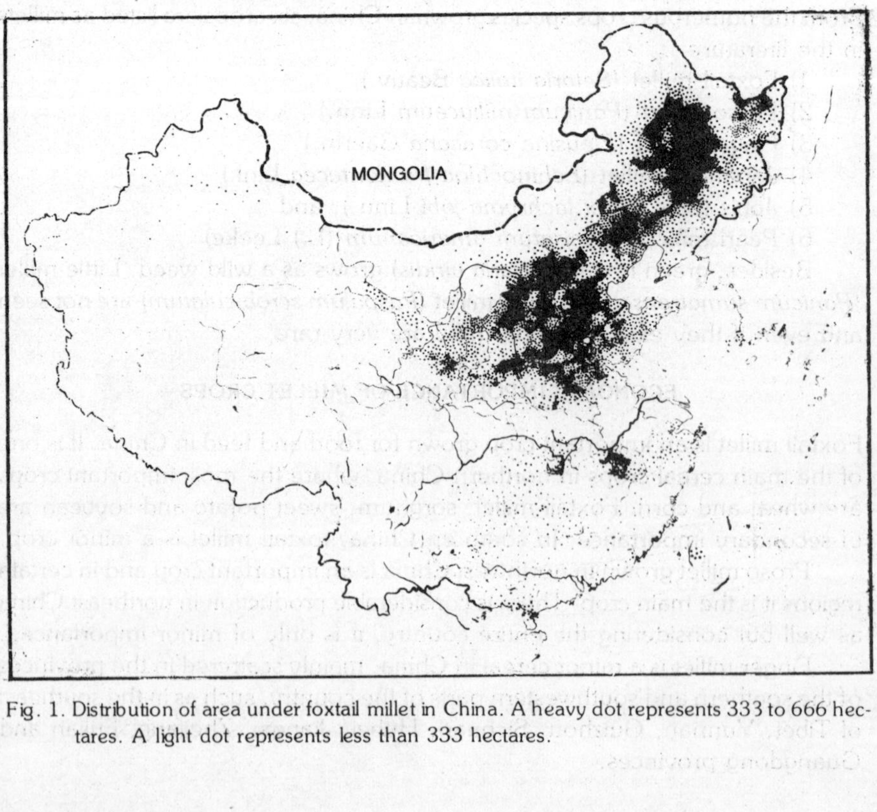

Fig. 1. Distribution of area under foxtail millet in China. A heavy dot represents 333 to 666 hectares. A light dot represents less than 333 hectares.

Figure 2 shows the distribution of seeding area in percentage of foxtail millet to other field crops. Foxtail millet generally occupies 10 per cent or more, in certain districts more than 20 per cent and in a few districts up to 30 per cent of the total seeded area.

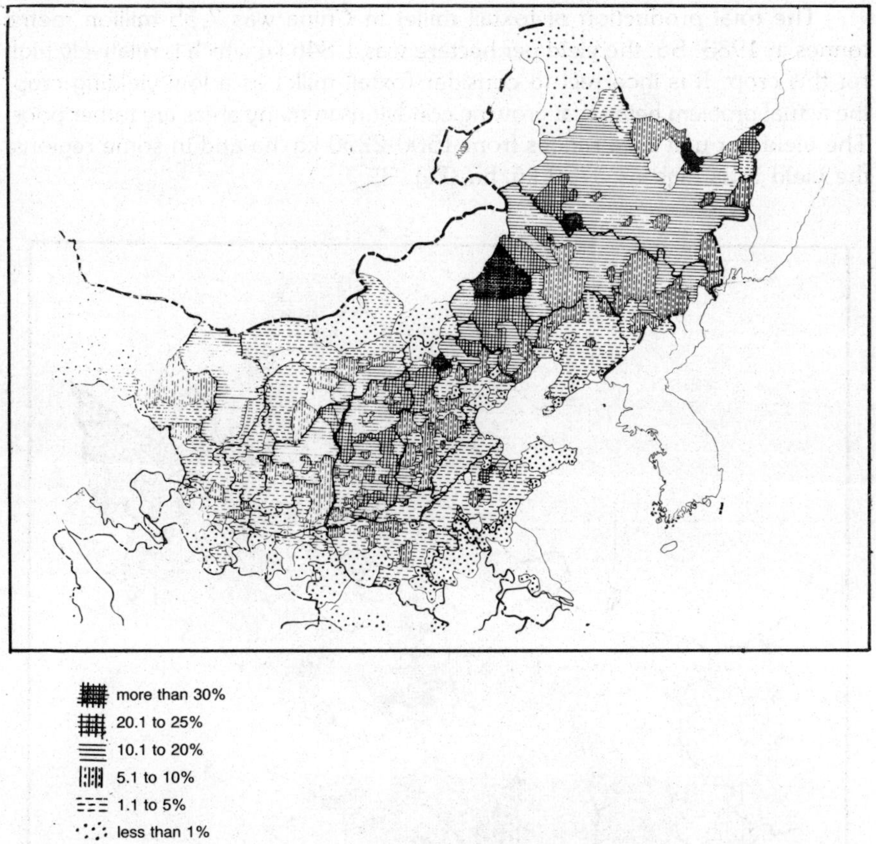

Fig. 2. Distribution of seeding area of foxtail millet in China, expressed as percentage of total field area.

As for the production figures, there were no published figures for many years. However, recent production figures since 1981 are given in the Chinese Yearbook. The cropped area in 1980 and 1983 under foxtail millet was 3.87 and 4.09 million ha, respectively. This accounted for 6.4 per cent of the food crops in the main growing region.

This figure is much lower than that of the 7 million ha during 1937-1945 and in 1936 it was 8.09 million ha. This reflects the decline in area in the past 30 years. After 1954, the change in the cropping system of agricultural cooperatives decreased the millet area. The promotion of highly productive and economic crops also replaced millet area to some extent. However, it is of importance to note that the cropped area has increased somewhat since 1982. Under the responsibility system, farmers can grow now whatever they want. It is estimated that 4 million ha area will be planted hereafter under foxtail millet.

The total production of foxtail millet in China was 7.55 million metric tonnes in 1983. So, the yield per hectare was 1,846 kg which is relatively high for this crop. It is incorrect to consider foxtail millet as a low yielding crop, the actual problem being that growing conditions in many areas are rather poor. The yield per unit area ranges from 1500-2250 kg/ha and in some regions, the yield is as high as 3750 kg/ha (Fig. 3).

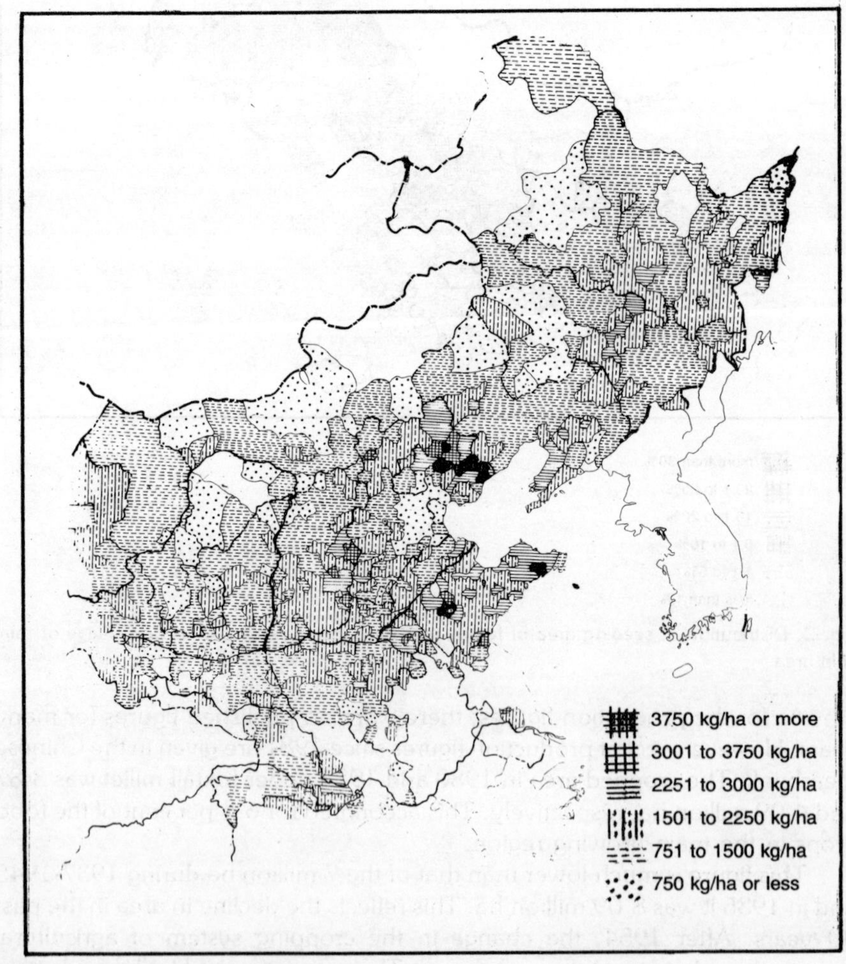

Fig. 3. Distribution of area under foxtail millet in China.

The climate in the principal growing regions of foxtail millet is characterized by moderate temperatures and low precipitation (400-800 mm). Much of the millet is grown on hilly and mountainous land and foxtail millet prefers warm and sunny conditions. Its growth period is rather short and it tolerates drought and low soil fertility. It requires water in the later stages of growth, but waterlogging is harmful. The environmental conditions in the principal growing regions are very suitable to foxtail millet cultivation.

PROSO MILLET

Production and distribution

Proso millet is the second important millet in China. Though precise statistical figures are not available, experts estimate that this crop is now grown on about 1.3 million hectares. However, it was approximately 2 million hectares in 1957. The crop is distributed over arid and semi-arid areas of north, northwest and northeast of China. The main growing regions are Inner Mongolia, Shanxi, Gansu, Ningxia and Heilongjing. There is only sparse cultivation in other areas.

Proso has an even a shorter growth period and is more tolerant to drought and salinity as it can grow under 0.35 per cent salt content. Under drought and poor soil conditions, proso millet gives a yield which surpasses the yield of all other crops.

PRESENT STATUS OF CONSERVATION OF GENETIC RESOURCES OF SMALL MILLETS IN CHINA

Small millets are very widely distributed in various ecological environments in China. During the long history of domestication and cultivation, small millets have derived a great diversity in genetic resources. In the late 1950's and early 1960's China began to collect plant genetic resources of field crops from all over the country. More than 40 kinds of crop plants totalling about 200 thousand accessions were collected, including foxtail millet, proso millet and other small millets.

Foxtail millet genetic resources

Collection

The collection of foxtail millet genetic resources began in the 1920s and large number of landraces were collected in 1950s. According to the statistics of Conference of Field Crops (1958), about 20,000 accessions have been collected, including a number of duplicates.

In 1954-55, the author was in charge of field observations, selection and elimination of duplicate materials collected from Hebei Province. During 1956, identification of 3500 accessions collected from Hebei and Shandong provinces were carried out at the Institute of Crop Breeding and Cultivation, CAAS. Similar collection works have been carried out by other scholars and these materials are generally maintained by provincial agricultural academies.

In the 1980s, several thousand accessions of foxtail millet were collected through large scale crop recollection in Shanxi, Yunnan, Guishou, Goangdong, Hunan, Hubei, Jiangxi and Xinjiang Provinces (Districts). A few foxtail millet landraces were also obtained from various collection activities. Another 40 accessions were collected from Tibet in 1984.

Up to now, 350 accessions of foxtail millet have been introduced to China from foreign countries. We welcome the exchange of germplasm with other countries and the Institute of Crop Germplasm Resources, CAAS is responsible for this task.

Conservation

Dr. Xu Yun-Tian reported that 3,226 land races are stored in the Institute of Crop Germplasm Resources. Besides, about 12,000 accessions of foxtail millet are held in 45 local agricultural institutes in the northern part of China. Seed viability is maintained by rotational planting in fields. Usually, seeds are stored in seed depots, where the temperature and moisture are low and regenerated at an interval of three to five years.

In Qinghai Province, because of its dry summer and cold winter, a medium-term storage seed depot was built using natural low temperature and moisture for conserving seed materials. Other methods like locating seed storage facilities in very dry places, or storing small quantities of seeds in moisture proof containers have also been put to use.

At present, about 2,000 accessions are stored in a medium term bank at the Institute of Crop Germplasm Resources, CAAS Beijing, at temperature −10°C, in tight cans. Above all, a new gene bank for long term storage is already functioning in Beijing, and will conserve about 15 thousand accessions of foxtail millet seeds in next five years (1986-1990) at temperature −18°C and relative humidity of 50 per cent in sealed vacuum cans.

Characterization and evaluation

Characterization for more than 20 agronomic and biological descriptors concerning plant morphology and yield characteristics has been underway for many years. More detailed observations on 60 or more descriptors were also made on 3,000 foxtail millet accessions and information on data cards are put in computerized CAAS ICGR Millet Germplasm Data Bank. A sample of this variation is shown in Fig. 4.

So far, individual research needs are met according to the selection and breeding requirements such as pathological resistance, stress resistance, blast, downy mildew and kernel smut resistance, drought tolerance, salt endurance and nutritional quality needs. The progress in this regard is encouraging and

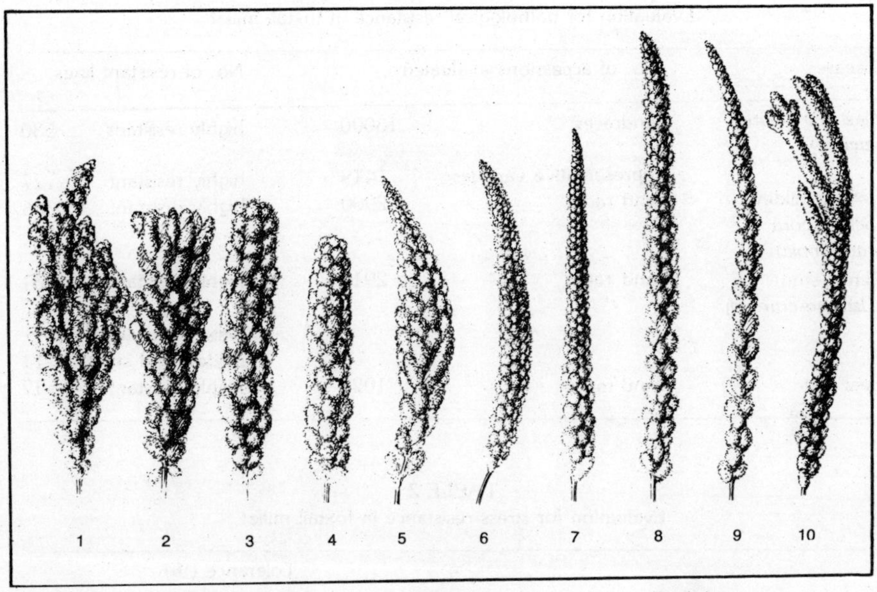

Sample Number	ICGR, CAAS Accession Number	Mandarin Name	English Translation
1	1039	Long Zhua Gu	Dragon's claw
2	1151	Ma Qing Miao	Green seedling
3	149	Diao Ba Qi	Handle
4	15	Huanong Si Hao	Huanong No. 4
5	359	Tie Ji Zui	Hen's beak
6	381	Jian tou Mao Yi dan	Tapered ear, long bristle
7	1138	Da Bai Gu	White grain, long duration
8	545	Qing ca Gen	Green stalk
9	73	Ma bian Zi	Whip
10	1012	Ma Zhua Huang Gru	Cat's paw, yellow grain.

Fig. 4. Sample of the variation in foxtail millet germplasm in China.

various resistant and stress tolerant materials have been identified (Tables 1 and 2).

Most studies on foxtail millet are described in the books *Manual of Cultivars of Chinese Foxtail Millet* and *Cultivation of Foxtail Millet in China*.

Proso millet genetic resources

Collection of proso millet started in 1940s and some 5000 accessions of proso millet were collected from 11 provinces in northern China. In these collected materials, 17 descriptors were studied and about 4200 accessions were

TABLE 1
Evaluation for pathological resistance in foxtail millet

Disease	No. of accessions evaluated		No. of resistant lines	
Blast (*Piricularia setariae*)	landraces	13000	highly resistant	530
	representative varieties	419	highly resistant	27
Downy mildew (*Sclerospora graminicola*)	land races	2900	highly resistant	366
Kernel smut (*Ustilago crameri*)	land races	2910	highly resistant	61
			resistant to downy mildew and smut	11
Leaf rust	land races	1021	highly resistant	17

TABLE 2
Evaluation for stress resistance in foxtail millet

Stress	No. of accessions or land races screened	Tolerance (%)			
		Rate I	Rate II	Rate III	Rate IV
Drought tolerance	1015 (Pot screening)	7	12		
	1008 (Field screening)	2.5	25		
Salt endurance	1929	0	0	9	9.8

described. Among them 53 per cent are non-glutinous and the remaining 47 per cent are glutinous varieties. In China the non-glutinous varieties of proso are called 'Ji' and glutinous varieties are called 'Shu' as they had long been recognized as different crops since ancient times.

Ecological studies, classification and descriptions of 400 varieties are given in the book *Manual of Cultivars of Proso Millet in China*.

10

BREEDING AND VARIETAL IMPROVEMENT OF FOXTAIL MILLET IN CHINA

Chen Jiaju

BACKGROUND

Recognizing the role of improved varieties in agricultural production, many regional breeding organizations were established during the 1950's.

Germplasm of foxtail millet, are diverse in morphology and adaptability. Since the 1920's land races were collected and selections were carried out; by the end of the 1940's, several varieties such as Yanjing No. 811, Huanong No. 4 and Biangu No. 1 were released for cultivation.

In the beginning of the 1950's many local varieties were evaluated and extended to farmers. Meanwhile, single-ear and bulk selection approaches were followed at many research organizations and from the later part of the 1950's to early 1960's, pedigree selection became a primary approach. In 1959, the first variety "Xinnongdong No. 2" was bred through hybridization. By the 1970's 50 per cent of the varieties used in production were derived from hybridization breeding which has been the main method of foxtail millet improvement in the country. Induced mutation breeding by irradiation also started in the 1970's. Achievements in developing male sterile lines in foxtail millet have been made, and a line with high male sterility (more than 95 per cent) was reported, and hybrid F_1 seeds could be produced by the two-line method. Most of the land races could restore fertility when crossed with the male sterile line. Since some 5 per cent seeds could be harvested from the sterile line under open pollination, no maintainer line is necessary.

BREEDING METHODS

Foxtail millet is essentially a self-pollinated crop and usual breeding methods are all applicable and effective. The breeding methods adapted in China are briefly described here:

1) Introduction

Varieties introduced from abroad mostly exhibited poor adaptability and were susceptible to diseases. Direct use of this exotic germplasm was rare. However, a few lines from Korea showed good adaptability and grew well. From these introductions the varieties such as Yonyi and Jigu No. 2 were selected for direct use. Many introductions have been used as parents in hybridization. Therefore, we are interested in extending the scope of introduction from foreign countries.

2) Selection

Single-ear selection was effective in the early stages, particularly in the landraces having many derivative forms. However, continuous selection was not so effective in pure lines or homogeneous landraces.

3) Hybridization

Since the floret of foxtail millet is very small, hand emasculation is very difficult and ineffective. During the 1950's hot water emasculation technique was developed and widely used. But emasculation with hot water or other chemicals can hardly be complete. So, normally a dominant marker character is used to identify true F_1s. In a hybridization programme, parents should be carefully chosen to produce genetically superior progenies. In special cases, interspecific hybridization may induce sterility, which is considered to be one of the ways of generating male sterile lines.

4) Rapid generation advancement

In China, rapid generation advancement (RGA) and overstepped advancement of selection are widely followed.

ACHIEVEMENTS

From 1949 to 1979, 158 varieties, developed by breeding, were grown in a total area of 4.5 million ha. Among the 422 varieties listed in the *Manual of Cultivars of Chinese Foxtail Millet*, 197 are derived from breeding. Table 1 shows the origin of these 197 bred varieties.

Four aspects have contributed to the increase in foxtail millet production.

1) Extensive use of improved varieties have greatly increased productivity of foxtail millet. Many varieties such as Longgu No. 23, Longgu No. 24, Suigu No. 1, Angu NO. 18, Hualian No. 1, Gonggu No. 6, Baisha No. 971, Baishazhan, Zhaogu No. 1, Meligu, Xinnong No. 724, Jingu No. 1 and Jingu No. 2 are grown in more than 67 thousand hectares (million mu in Chinese system). Improved varieties are popular in all areas except in mountainous regions. In certain locations, varieties were renewed two to three times.

TABLE 1
Methods adopted in bred varieties

Method	No. of bred varieties	Percentage
Pure line selection	134	68.1
Hybridization breeding	57	28.9
Artificial induction	6	3.0
Total	197	100.00

2) The level of resistance to lodging, diseases, insects and various forms of stress in improved varieties have been increased. Jigu No. 6 and Yugu No. 1 are resistant to lodging. Jigu No. 1, Jinggu No. 1 and Minquanginggu are resistant to blast, Lujin No. 3, Beihuang No. 3 and Zhenggu No. 2 are resistant to green ear disease (downy mildew). Some of these varieties are grown in more than one hundred thousand hectares.

3) Development of early maturing varieties has facilitated the improvement in cropping systems. Early varieties can be sown after the harvest of winter wheat and they mature before the onset of sowing time for the next crop of winter wheat. In northeast China, early varieties can be grown in the region north of latitude 50°N.

4) Besides ordinary types, special types with glutinous, or white or green or grey endosperm and with good seed quality were also developed.

REGIONAL TESTS FOR VARIETAL EVALUATION

Three major locations for the regional test of foxtail millet cultivars were established namely highland spring sowing region, north China plain summer sowing region, and northeast China spring sowing region. In provinces, there are provincial, districts and country regional tests sponsored by local evaluation committees. The national evaluation committee for regional tests is responsible for the evaluation of eight crops including foxtail millet.

PRESENT STATUS OF FOXTAIL MILLET IMPROVEMENT

In north China, there are 12 provinces and districts where breeding units for foxtail millet are located. In them, there are several provincial academies of agricultural sciences that have undertaken primary work on foxtail millet breeding in Hebei province, Shanxi province, Heilongjiang province, Jilin province, Inner Mongolia Autonomous Region, Shandong province and Henan province. The present breeding goal is to develop varieties that are suitable for spring and summer sowing, have high yield potential, good stress resistance and good eating quality. Specific varieties are also bred for areas with one or

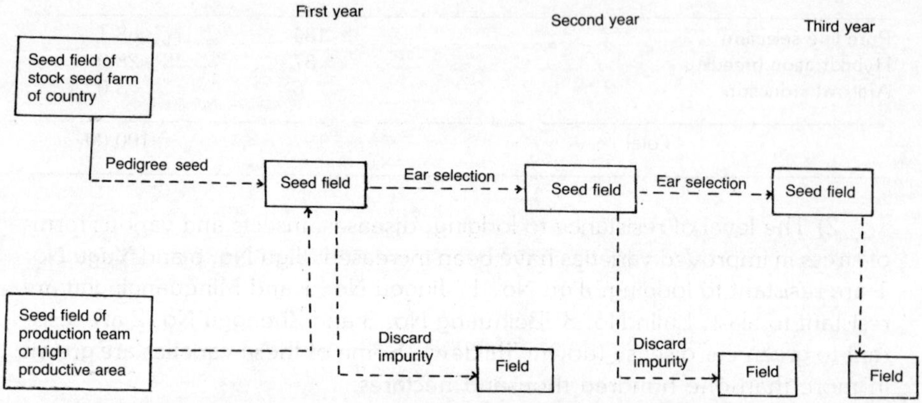

Fig. 1. Procedure of seed production in one-step seed field

two major problems, like drought or/and cold weather, etc. Utilization of heterosis in foxtail millet has been included as one of the breeding goals.

MULTIPLICATION AND PROMOTION OF IMPROVED VARIETIES

Establishment of national seed companies was necessary to cope up with the multiplication and promotion of improved varieties. Localization of certain varieties in some areas necessitate seed production, standardization and mechanization of foxtail millet cultivation. More than 2,300 seed companies have been established at various levels of local government provincial, district and country level. Multiplication of seed is undertaken according to the national seed production procedures (Fig. 1). Since foxtail millet has a high reproductive rate, field selections, which is a traditional method, can still be used by farmers who might probably develop certain special types for steadying the foxtail millet production in their areas.

11

BREEDING OF PROSO MILLET (*Panicum miliaceum* L.) IN VOLGA REGION OF USSR

V.A. Ilyin and E.N. Zolotukhin

The main millet growing areas of USSR are the Volga region and parts of Kazakhastan, which are characteristic of semi-arid conditions, with an annual rainfall of 250-400 mm. As only 35 to 40 per cent of the precipitation is received during the crop growth period, the breeding strategies are to evolve suitable varieties for these conditions.

Many improved varieties have been developed at the Millet Breeding Department of the Agricultural Research Institute for Southeast Regions located at Saratov and these varieties have occupied more than 50 per cent of the millet growing area in the U.S.S.R.

The present breeding strategy is to develop varieties with the following agronomic characteristics:

1) Drought tolerance
2) Early maturity
3) High crop growth rate and photosynthetic efficiency
4) Bigger panicle size and higher grain number
5) Resistance to diseases and pests
6) High quality of the grains
7) Wider adaptability to suit the local agroclimatic conditions.

With the above objectives in mind, breeding work is being carried out at this institute.

An important subspecies of the proso millet (*Sanguineum*) with red kernels is widespread in the dry southeast areas. This subspecies is fast growing with vertical development of the root system, a high hydraulic conductivity, compact dense panicles and glumes. The important feature of the seed is its maintenance of bright yellow colour till harvest and during storage. We are carrying out our breeding programme with the existing variability in this subspecies.

The main breeding method adopted in proso millet is intraspecific hybridization. We have been consistently improving the methodology of crossing to suit the local climatic conditions. Crossing is done by squeezing the style and stigma and it does not warrant opening of florets again for pollination. The best results are obtained when pollination is carried out by dry emasculation as per the method of Borluag (twirl method). Often as high as 70 per cent success is achieved in obtaining hybrid seeds which otherwise is not possible by adopting other methods.

In our laboratory, considerable work is being carried out on the theoretical genetic aspects. An attempt was made to study the genetics of grain colour. The grain colour is determined by at least four genes, but these are not completely dominant. One of the genes seems to code for white colour, which acts as an inhibitor. The dominance of colours follows order of: white, brown, grey, cream and red colour.

With regard to seed size, we have established that the lower test weight is dominant over the higher test weight. By the hybridization programme, it has been possible to increase the range in 1000 seed weight from 5.5 to 6.7 g which was prevailing in varieties grown in the 1930's to 7.4 to 8.19 g in the present-day varieties. The local land races possess the dominant genes and hence their grain weight is always low. They might have inherited this character from their wild ancestors and through natural selection. This resulted in smaller grains and more grain number per panicle.

The variety Saratov 853 was developed with increased test weight of 7.5 g. However, despite considerable selection work carried out between 1938 and 1968, it was not possible to increase the seed weight in the local varieties. On the contrary through the hybridization programme, recessive genes were combined and significant improvement was made in increasing the 1000 seed weight. Some of the newly developed cultures have a test weight of even 10 g. However, it is necessary to mention here that although test weight was increased, the seed viability seems to have decreased. Further, it can be concluded that it may not be possible to increase the test weight beyond 9.0 g with longer viability.

The inheritance pattern with regard to seed number per panicle has been extensively worked out. The same plant material which was used for increasing test weight was used for this purpose also. In the F_2 generation, seed number is super-dominant, consequently resulting in higher seed number than the parental lines. Breeding for high seed number is relatively difficult, irrespective of its dominance nature and this character is always associated with a few other important characters which determine productivity. It is difficult to achieve high seed number through mutation breeding. With an objective to evolve a suitable variety with high seed number, we adopted the following approaches:

1) The parental lines selected for the breeding programme were predominantly varieties with high yield per unit area through high seed number

per panicle. This facilitates considerable variability in the subsequent generations.

2) Selection was carried out under conditions which facilitated the development of more seed number per panicle.

3) Secondary selection pressures were applied to identify lines with high seed number.

4) During the process of selection, importance was given to synchrony in flowering, complete grain filling from the lower parts of the panicle with high density and high flower to seed ratio or seed fertility.

The main emphasis in proso millet improvement in our institute has been to develop varieties with high productivity, good grain quality and disease resistance particularly for grain smut.

BREEDING FOR PRODUCTIVITY

Selection of parent materials

One group of parents consisted predominantly of high yielding local cultivars having local adaptation with high plasticity. The other group was selected taking into consideration the ecological and geographical classification of proso millet developed by Lisov.

The best results were obtained by crossing cultivars from the dry plains of the Volga, Ukranian and Kazakhastan regions. For instance the variety Scorospeloe 66 was developed by crossing Aureum 1113 (Kazakhastan group) with Saratov 853 (Volga group). Similarly, the variety Volga 3 was evolved by crossing hybrid populations of Victoria 48 (Khazakhastan) with the hybrid populations obtained from, Orenburgskoe 42 and Sanquineum 75 (Volga group).

Whenever diverse types were used in hybridization we adopted a multistep crossing programme, first developing stable lines and further crossing these lines to develop the required cultivars. For example the variety Start was developed by this method. Hybridization was started in 1956. The female parent was obtained by multiple crossing and through selection in F_3 generation. From this multistep programme a uniform population was developed known as Cremy-15. The male parent Sanguineum-7 was developed in the sixth generation after crossing and in 1969, these two lines were crossed which resulted in the development of the variety start which is widely cultivated now. The salient features of such a multistep programme are:

1) Involvement of diverse ecological and geographical groups;

2) Repeated hybridization with samples from the Volga group;

3) Continuous directed selection;

4) Repeated involvement of local high yielding cultivars with high grain quality, and

5) Appearance of transgression in a number of characteristics.

The investigations carried out led to the conclusion that there are two distinct biotypes with two distinct yield attributes. The contribution of these factors were assessed by using a row of formulae which helped us to identify characters associated with higher productivity.

The local varieties are generally early maturing, but with small seeds. Further, they are also drought resistant with high survival under stress. Further development of this biotype resulted in evolving a variety, Scorospeloe 66 in 1962. Similarly a few other varieties like Saratov-3 and Saratov-6 were developed during the later years. The main characters associated with productivity in these two lines were higher survival under stress and higher plant density per unit land area.

Working on various aspects of improving productivity Arnold selected from the locals a few late maturing lines from which the variety Saratov 853 was selected later on. From this biotype a few other varieties were developed like Volga-3, Saratov-3 and Start. These varieties are relatively long duration types with bigger panicles and high grain number. In these varieties the grain number per unit land area was high and the yield increase was also due to higher seed weight. The biotype-I is more drought resistant and makes best utilization of winter and spring precipitation whereas the biotype II is more productive and more responsive to soil fertility and precipitation during the later half of the summer season. The relative performances of these varieties are given in Table 1.

TABLE 1

Comparative yields of varieties of two different biotypes

Variety	Biotype	Yield (q/ha) 1983	1984	1985	Mean	Yield increase over Saratov-853 (q/ha)	Duration (days)
Saratov-3	I	20.2	11.0	23.4	18.2	2.1	84
Saratov-6	II	21.4	13.1	22.6	19.0	2.9	84
Saratov-853	II	18.2	11.2	19.0	16.1	0.0	84
Volga-3	II	20.7	13.2	22.2	18.7	2.6	87
Start	II	24.5	14.5	23.7	20.9	4.8	92
(Sanguineum-7 × Brown) × Start	II	25.4	16.1	23.8	21.8	5.7	90
(Start × Scorospeloe-66) × Orel-92	II	25.1	16.0	25.5	22.2	6.1	92
CD at 5%		2.7	2.2	2.0	—	—	—

The new lines of biotype-1 gave higher yield over the check variety Volga-3 by about 3.1 to 6.9 q/ha. This indicates that by increasing the duration up to a reasonable limit, productivity could be enhanced.

Selection for grain quality

The work in this area is directed to improve technological and consumer qualities of the seeds of the varieties. Among the varieties developed in our laboratory, variety Saratov-2 could not meet the consumer requirements.

As regards quality parameters, the selection work is carried out for bigger seed size, bright endosperm colour and density, and resistance to Melanos disease. In all such studies, the standard variety for quality comparison has been Saratov 853 although in recent years there are many varieties with better grain quality than Saratov 853. Another variety Saratov 6 is more stable in grain quality across ecological zones.

Among many grain quality characters, the brightness of the endosperm which contains carotenoids and resistance to Melanos disease are most important. Varieties with bright yellow endosperm are preferred by consumers. The brightness of the endosperm is directly related to the content of the carotenoids. Hence brighter the yellow colour greater the carotenoid content. Such types are not only preferred by consumers but also are more nutritious. However, this character is unstable and hence needs careful assessment. Red-coated varieties are preferred mainly because they offer protection from light degradation of carotenoids under field conditions.

To improve seed quality characters, variety Sanguineum-7 is often used in the crossing programme. This variety was developed as a result of transgression and possess bright endosperm with more carotenoid content and also high test weight (Table 2).

In some years due to the attack of Melanos disease the endosperm becomes black. To some extent the varieties with big seed and inadequate compactness of the glumes are susceptible to this disease. Attempts have been

TABLE 2
Grain quality of Sanguineum-7 and its parent forms, 1977-79

Variety	1000 seed weight (g)	Brightness of endosperm (index)	Extent of grain filling (index)	Carotenoids (mg/kg)	Infected seeds (%)	Tests of prepared produce (index)
Sangineum-75	7.9	2.9	3.0	9.8	1.2	4.4
Aureum Chakinskoe × Aureum-38	8.2	2.2	2.2	8.9	7.4	3.6
Sanguineum-7	8.5	3.9	3.9	12.4	2.5	4.5
CD at 5%	0.2	0.3	0.3	1.1	1.3	0.5

made to transfer resistance to this disease by crossing resistant genotypes from world germplasm but have not met with any success. There is always a negative relationship between the seed size and the extent of disease-affected seeds.

The best approach to identify types having resistance to Melanos has been through the adoption of suitable breeding techniques. This approach only ultimately led to the development of resistant types with big seed size like Saratov-6.

Breeding for smut resistance

Smut is the most widespread disease on proso millet and resistance is controlled by a single gene. A selection procedure has been developed to identify more resistant varieties. The main feature of this selection process is to adopt vigorous roguing at different stages of breeding programmes right from the beginning.

Breeding for resistance is a difficult and slow process because most of the resistant donors are poorly adapted to the local conditions. The bulk of the breeding material has been discarded purely based on one character, susceptibility to smut. We feel that the selection of resistant types in infected sick plots is not always advantageous, besides it is often difficult to combine high resistance with other desirable plant characters.

One of the important aspects of this programme was to identify the suitable parent material. Initially, we used the varieties VNIS 29 and VNIS 223 × 1843 as resistant donors. However, these types did not possess the required drought resistance, or other agronomic characters; they were of long duration and seed quality was poor. So, we adopted a multistep selection procedure by crossing types of different ecological and geographical forms. This resulted in the development of the variety Saratov-2, the first resistant variety which is widely cultivated in the drought-prone areas of southeastern U.S.S.R.

From 1967 onwards we started adopting the backcross breeding technique to obtain better results. By this method it was possible to develop Saratov-3 resistant to smut. This variety possesses all other economical and biological properties as that of Scorospeloe 66 from which it was developed. They are isogenic except for resistance to smut. For the transfer of resistance, line no. 356 has been used, which is acclimatized to local conditions. Further breeding programme for smut resistance was carried out by using this material which ultimately resulted in the evolution of varieties, Sanguineum-7, Saratov-3 and Saratov-6.

These new varieties were evaluated against the local races of smut populations. In these resistant lines, the gene responsible for resistance was obtained predominantly from a population 1843 (K-8753) and was named as Sph-1. As we started growing relatively resistant types under field conditions new virulent smut races emerged and these resistant varieties became susceptible. The new strain of smut developed in Saratov region was named No. 2 and the strain which was developed in the Ukraine region was named No. 3. Work is now in progress to identify new donors from the national collection. Our studies have shown that the character is controlled by a single dominant gene.

From the national collection it was possible to identify lines which contained a resistant gene which is named Sph-II. This possesses resistance to race 1 and 2 of smut. Similarly we have identified lines containing gene named Sph-III, which is resistant to race 3. We have already developed resistant populations with high productivity and grain quality on par with the standard variety.

Our results have shown that breeding for smut resistance is more complicated than we thought earlier. In this connection we feel that it is necessary to develop multiline varieties, to adopt phytosanitary measures and frequent change of the varieties in order to maintain the level of resistance for a long period.

III

IMPORTANCE, GERMPLASM AND VARIETAL IMPROVEMENT IN AFRICA

12

FINGER MILLET RESEARCH IN THE SADCC (SOUTHERN AFRICAN) REGION

S.C. Gupta, S. Appa Rao and L.R. House

INTRODUCTION

Among the millets of the world, finger millet ranks fourth after pearl millet (*Pennisetum americanum* L.), foxtail millet (*Setaria italica*), and proso millet (*Panicum miliaceum*). There are two regions of the world where finger millet is most intensively grown; immediately surrounding Lake Victoria in East Africa, the south-eastern parts of Karnataka and parts of Tamil Nadu and Andhra Pradesh in southern India. These regions account for nearly 75 per cent of the world's production of this cereal (Rachie, 1975).

In Africa, finger millet is produced principally in Uganda, Tanzania, Rwanda, Burundi, Eastern Zaire, Kenya and to a lesser extent in Ethiopia, Sudan and Somalia. It is also grown in Zimbabwe, Malawi, Zambia, Tanzania, Botswana and Madagaskar. In central and western Africa, it is grown to a limited extent in central Africa, southern Chad and northeastern Nigeria. In Uganda, finger millet is the most important cereal crop equalling the acreage and production of other cereals combined. It is really the only other millet of consequence after pearl millet throughout Africa (Leonard and Martin, 1963; Rachie, 1975). The most important region is in the vicinity of Lake Victoria and Lake Kyoga and between Lake Tanganyika and Lake Victoria (Johnson and Raymond, 1964).

The Southern African Development Coordination Conference (SADCC) Region consists of nine countries: Angola, Botswana, Lesotho, Malawi, Mozambique, Swaziland, Tanzania, Zambia, and Zimbabwe. All these countries are south of the equator. Finger millet is the most important small millet grown in this region. Teff is planted in certain areas of Swaziland but no scientific information is available on this crop. Some research work and germplasm collections of finger millet were carried out in the last few years in several SADCC countries. A systematic regional finger millet programme was started in 1985 by SADCC/ICRISAT (International Crops Research Institute for the Semi-Arid Tropics) Sorghum Millet Improvement Project at Matopos (Zimbabwe) for the SADCC region. In this paper, efforts are made to review the knowledge available on finger millet in the region.

AREA AND PRODUCTION

The area and production of finger millet is often combined with pearl millet and in some countries such as Angola and Malawi, the statistical division reports millets and sorghum combined together. The figures vary greatly from source to source and year to year. However, the area and production data on finger millet is available for Zambia and Zimbabwe which is presented in Tables 1 and 2 respectively. Over 100,000 ha of finger millet is grown in each of these two countries with an average yield of 500 kg/ha. Tanzania is the other important country where over 100,000 ha is grown. In Malawi, Mozambique and Botswana, finger millet is grown in patches. The crop is not reported in Lesotho and Swaziland.

GERMPLASM COLLECTION AND MAINTENANCE

Finger millet germplasm has been collected in certain regions of Zimbabwe, Zambia, Malawi, Mozambique, Tanzania, and Botswana during the past few years. The present status of germplasm collections country-wise is described here.

TABLE 1
Area and distribution of finger millet and sorghum in Zambia (1969-70)

Province	Finger millet		Sorghum	
	Area ('000 ha)	Production (tonnes)	Area ('000 ha)	Production (tonnes)
Central	7.4	4100	13.1	6500
Copper belt	0.8	300	8.8	13900
Eastern	11.9	6000	5.4	1000
Luapule	9.6	7600	0.3	50
Northern	43.6	38700	4.0	4500
N. Western	9.4	6500	11.7	6000
Southern	7.1	2200	15.0	8400
Western	40.7	19500	19.3	11100
	130.5	84900	77.6	51450

Source: World Bank Report No. 841-ZA, Agricultural and Rural Survey Coll-III, October, 1975.

TABLE 2

Area in Zimbabwe planted (in hectares) to maize, sorghum, finger millet and pearl millet in the 1980-81 summer season

Province	Maize	Sorghum	Finger millet	Pearl millet
Mashonaland West	66,534	1,910	2,755	151
Mashonaland East	85,960	1,507	4,833	2,560
Mashonaland North	44,696	5,663	3,124	3,260
Matebeleland South	92,770	59,322	6,735	97,621
Midlands	178,592	12,984	26,984	72,479
Victoria	275,572	32,775	53,244	40,938
Manicaland	88,220	44,818	21,931	26,162
Total	875,859	185,748	118,784	271,750

Source: Unpublished data from Agricultural Technical and Extension Services (Agritex), Zimbabwe.

Botswana

During 1985, the International Board for Plant Genetic Resources, (IBPGR)/ICRISAT in collaboration with the Department of Agricultural Research (DAR), Botswana, collected six cultivated samples of finger millet from the northern province (Appa Rao *et al.*, 1986a).

Malawi

During March-April, 1979, ICRISAT/IBPGR organized germplasm collecting expedition to Malawi in collaboration with the Ministry of Agriculture and Natural Resources of the Government of Malawi. A total of 1106 accessions were collected including 190 of finger millet (Appa Rao, 1979). These accessions were deposited at the Chitedze Agricultural Research Station. The collecting mission moved from Shire Valley in the extreme south to north through the Lakeshore, Lilongwe, Kasungu and up to Rumphi. The local name of finger millet is "Mawere" in the south, "Kapuku" in the central region and "Lupoko" in the northern region. It is extensively grown in the northern region around Mzimba and Karonga. There are two distinct types based on the head characters. The open type has long thin ribbon like fingers that open outward. It is called 'Phazi-la-ngobuu' which means elephant foot. The fist type in which short fingers fold inwards, is called 'Fumbata'. The landraces are classified into two maturity types. The early type which matures in about three months is called 'Nthanga' which is common around Mzimba. It has incurved heads and the seed is very small. The late maturing types which take more than 120 days to mature produce large heads and bold grain (Appa Rao, 1979).

Mozambique

During 1981, IBPGR/ICRISAT in collaboration with University of Edurado, Mondlane (UEM) and National Institute of Research in Agronomy

(INIA), Maputo organized a germplasm collection mission in Mozambique. This mission was mainly aimed at collecting groundnuts along with other crops which matured during the collection period. Two samples of finger millet were collected (Rao, 1981). Locally finger millet is known as 'Marupi'. It was suggested that an expedition timed during May-June should recover much variability in sorghum, pigeonpea, pearl millet and finger millet from most of the regions of Mozambique.

Tanzania

Two collecting missions were organized by IBPGR/ICRISAT in July 1978 and in May 1979. The regions covered were Morogoro, Dodoma, Singida, Shinyanga, Mwanza, Musoma, Babati, Kondoa, and Iringa. From these regions five finger millet accessions were collected. The important regions for finger millet are Rukuwa, Mbeya, Dodoma, Ausha, and Moshi where collection trip should be organized in near future. *E. indica* and *E. multiflesa* are the wild types which are commonly found in Tanzania (Rao and Mengesha, 1980).

Zambia

From 1980 to 1984, four collecting missions were organized by the IBPGR in collaboration with ICRISAT and other organizations in which a wide range of crops and diversity were collected from all parts of the country. This included 273 samples of finger millet (Attere, 1985). Finger millet is locally known as 'Luku Kachiaye', 'Katombela Massa'. Attere (1985) has indicated that the crop genetic resources in Zambia have been well collected. To conserve this germplasm, the Zambian Government is establishing a cold storage facility with assistance from the Swedish Government at the Mount Makulu Research Station. Zambia has been identified as a possible site for the SADCC centre for the mobilization of plant genetic resources.

Zimbabwe

Finger millet is the third important crop next to maize and sorghum in the communal areas of Mashonaland, Midlands and Manicaland provinces. It is referred by one of its vernacular names, such as 'Rupoko', 'Rukweza', 'Njera' or 'Zviyo'. Two germplasm collecting expeditions were launched in Zimbabwe during 1982 and 1985 jointly by IBPGR and ICRISAT in collaboration with the Department of Agriculture of Zimbabwe (Appa Rao and Mengesha, 1982 and Appa Rao *et al.*, 1986b). Variations existed for head compactness, number, length and type of finger, and grain colour. In some samples the fingers were stiff, erect and more or less divergent while in others they tend to curve inwards at the top. At several locations a mixture of wild and weedy types were found along with the cultivated types. In addition to cultivated finger millet, six wild accessions were also collected (Appa Rao and Mengesha, 1982).

VARIETAL IMPROVEMENT

In Southern Africa, breeding work on finger millet has been carried out in Zambia and Zimbabwe. Recently in other countries, evaluation of local accessions has started.

Zambia

At Mount Makulu Research Station in Zambia, breeding work started in the early sixties. This work produced a high yielding variety called M 144. This variety was very susceptible to lodging and had a less than desirable grain colour. Selections were made from natural outcrosses of M 144 for yielding ability and lodging resistance. During the course of breeding, the single plant selection method was used initially mainly to obtain pure strains. Later hybrid plants were selected and further developed by using the pedigree method. Varieties and selections were tested in two groups of variety trials in the main finger millet growing areas. Mean yields of the experiments were never under 2000 kg/ha while the best yielding strains often exceeded 4000 kg/ha level (Sarmezy, 1978).

From this programme a new non-lodging variety named Steadfast was selected. This variety was derived from an outcross of M 144 × Line 197. The grain colour was light brown and more acceptable than M 144. The variety Steadfast is recommended for all areas, especially where management levels are good. Attempts to produce improved strains from M 144 × *E. africana* outcross failed, and the selections were discarded after five generations (Sarmezy, 1978).

The work was restarted in 1984 by evaluating 175 local collections and the introduced genotypes. Twenty lines were retained for retesting during this year which were superior to Steadfast.

Zimbabwe

A brief reference to the selection of high yielding strains of *E. coracana* in Zimbabwe was made by Hill (1945) and the author concludes that slow-growing millets like *Eleusines* and *Pennisetums* are considerably lower yielding than sorghum and are therefore of lesser value for food consumption. He indicated that the chief advantage is their immunity against stem borer under Zimbabwe conditions.

Breeding of finger millet was initiated in as early as 1968 at Chibero Agricultural College, where students from African areas throughout Zimbabwe were asked to bring in unthreshed heads from standing plants of genetically mixed populations. Sixty-four pure lines were obtained by threshing single plants. These lines were evaluated for four years. The main conclusions were: there were significant differences among pure lines for yield, threshing percentage, grain weight per head, grain size, days to maturity and plant height. Grain yield was highly significant and positively correlated with grain weight per head, seed number per head, and threshing percentage. There was positive correlation between grain weight per head and the number of seeds produced per head. It appears that under local conditions the important characteristics of high yielding pure lines are high grain weight per head, which is brought about by high number of seeds per head, high threshing percentage and early maturity (M'shonga, 1974).

Systematic efforts were made in 1982 in the research project that was initiated by the Crop Breeding Institute of the Research and Specialist Services. The project started with the collection of different varieties grown by communal farmers throughout the country. All the samples collected were evaluated at Gweti variety testing centre and Panmure Experiment Station during the 1984-1985 rainy season. Promising lines were further evaluated during the 1985-86 season. The selected lines will be used to make crosses among themselves and with the introduced lines from elsewhere to produce high yielding varieties.

The reports on a variety trial of finger millet at the Lilongwe station near Mzimba was reported in Nyasaland quarterly (Nyasaland, D.A. 1952). The varieties, Phagalala and Fumbata gave the highest grain yields of 1937 and 1880 kg/ha respectively at this location. Some work was carried out in the seventies and four lines were identified.

SADCC programme

SADCC/ICRISAT regional program has evaluated 394 accessions collected from Zimbabwe (374), Zambia (14), Malawi (4) and Tanzania (2) during the 1985-86 rainy season. Sixty-seven accessions were retained for further testing. A total of 1285 accessions have been obtained (Table 3) and the seed is being multiplied. The selected accessions from the previous season and the newly introduced accessions will be evaluated at four or five locations during the coming rainy season. The selected entries will be utilized in a breeding programme.

CROPPING SYSTEMS

A kind of primitive agriculture is practised in Tanzania as described by Lunan (1950). This consists of turning over grassy chunks of sod to form mounds about 1 m in diameter and 60 to 75 cm high. Sometimes trees and bushes may be incorporated with the mound to be burnt during the dry season. The first crops to be planted are beans, cassava, sweet potatoes, chickpeas and wheat during the first season from February to April. Following summer rains, the mounds are broken and spread in November and finger millet or maize are sown. After harvesting in July, the mounds are again formed to cover the weed patches, resown with finger millet, and harvested the following June.

TABLE 3
Finger millet germplasm accessions available at SADCC/ICRISAT Regional Centre at Matopos (as on 20 September 1986)

Country of origin	Number of accessions NP[a]	PCQ[b]	Total
Zimbabwe	376	1	377
Zambia	14	0	14
Malawi	194	0	194
Tanzania	2	1	3
Uganda	27	6	33
Kenya	0	3	3
Ethiopia	0	3	3
Zaire	0	3	3
Mali	5	0	5
South Africa	0	1	1
India	0	649	649
Total	618	667	1285

a = Accessions contributed by national programmes
b = Accession contributed by Plant Germplasm Quarantine Centre, Beltsville, U.S.A.

Ir Kenya and Uganda, a similar practice is followed known as Chitemane cultivation: This consists of chopping down trees or lopping tree branches from a wide area, piling them over a much smaller area, burning and then sowing millet in the ashes at the beginning of the rains and without any further cultivation.

In Zimbabwe, finger millet seed is broadcast and brushwood is drawn over immediately or the field is trampled with animals. Gaps in the fields are filled by transplanting after thinning when the soil is moist. Weeding is commonly done by hand and fertilizers are seldom applied to the crop except by a few farmers around urban areas. The dried heads are stored without threshing. The crop matures in three to five months depending on the variety and the temperature. Finger millet is often mixed with maize, sorghum, groundnut, beans or cowpeas. However, it is also grown as a sole crop especially on virgin land after slashing and burning the bush. The most common rotation is groundnut, finger millet, and maize. Cucurbits are also grown along with finger millet, pearl millet and groundnut.

PRODUCTION TECHNOLOGY

Fertilizer use

In Malawi, application of kraal manure to finger millet increased yields by 400 per cent at Zomba in 1950 (Nyasaland, D.A. 1952). Similar experiments carried out at Lusaka (Zambia) in 1954, resulted in yields of approximately 1600, 500, and 350 kg of grain per hectare from manurial treatment of 6, 2 and 0 tons of farmyard manure per hectare (Rhodesia, C.A. 1955).

Agronomic studies during 1960 to 1971 in Zambia showed that finger millet respond well to ammonium sulphate at the rate of 200 kg/ha or, more preferably, in the form of top dressing applied six weeks after planting. The response to phosphorus, in the form of single superphosphate, was less marked and gave significant results only on virgin land which had previously received no fertilizer. The residual response to phosphorus was good when finger millet followed fertilized groundnut or soybean (Sarmezey, 1978).

Dhliwayo and Whingwiri (1984) studied the nitrogen and phosphorus response to finger millet at Makoholi Experiment Station of Zimbabwe. The responses of the crop to different nitrogen levels suggest 100 kg N/ha as the optimum level during 1980-1981 season. Inadequate nitrogen can limit yield in finger millet through a reduction in number of heads/m^2. This study also suggested that where the soils had moderate level of phosphorus, no fertilizer response was observed.

Seeding experiments

A factorial trial conducted at Baka, Karonga, in Malawi showed December sowing to be superior to February sowings; drilling on 45 cm row on the flat was superior either to broadcasting on the flat or to sowing two rows of hills on ridges 90 cm apart (Nyasaland, D.A. 1960).

Spacing and drilling experiments demonstrated that broadcasting was not the best or only method, as weed control and harvesting were easier to manage with drilled crop (Sarmezey, 1978).

Weeding and tillage practices

Finger millet is very labour-intensive at weeding time. The common practice of hand weeding in broadcast crops is time consuming and it is often difficult to distinguish rapoko grass (*E. africana*) from finger millet in the early growth stages (Howden, 1965). The crop cannot stand much weed competition in the early stages of growth.

Few generalizations can be made about tillage practices, planting patterns or land use patterns. Ploughing with animal traction, particularly oxen, is almost universally used in some countries such as Botswana, Lesotho and Swaziland, while hand tillage continues in some or most of the farming systems in the other SADCC countries. The majority of farmers plant crop mixtures but sole cropping has become more common in Swaziland and Lesotho. Historically important shifting cultivation and land use systems such as the Chitmene ash based system have given way to other systems in SADCC countries (Norman *et al.*, 1984).

PLANT PROTECTION

Diseases

Blast (*Pyricularia* sp.) has been recorded on finger millet in almost all regions where finger millet is grown . It is common in Uganda, and was particularly severe in very wet years such as 1961 and 1962 (Dunbar, 1969). The varieties Mozambique 359 was used as a source of resistance to *Pyricularia* in a programme to transfer its resistance to local Uganda strain (Uganda, D.A. 1958).

Phyllachora eleusinis Spet. has been reported from Uganda and Tanzania and causes tar spot on finger millet leaves (Small, 1922). *Cercospora fusimaculans* Atk. has been reported from Tanzania (Wallace and Wallace, 1947). *Gloeocercospora* sp. has caused severe blight of leaves in Malawi (Weihe, 1950).

Tar spot (*Phyllachora eleusines*) causes small, jet black, brown and slightly water-soaked spots or lesions which are irregularly distributed on both sides of the leaves. This disease usually appears late as the crop approaches maturity (Dunbar, 1969).

Striga

Striga hermonthica is an important pest of finger millet in east Africa. Control measures involve uprooting before seeding, crop rotation and the possible use of chemical herbicides.

Insects and animal pests: Common insects and pests are locusts and grasshoppers, stem borers and foliage caterpillars.

FOOD USES

Finger millet is frequently an important constituent in local beer making. Sometimes it is the major constituent, but frequently it is added to sorghum and other carbohydrate sources. The finger millet enzyme is reported to have a saccharifying power greater than the corresponding enzymes from sorghum or maize malt, but less than that of barley malt amylases (Patwardhan and Narayana, 1930).

In Africa, finger millet is mainly consumed either cooked as porridge or used in brewing (Dunbar, 1969). White finger millet is particularly well-suited for making porridge, with the dark brown grain colour preferred for making beer.

In much of central and southern Africa and India, finger millet is consumed as a thick porridge. It is called sadza in Zimbabwe, nsima in Malawi, ugali in Uganda or mudde or sankati in India. The porridge is prepared by adding the flour to boiling water with constant stirring to obtain the desired consistency. At times, dry cassava tubers are added to improve the texture of the porridge.

A fermented thin porridge called ambali is also consumed as food during lunch and breakfast. The flour is mixed with water. A small quantity of fermented flour is added as a starter and kept in a warm place for a day. This is then added to boiling water with constant stirring to obtain a porridge of free-flowing creamy consistency. It is cooled and eaten.

A particularly important feature in the humid tropics is the excellent keeping quality of finger millet grain which is the best of all the cereals.

As shown in Table 4, finger millet from Zimbabwe has a lower protein content than maize or the other tropical small grains. It is however, much richer in calcium than most of the other cereal grains (Johnson, 1968).

TABLE 4
Comparative analysis of Zimbabwe grains: percentage by weight

	Finger millet	Pearl millet	Sorghum	Maize
Dry matter	87.0	89.0	90.0	90.0
Carbohydrate	72.0	70.0	72.0	74.0
Crude protein	7.0	10.0	10.0	8.5
Fat	1.3	4.4	2.7	4.3
Fibre	3.4	1.8	2.1	2.1
Ash	4.0	2.6	1.7	1.5
Digestible protein	5.5	8.0	7.2	6.8
Total digestible nutrients	72.0	80.0	78.0	80.0

Feed uses

In a feeding study made in Zimbabwe, finger millet was fed to fatten pigs in combination with maize (75 per cent millet + 25 per cent maize) or pollards (70 per cent millet + 30 per cent pollards) and compared with maize; pollard blend (60:40). The protein contents in the above feeds were 15.5, 18.4 and 17.2 per cent, respectively (Calder, 1960). When 25 per cent of maize was replaced by millet, there was a small improvement in efficiency of feed conversion over maize alone or maize + pollards. However, the millet + pollards combination was less effeciently used than maize alone.

Grinding

Grinding of finger millet is considerably simpler than that of maize or kafir corn. Water is not used and the grain is merely ground in the pestle and winnowed, or sometimes sifted, winnowed and reground. This process resulted in meal recovery of 80 per cent with 18 per cent ofals and 2 per cent waste. It should be pointed out that rapoko is used commonly for brewing and is not commonly ground to a meal to make thick porridge. It was reported that the extraction rate by traditional method of grinding was 80 per cent for finger millet whereas it was poor for maize (55 per cent), sorghum (65 per cent) and pearl millet (75 per cent). Loss of vitamins (Thiamine and Riboflavin) was negligible for finger millet whereas it was up to 90 per cent in maize (Carr, 1961).

SUMMARY

Finger millet ranks fourth in area and production among cereals next to maize, sorghum, and pearl millet in SADCC countries. This is a poor man's crop and grown by small farmers. Very little research work has been carried out on this crop. Germplasm has been collected from Botswana, Zambia, Zimbabwe, Malawi, and parts of Tanzania. There is a need to collect germplasm from Mozambique, Angola and Tanzania. Finger millet is used as food known as sodza, and nsima (thick porridge). It is also used for brewing. The beer from finger millet is preferred over the beer made from sorghum, pearl millet and maize. The protein content of finger millet is inferior to many cereals such as maize, sorghum, pearl millet and wheat but it is very rich in ash and fibre content.

A systematic research programme for the improvement of finger millet for grain yield and quality was initiated in 1985 at Matopos (Zimbabwe) by SADCC/ICRISAT sorghum and millet improvement project. A total of 1285 germplasm accessions have been collected locally or introduced from elsewhere for evaluation at four to five locations in different SADCC countries during rainy season of 1986-87. The selected lines will be used in finger millet improvement programmes.

REFERENCES

Appa Rao, S. 1979. Germplasm collection mission to Malawi. ICRISAT, pp. 1-25.

Appa Rao, S. and M.H. Mengesha. 1982. Germplasm collection in Zimbabwe. Genetic Resources, Progress Report 41, ICRISAT, pp. 1-16.

Appa Rao, S., L.M. Mazhani and A.F.Y. Attere. 1986a. Collecting in Botswana. FAO/IBPGR *Plant Genetic Resources Newsletter* **68:** 27-28.

Appa Rao, S., J.N. Mushonga and R.F. Muza. 1986b. Collecting in Zimbabwe. FAO/IBPGR *Plant Genetic Resources Newsletter* **67:** 41-43.

Attere, A.F.Y. 1985. Germplasm Collection in Zambia. *Plant Genetic Resources Newsletter,* **62:** 11-14.

Calder, A. 1960. The value of Rupoko millet for pig feeding, *Rhodesia Agric. J.* **57:** 116-119.

Carr, W.R. 1961. Observations on the nutritive value of traditionally ground cereals in Southern Rhodesia. *Brit. J. Nutr.* **15:** 339-347.

Dhliwayo, H.H. and E.E. Whingwiri. 1984. Effect of nitrogen and phosphorus on finger millet. *Zimbabwe Agric. J.* **81:** 115-118.

Dunbar, A.R., 1969. *The Annual Crops of Uganda.* East African Literature Bureau, Dar es Salaam, Nairobi, Kampala.

Hill, A.G. 1945. The improvement of native food crops. A precis of the more important work done in East Africa during 1944 (Millets). E. Afr. Agr. Res. Inst. Amani No. DF/5/2: 15.

Howden, R.H.G. 1965. Rapoko. Extension officers Handbook. Government Printer, Salisbury.
Johnson, D.T. 1968. Finger millet—Rhodesia's neglected crop. *Rhodesia Agric. J.* **65:** 87-97.
Johnson, R.M. and W.D. Raymond. 1964. The chemical composition of some tropical food plants. 1. Finger and bulrush millets (*E. coracana* and *P. typhoides*). *Trop. Sci.* **6:** 6-11.
Leonard, W.H. and J.H. Martin. 1963. *Cereal Crops.* The McMillan Co., New York, 824 p.
Lunan, M. 1950. Mound cultivation in viira, Tanganyika, *East Afr. Agri. J.* **16:** 88-89.
M'shonga, S. 1974. Finger millet (*Eleusine coracana*) seed selection programme at Chibero Agricultural College. *Rhodesia Agric. J.* **71:** 51-55.
Norman, D., D. Baker and J. Siebert. 1984. The challenge of developing agriculture in the 400-600 mm rainfall zone within the SADCC countries. *Zimbabwe Agric. J.* **81:** 205-214.
Nyasaland Department of Agriculture. 1952. Report of the year 1950. Part II. Zomba 30-36.
Nyasaland Department of Agriculture. 1960. Report of the Department of Agriculture of Nyasaland for the period, 1959-60.
Patwardhan, V.N. and N. Narayana. 1930. Amylase from ragi (*Eleusine coracana*). *J. Indian Inst. Sci.* **13:** 38-41.
Rachie, K.O. 1975. The millets: Importance, utilization and outlook. Tech. Bull. 63 p. ICRISAT, Hyderabad, India.
Rao, V.R. 1981. Germplasm collection mission to Mozambique. Genetic Resources, Progress Report-35. ICRISAT, pp. 1-31.
Rao, K.E.P. and M.H. Mengesha. 1980. Sorghum and millets in Tanzania. Plant Genetic Resources Newsletter-180. pp. 21-23.
Rhodesia Department of Agriculture, 1955. Annual report for the year, 1954: 28.
Sarmezey, A.A.V. 1978. Finger millet Review of Agronomy and Breeding 1960-1977. Research Memorandum No. 24. Zambia, pp. 1-42.
Small, W. 1922. Diseases of cereals in Uganda. *Cir. Dept. Agr.* Uganda 8: 19.
Uganda, Department of Agriculture. 1958. Annual report for the year ended 31 December 1958. Uganda Dept. Agr. Ann. Rept., 75.
Wallace, G.S. and M. Wallace. 1947. A second supplement to the revised list of plant diseases in Tanganyika Territory. *E. Afr. Agri. J.* **13:** 61-64.
Weihe, P.O. 1950. The plant diseases of Nyasaland. Mycol. Pap. 53 Commonwealth Mycological Inst., KEW.

13

SMALL MILLETS IN UGANDA AGRICULTURE

Bill Williams Khizzah

INTRODUCTION

Uganda lies astride the equator, enclosed by latitudes 4° 12′ north and 1° 29′ south, longitudes 29° 34′ east and 35° 0′ west. Rainfall is bimodal and well distributed along the northern and northwestern shores of Lake Victoria. Away from the shore, northwards it becomes more monomodal with the peak coming in the month of June. Here, the annual rainfall varies between 120 and 180 cm and conditions are more of the equatorial type. Northeastwards the dry seasons are longer and much more severe. Over 80 per cent of the country lies between 900 and 1500 metres above sea level, with the daily temperature ranging between 30°C and 15°C.

Uganda is dominated by agriculture with over 80 per cent of the 15.3 million people involved in subsistence farming. The most important cereals are maize, finger millet and sorghum in that order. Other small millets such as foxtail, barnyard, kodo and teff are not grown at all. This paper therefore emphasises on finger millet as it is the only important small millet in Uganda agriculture.

IMPORTANCE

Finger millet occupies over 500,000 hectares annually and produces over 500,000 metric tonnes of grain. By area, the relative importance of finger millet, sorghum and bulrush millet are 55, 42 and 3 per cent respectively (Zake, 1985). The national production and area under finger millet, maize and sorghum for the years 1981-1985 are shown in Table 1. The figures for finger millet showed an upward trend in both area and grain production until 1983, and in 1985 the area declined to 312,000 ha. During the same period production also declined from 545,000 metric tonnes in 1983 to 210,000 metric tonnes in

TABLE 1

Area ('000 ha) and production ('000 metric tonnes) of major cereals in Uganda (1981-1985)

Crops	1981		1982		1983		1984		1985	
	Area	Production	Area	Production	Area	Production	Area	Production	Area	Production
Finger millet	300	480	330	528	341	545	332	223	312	210
Maize	260	342	285	393	295	413	347	281	271	251
Sorghum	170	320	200	256	207	407	206	164	186	148

Source: Ministry of Agriculture and Forestry, Uganda.

1985. Finger millet is the staple food for the Nilotic and the Nilo-Hamitic tribes of the interior plateau. It is an important item of diet for the Bantu people in the south and southwestern region and Sudanic tribes in the northwest on either side of the River Nile (Thomas, 1970). Because of its good storage ability and absence of major storage pests, finger millet is often kept aside as an insurance against famine, drought or lean periods.

The grain can easily be ground with simple tools (grinding stone) possessed by peasant farmers. The flour mixes well with dry cassava or sorghum to make excellent food *ugali* (stiff porridge). Finger millet is increasingly becoming a major cash crop. The grain can be sold directly for cash at local markets or shops soon after harvest or may be stored until the market conditions are favourable. Often grain is brewed and the beer is sold for cash. The grain may be used as a means of payment for labour wages either directly or in the form of beer or used in barter exchange for other commodities like meat, livestock or chicken.

Millet porridge is considered an important means of nutrition for expectant or breast-feeding mothers. The byproduct of millet beer is usually fed to chickens, pigs or goats. The use of finger millet straw as fodder in Uganda is not in vogue. Perhaps it is because there is always plenty of fresh grass available all the year round.

However, grazing in the field after harvesting finger millet is common, and with increasing pressure on land, the use of stalk as fodder needs to be investigated. There is a strong belief that direct grazing of stalks in the field causes abortion in cows, though reasons for this belief is not quite clear. In many parts the dry stalk is used as roofing material especially for storage structures.

DISTRIBUTION AND PRODUCTION STATISTICS

Finger millet is grown in almost all ecological areas of the country. Its distribution is closely related to the different tribal groups, their social history, background where they live, soil type, rainfall pattern, altitude and the yield potential of the varieties. On the basis of the above, five agricultural farming systems are identified.

1) The banana-coffee and banana-millet-cotton region

This is the fertile lake crescent area with high humidity throughout the year. It is difficult to dry the finger millet and there is no need to store reserve foodgrain as there is assured and continuous supply of fresh banana, maize and sweet potato. Therefore, finger millet is grown only in small patches for beer making. The crop becomes important between this area and the shorter grass area where rainfall is too low to support banana cultivation. This zone can be referred to as the banana-millet-cotton region.

2) The northern system

This is a system practised in the short grass or the millet areas of Acholi, Lango and West Nile districts. The soils here are more fertile and less heavily populated. The monomodal rainfall with a long dry season (December-February) makes finger millet the most favoured cereal. In West Nile, however, cassava tends to be more important. Throughout the district, finger millet is being progressively replaced by cassava though the former is still being extensively cultivated. Also in the northeast towards Karamoja where conditions are more severe and dry, finger millet gives way to sorghum and bulrush millet. This is mostly due to rainfall uncertainties and low soil fertility; with the proportion of each cereal varying in transitional areas from the pure finger millet stand to the pure sorghum or bulrush stand.

3) The eastern or teso system

This environment is found predominantly in the districts of Kumi, Soroti, parts of Tororo and Pallisa. It is characterized by light infertile soils, heavy bimodal rains and fairly prolonged dry season (December-March). Cattle care is an integral part of this system. Up to the mid-1970's cotton was the main cash crop. The situation has changed since then with finger millet and cassava becoming cash crops.

4) The mountain system

This region is similar to the banana-coffee region except for higher altitudes and more dense population. Finger millet is found in Kigezi and Rwenzori range in the south, and Mbale and Kapchorwa in the hills of Mount Elgon in the east. In Kigezi, sorghum has been substituted for finger millet. Factors responsible for this change may be high population increase and the need for more quantity of food even if less palatable, low rainfall, and probably falling soil fertility levels. Finger millet is grown in higher elevations in small plots year after year mainly for beer making. In Kigezi and the Rwenzori range, it occupies an important place and grains are used both for beer making and as food.

5) The pastoral system

This is a system which, since early 1960s was basically pastoral and agriculture was insignificant. The system is dominated by low and unreliable rains (70 cm/year) and extended dry season (November-April). Cattle farming is the most important occupation. The Suk and the Karamejong tribes in the east do cultivate sorghum, maize, bulrush mixtures in the drier parts and finger millet in the valleys and wetter areas of the west normally around permanent settlements. In Kabarole and Mbarara in the west, finger millet is grown in swamps and valley bottoms.

Finger millet area and production in different regions during the years 1981-1985 is given in Table 2. The data indicate the northern region as the

TABLE 2

Area and production of finger millet in different regions: 1981-1985

Regions	1981		1982		1983		1984		1985	
	Area (ha)	Production (t)	Area (ha)	Production (t)	Area (ha)	Production (t)	Area (ha)	Production (t)	Area (ha)	Production (t)
1. Banana, coffee, banana, millet and cotton	67341	109610	72652	88297	75007	120011	74263	58156	—	—
2. Northern region	112786	194482	111184	261980	114793	183567	121649	97180	—	—
3. Toso/Tororo region	70048	114621	67253	81737	69437	111099	80491	64392	—	—
4. The montane region	22162	36296	49526	60226	51164	81861	39477	31677		
5. The pastoral region	21667	24265	29425	35762	30379	48607	16122	12585		
Total	299995	480001	330120	528007	340780	545245	332002	223206	312000	210000

Source: Ministry of Agriculture and Forestry, Uganda.

largest finger millet producer accounting for about 40 per cent of the total national grain production, followed by the eastern region with 21 per cent. Production in the banana-coffee and the mountain regions has tended to remain constant. This is attributed to population pressure and availability of other fresh foods. In the pastoral system, production showed a steady increase, because, as more nomads settle in permanent homes, agriculture assumes importance. In Figs. 1 and 2 production trends and area for finger millet, maize and sorghum for the period 1981-1985 are shown. Production of finger millet and maize increased steadily from 1981-1983. Production of all cereals dropped sharply in 1984 and continued to drop in 1985 most probably due to political instability, increased insecurity followed by wars. This drop is also experienced in area for these crops (Fig. 2). However, the Government's commitment to increase grain production to meet the ever-increasing national demand is evident through its ten-year production projections from 1981 to 1990 (Fig. 3). As per these projections, the grain production by the year 1990 is expected to be 1.3, 1.2 and 0.92 million metric tonnes for finger millet, maize and sorghum respectively.

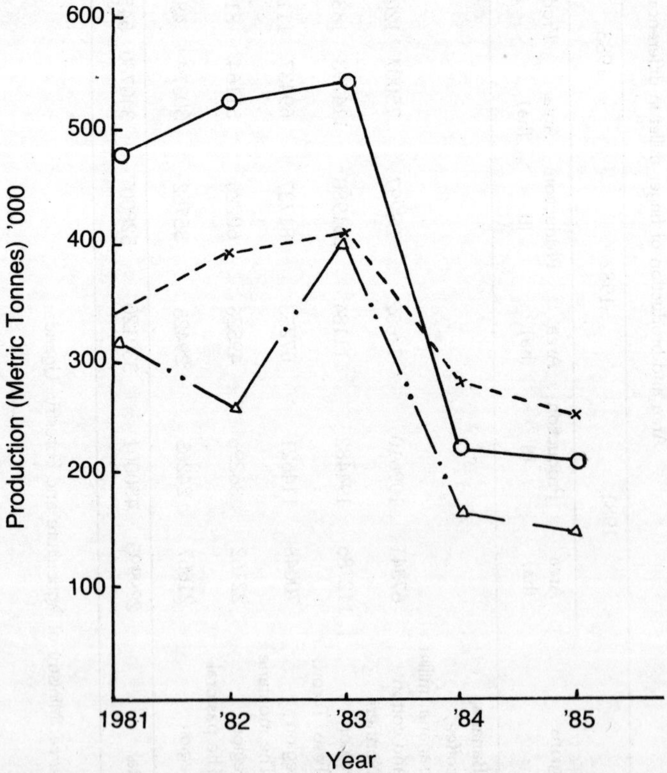

Fig. 1. National production trend for finger millet, maize and sorghum for 1981-1985.

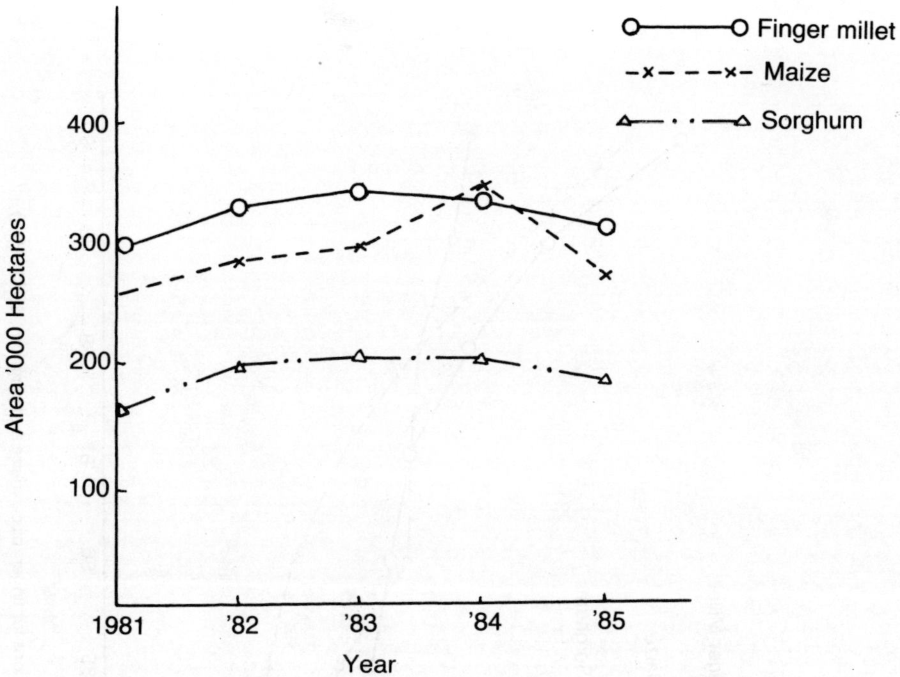

Fig. 2. Area '000 ha for finger millet, maize and sorghum for 1981-1985.

Local varieties

There are a number of well-known local varieties of finger millet grown throughout in different regions of Uganda. The most common ones in the northern region are: Odyera, Kalachol, Nalyongolyongo, Lawatmio, Embalasasa, Nyaracholi, Agwe and Adyela. In the western region: Kaguma, Mate, Kyomiguru, Omusara and Katomi are most common. In the eastern region Engeny, Okiring, Eding, Emiroit and Emoru are grown. Most of these varieties are identifiable at maturity by characters such as plant height, maturity period, panicle shape and size and grain colour. More than one type is often grown in the same plot. The recommended varieties like Engeny, Gulu E, Serere I and P 224 have not been largely adopted by the farmers due to the insufficient seed production and distribution system and poor extension service. However, the need for supply of improved seeds to farmers has been recognized at the station as well as at the national level (Zake, 1985). To help and solve these problems, "on the farm and adaptive research" programmes have been launched to understand the local farming systems, to assess the farmers' readiness to adapt the new technology, and to find out the degree to which such a change will affect farmers technical, economic and social status.

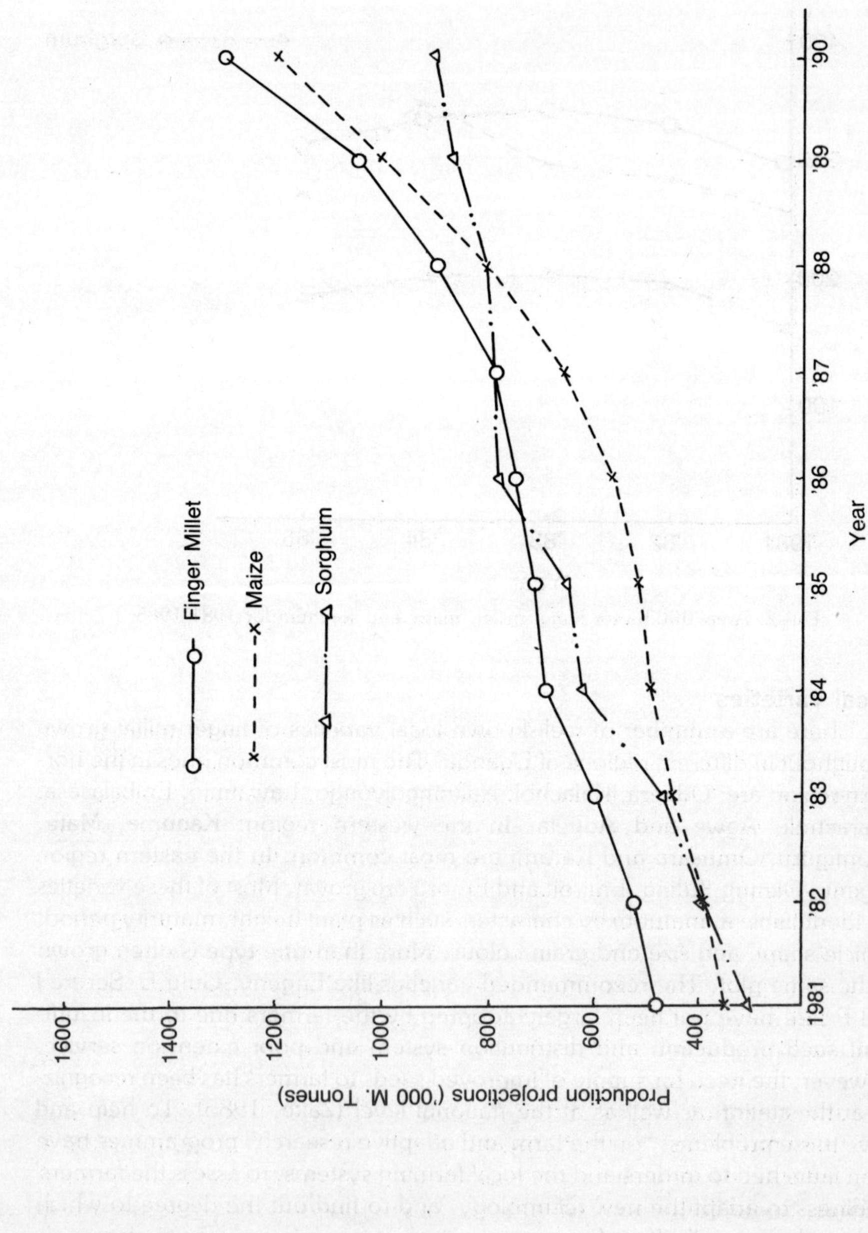

Fig. 3. Production ('000 metric tonnes projections for finger millet, maize and sorghum, 1981-1990.

GERMPLASM RESOURCES

Uganda is regarded as the centre of origin of finger millet because of its long traditions of religious and other ceremonies connected with its cultivation (Hulse, Liang and Pearson, 1980). There is extensive variability in the local varieties and in their wild relatives. Generally all varieties grown in the country have exposed grains which are known by local vernacular names. Wild species *Eleusine indica* (L.) Geartn. and *Eleusine africana* are common aggressive weeds; the latter being more aggressive because it matures, shed seeds and establishes seedlings much earlier than finger millet. The present finger millet and small millets germplasm activities at Serere includes efforts to assemble, characterize, evaluate, document, utilize, distribute and maintain the collections. Uptil the beginning for 1985, with the assistance of the International Board of Plant Genetic Resources (IBPGR); three collecting missions were conducted in Northern, Eastern and Western regions of Uganda. The existing finger millet and other small millets collection consists of materials as indicated in Table 3.

TABLE 3
Collections of small millets germplasm in Uganda

Crop	Source		No. of accessions	Total
Finger millet	Indigenous	Uganda		
		Northern	336	
		Eastern	27	
		Western	238	
		Others	177	778
	Exotic:	India	100	
		Kenya	26	
		Malawi	20	
		Nepal	10	
		Tanzania	3	159
Total finger millet accessions				937

Small millets	Seed source	No. of accession
Foxtail millet	India	83
Common millet	India	93
Kodo millet	India	84
Little millet	India	88
Total small millets		348

CONCLUSIONS

Finger millet is the most important cereal in Uganda exceeding maize and sorghum both in area and production. It is the staple food for over 50 per cent of the country's 15.3 million people and increasingly a major source of income.

It grows in all ecological areas of the country. Its preference for food is related to tribal and social groupings and ready available supply of other foods like banana, sweet potatoes, maize or cassava. The areas where finger millet can grow is also determined by rainfall and soils.

Much of the finger millet technologies have not been readily accepted because farmers do not consider them economical under the prevailing social conditions. On-the-farm and adaptive research approach have currently been introduced to help solve these problems.

Wider utilization of finger millet grains in livestock feeds, beer, biscuits and in making other finished products will help in widening its consumption and production. Its utility as fodder still needs to be investigated in Uganda.

BIBLIOGRAPHY

Hulse, J.H., E.M. Liang and O.E. Pearson. 1980. Sorghum and the millets: Their composition and nutritive value. Academic Press, London.

Thomas, D.G. 1970. Finger millet (*Eleusine coracana* (L.)) Gaertn. In: J.D. Jameson (ed.), *Agriculture in Uganda*. Second edition. Oxford University Press. 127-138.

Maxted, *et al.* 1985. Collection of sorghum and millets in Uganda. *Plant Genetic Resources Newsletter* **64:** 21-22.

Zake, V.M. 1985. Overview of sorghum and millets research in Uganda. In: Proceedings of the Regional Workshop on sorghum and millet improvement in Eastern Africa, Soroti, Uganda, 22-26 July 1985. SAFGRAD/ICRISAT, Eastern Africa Programme, pp. 19-21.

Oryokot, Joseph, O.E. 1985. Adoption of sorghum and finger millet technology in Uganda. In: Proceedings of the regional workshop on sorghum and millet improvement in East Africa, Soroti, Uganda, 22-26 July 1985. SAFGRAD/ICRISAT, Eastern Africa Programme, pp. 52-58.

14

FINGER MILLET IMPROVEMENT IN UGANDA

Vincent Makumbi Zake and Bill Williams Khizzah

INTRODUCTION

Uganda is regarded as the centre of origin of finger millet *Eleusine coracana* (L.) Gaertn. This is indicated by the long tradition religious and other local ceremonies connected with its cultivation and utilization (Hulse *et al.*, 1980). Finger millet has been grown in Uganda for over 5,000 years, and was probably introduced to India more than 3,000 years ago. Though found in many other tropical countries, it has gained little importance outside Africa and India. Its improvement in Uganda is perhaps as old as the crop itself. The practice, by farmers of selecting and preserving attractive panicles to form the next years seeds, is the first step towards the improvement of the local varieties. This practice is still common (Khizzah, 1985; Jameson, 1970).

In 1965, serious breeding work was started at the Serere Research Station. Prior to 1965, breeders at Serere had introduced and tested several local varieties for yield, disease resistance and other desirable characteristics. Finger millet breeding was aimed at developing varieties with high yield, early maturity for marginal rainfall areas of the country, lodging resistance, blast and virus disease resistance, and good grain quality. There is still a great need in Uganda to increase the national finger millet grain production to meet the ever-growing food demands.

The broad objectives of the finger millet improvement programme are:

1) To develop varieties of finger millet that are resistant to pests and diseases, acceptable to farmers and give consistent and persistent high yields over a wide range of environments.

2) To continue to evaluate the performance and adaptability of improved entries screened from existing programmes in as many locations or environments as possible.

3) To study the biology and epidemiology of blast disease and develop effective blast resistant varieties for utilization in resistance breeding.

4) To assemble and establish a large finger millet germplasm and their wild relatives and populations from as many locations as possible particularly within Uganda.

5) To continue to select for lodging resistance.

6) To select for high yielding, early maturing varieties for low rainfall areas of the country.

EMASCULATION AND HYBRIDIZATION TECHNIQUES

One of the problems of working with finger millet is the difficulty of making crosses. The traditional finger millet improvement through hybridization has only given modest increases in yield over local cultivars. The method used for emasculation is by hot water, which before 1974 was heated to 47° C and the heads immersed for ten minutes. In a recent modification the temperature has been increased to 52°C, and heads are immersed for only two-and-a-half minutes. Although this temperature gives a good pollen killing, seed set is low. Nonetheless one successful cross (F_1) gives several thousand seeds.

Hand pollination using special paper bags for collecting and dusting pollen on the female panicles is carried out between 7.00 and 9.00 a.m., four days after emasculation. Through traditional breeding methods, several white-seeded varieties were developed. These include WC 65 × Serere 1/1/78 and WC 65 × Engeny 1/1 which have given consistent satisfactory yields, though the yields are not significantly different than the best brown-seeded varieties, Engeny and Serere 1. Traditional breeding methods have been successful in producing varieties with stable, good yield across sites and years, yet no breakthrough in yield has been achieved. This may be due to lack of heterotic response or possibly limited genetic diversity. Single, double and triple hybrid crosses are usually done in the hybridization programme. However, due to the small amount of F_1 seed, no study has been possible to assess the heterotic response among hybrid crosses.

INDUCTION OF MALE STERILITY BY MUTATION BREEDING

An attempt to induce sterility by fast neutrons in several varieties of finger millet was made in 1971. Dosage rates of 2 Kr, 4 Kr, 6 Kr and 8 Kr were applied to each of the 32 varieties. No male sterility was identified in M_0 and M_1 generations (Mukuru *et al.*, 1974). However, it was observed that Gulu E treated with 4 Kr, had ten plants which failed to set seed in M_3 generation. These plants were suspected to be carrying a male sterility gene. The anthers were observed to be small, almost white to very light yellow and were devoid of pollen and were incapable of dehiscing or shedding pollen.

During the following seasons, further studies were carried out by crossing the identified male sterile plants in Gulu E variety with other varieties. All F_1 plants were fertile, however a segregation ratio of 3:1 fertile to sterile plants was not obtained in 10 crosses (Mukuru *et al.*, 1975 and Mukuru *et al.*, 1976). Studies indicate that while a ratio of 3:1 fertile to sterile could be obtained depending on the variety and season (Mukuru *et al.*, 1974) and that while other ratios of 5:1, 2:1 and 1:1 fertile to sterile could also be obtained, the existing variation in male sterility expression was influenced by modifier genes and the environment (Zake *et al.*, 1984). The male sterile trait has been introduced into more than four hundred lines. These have been pooled to form populations.

Okiror, Mukuru and Zake (1976) reported optimum seed set in male sterile plants if pollinated four days after flowering at 8 a.m. These authors reported significant differences in plant height and grain yield per plant between fertile and male sterile plants, but found no significant differences in tillering capacity and days to flower. Grain yield per plant for fertile was almost double that of sterile plants. Low seed set on sterile plants was attributed to insufficient pollen in the air at the time of optimum stigma receptivity, since pollination by hand resulted in good seed set. They further reported yield per plant to be significantly correlated with tillering capacity, days to flower and plant height for the fertile and male sterile plants in the population they studied. The results indicate that male sterility can reliably provide a mechanism to enforce outcrossing on a large scale and allow new selection schemes to sort out superior genotypes in finger millet. The hereditary control mechanisms of male sterility still need to be fully understood.

The use of genetic male sterility in finger millet has, to a great degree, enhanced outcrossing and widened the gene pool. In this connection four populations have been formed for early maturity, lodging resistance, pests and disease resistance and grain yield and agronomic performance *per se*. All populations were random mated for three seasons in isolation. For some populations, S_1 and sibs have been tested. Direct selections from the population to form varieties has also been done to advantage. However, testing for genetic gain and genetic advance among cycles of the populations was not done for some years, as remnant seed was lost in some cycles of the populations. Nonetheless, testing of cycles in all populations was carried out last year. Due to late planting and poor performance it was thought best to repeat the trial this year. F_1 crosses have also been made with known varieties and were yield tested with their parents for hybrid performance in 7×7 triple lattice design during the 1985 first rains. However due to poor performance, the trial has been repeated in 1986.

ONGOING RESEARCH HIGHLIGHTS

Early maturing finger millet

Early maturing finger millet varieties have high potential in areas of low and uncertain rainfall. They can also be grown during shorter, lighter second rains in certain parts of the country.

Attempts are being made to develop varieties with short maturity periods and high basal tillering capacities. Twenty-six early maturing populations were planted with 95 selections in 11×11 triple lattice design in a screening trial during first rains, 1985. The results are given in Table 1. Most varieties flowered under 70 days. Population derivatives P_2C_3G, and FAO 49380-012 flowered earlier than most entries and produced yields equal or greater than longer maturing recommended varieties. Several other early entries produced grain yields in excess of 25 q/ha.

TABLE 1

Grain yield (q/ha) and days to flower of early maturing finger millet screening trial grown at Serere 1st rains, 1985

Varieties	Days to flower	Yield (q/ha)
P_2C_3G	55	24.0
IE 718	72	16.2
IE 673	62	18.5
IE 601	66	9.7
IE 42	63	23.0
IE 501	63	13.1
IE 84	78	12.5
P_1C_1P	71	19.3
IE 600	63	15.0
P_2C_3P	65	25.4
PR 1091	63	12.1
P_4C_4G	73	19.1
IE 8	64	15.2
P_4C_3P	68	23.4
IE 84	70	11.1
HR-231-1	71	15.4
KM-1	62	21.3
P_1C_4G	68	23.8
IE 909	63	17.4
TAH 657	71	20.5
U1	78	9.0
P_2C_2P	66	22.3
P_2C_6P	56	18.4
Gulu E	64	25.2
IE 43	61	10.7
IE 48	56	10.5
CO 11	71	15.5
Ckiring	68	25.2
FAO D-005	70	13.5
EC 131784-J-098	65	3.1

Table 1 (contd.)

Varieties	Days to flower	Yield (q/ha)
HRG 91 B	60	19.3
WC 707	65	15.8
P_1C_5P	71	26.9
IE 798	62	10.0
P_2C_5G	72	22.8
P_2C_1G	64	20.1
SAD 149	62	21.5
Serere Cross 10	68	30.5
IE 94	65	24.2
HP-B-7-6	75	21.9
P_4C_5G	71	16.8
IE 805	72	12.3
P_1C_2G	71	16.4
U 30	66	15.0
IE 596	62	20.3
P_1C_4P	71	22.6
IE 40	73	12.5
IE 46	61	12.5
EC 131783-034J	65	9.6
PES 19	64	14.4
IE 11	65	10.7
HPB-83-4	73	10.3
P_670	63	12.3
IE 85	64	13.7
FAO 007	68	20.3
WC 445	68	27.9
IE 773	62	15.0
P 277	72	26.2
IE 413	65	21.5
IE 700	71	20.3
IE 588	62	8.2
FAO 008	69	23.2
WC 293	66	19.3
P_3C_4P	65	23.4
P 318	72	25.2
Serere-1	64	31.2
IE 580	71	10.3
IE 802	63	15.0
P_2C_2G	70	19.1
P 224	71	13.7
IE 602	62	11.7
Hamsa (red & white)	66	13.1
Eding	63	25.8
JNR-3B-008	74	11 9
FAO 004	59	16.2
P_251	68	32.0
IE 947	67	18.9
EC 132100	58	15.2
IE 672	67	18.2

Table 1 (contd.)

Varieties	Days to flower	Yield (q/ha)
IE 710	58	11.3
IE 14	66	10.3
IE 91	71	13.0
UR-403	71	17.8
P_3C_3P	69	17.4
IE 7	62	12.1
IE 685	64	19.7
U 15	73	18.2
P_3C_5P	66	17.6
P_3C_3G	67	20.5
WC 274	71	17.2
IE 713	65	10.3
P_2C_4P	73	17.6
U 10	70	26.0
E-KEP-4	72	11.7
IE 740	68	10.3
P_2C_5P	70	21.7
Engeny	67	22.3
IE 97	60	13.3
IE 508	62	16.4
IE 900	67	17.8
PIC3G	67	22.8
KEP 9	71	9.0
Illuvalai	64	12.1
IE 680	65	11.7
IE 593	62	11.7
FAO 49380-012	59	22.6
P_1C_2P	71	25.6
IE 13	63	12.5
CO-10	78	10.7
IE 170	72	22.8
P_4C_2P	67	5.7
P_4C_4P	66	14.8
P_4C_2G	72	13.7
PR 202	79	10.0
P_3C_4G	68	24.6
M-I-302	78	17.8
IE 946	67	18.9
IE 712	64	14.1
Mean	67	17.2
L.S.D. (t 0.05)	10	8
C.V. %	10.8	46.6

Local Check: Serere 1, Gulu E, Engeny and P 224

Finger millet screening trial

Twenty-nine popular local varieties were compared with 71 selections from exotic introductions. The results reported in Table 2 indicate that farmers varieties yield better than exotic introductions. More attention is to be focussed on the improvement of existing varieties in future.

TABLE 2
Grain yield (q/ha), plant height (cm) and days to flower of local varieties and exotic selections grown in 10 × 10 screening trial at Serere, first rains, 1985

Entry name	q/ha	Days to flower	Height (cm)
IE 945	12.4	72	82.3
Lawatmio	15.2	68	96.0
FAO 4930-002	9.1	70	97.3
SE × 33 × Emiroit-P4	4.8	70	96.7
JNR-3-008	13.3	71	89.3
FAO 49380	9.1	61	88.7
UR-403	7.0	70	92.0
IE 782	13.3	66	91.0
Serere-1	12.4	65	89.3
Adiang	15.8	70	100.0
Gurbati	20.6	70	96.0
Hamsa	10.0	76	80.7
FAO 49373	12.1	68	99.3
T 249	21.8	70	93.7
FAO 49378	9.7	67	83.3
Lopus-1	10.6	77	113.7
SAD 167	8.8	63	104.7
FAO-J-29	16.7	68	100.3
IE 588	2.7	58	73.7
Omuga	21.8	71	103.3
SAD 149	9.4	63	94.0
HR 344	8.2	69	96.0
EC 131647	8.5	84	87.3
P 224	14.8	71	104.3
ADOKE	13.9	74	95.0
P 277	18.5	71	99.0
FAO 49385D	13.0	71	89.0
FAO-J-27	10.3	72	104.3
SE × 46 × 102	10.9	71	89.3
IE 1037	18.5	70	88.7
U 15	10.9	61	90.3
IE 911	7.6	69	93.7
Okiring	18.2	68	96.3
FAO 007	8.2	69	102.3
PR 1091	5.2	67	79.3
IE 894	10.9	69	88.3
Otuka	11.2	70	99.7

Table 2 (contd.)

Entry name	q/ha	Days to flower	Height (cm)
Lajok-okwero	21.5	73	100.0
EC 131785	3.6	73	75.3
IE 897	4.8	70	93.7
P 278	16.1	69	102.3
P 318	16.7	72	99.3
IE 904	14.8	69	99.7
Nalyongo luongo	7.9	83	82.0
FAO-J-31	9.4	73	87.3
Oyoke	16.6	71	110.3
EC 1316 48	12.4	71	78.3
FAO-J-30	6.1	70	95.0
FAO 009	3.9	74	97.7
P 251	19.1	67	102.7
FAO 386	17.0	70	99.7
FAO 0003 C	11.5	68	91.0
IE 90	7.3	76	91.7
P 283	10.0	61	96.7
Lawiliwili	14.2	69	100.3
FAO D 00/OC	4.5	70	89.3
OKE 110	16.1	64	89.7
Eding	13.6	60	85.3
Egeta	20.6	67	100.7
Palake	6.4	72	88.3
FAO 4937-002	6.7	69	93.0
FAO D 00C	7.0	73	87.3
Todyang	19.7	68	101.0
FAO-J-25	9.4	73	88.7
IE 714	14.5	71	99.7
Aremo	18.2	67	100.7
P 211	20.6	64	92.7
IE 982	13.9	64	88.7
Adex Okwok-1	20.3	73	94.3
Erute	13.6	68	94.0
IE 413	13.0	66	100.3
IE 927	13.0	74	85.0
IE 891	12.1	70	99.3
PES 1A	3.0	60	74.0
Pajimo	18.2	64	92.0
Lajok Petalo	19.2	70	105.7
FAO D 006C	16.1	66	98.0
Indaf 6	7.0	71	89.3
Awiye Apam	17.6	64	83.0
FAO 49375	13.3	69	91.7
FAO 49387	18.3	68	94.0
Okama	20.0	65	91.7
Gulu E	12.1	67	100.3
CO-10	4.0	77	88.7

Table 2 (contd.)

Entry name	q/ha	Days to flower	Height (cm)
FAO-J-26	10.0	71	98.7
Engeny	14.5	66	92.7
FAO 49372	11.5	71	94.7
EC 131647	10.9	84	89.3
IE 902	6.1	74	91.7
Amola-1	12.4	67	101.7
Agang	9.1	73	102.0
FAO D 008C	11.2	70	98.7
FAO 005	5.8	69	104.7
Odyera	13.6	76	84.7
Lopus-2	20.0	74	92.0
Okuruwiye	18.8	68	95.7
IE 945	7.0	68	86.0
U 10	14.2	65	87.3
Indaf 5	10.9	76	78.3
FAO 49389	11.8	67	103.0
Mean	12.4	69	93.7
L.S.D. (P = 0.005)	0.27	3.5	14.7
C.V. %	53.7	7.0	11.5

Local Check: Serere 1, P 22, Gulu E, and Engeny

District variety trials

To develop varieties to suit the farmers environment, the Department of Agriculture has divided Uganda into eleven agricultural zones, determined on rain seasons, amount of rainfall per annum, altitude, soils and the basic agricultural and livestock management activities in such areas. In these zones, 64 district variety trial centres covering all districts, are established. The centres, which are run by Variety Trial Officers, are used for screening breeders materials, testing of new technology, and serve as demonstration sites for farmers. The use of these centres is open to all, including Makerere University staff and private organizations. The finger millet programme makes use of the multilocational testing sites to identify suitable varieties for farmers.

The summary of the performance of some of the entries tested in these centres is indicated in Table 3. It will be observed that Serere cross 10, P 277 and U 10 are identified to be doing considerably well compared to the recommended varieties, Engeny, Serere 1 and Gulu E. The mean yield across sites of Serere cross 10 exceeded 50 q/ha. Most entries screened had satisfactory levels of resistance to neck blast and lodging.

TABLE 3
Mean grain yields in (q/ha) of 1983, 1984 and 1985 finger millet variety trials grown at locations in Uganda during first rains

Entry	Years			Mean	Rank
	1983	1984	1985		
Serere Cross 10	28.8	37.9	20.4	29.0	1
P 211	24.4	31.2	21.1	25.6	8
P 244	24.8	32.0	20.2	25.6	8
P 318	26.8	33.7	17.6	26.0	7
Eding	24.4	32.0	19.9	25.4	11
P 251	26.6	29.5	18.1	24.7	13
P 231	23.9	32.0	19.5	25.1	12
U 10	28.8	35.4	17.6	27.3	3
Serere 1	23.7	32.8	20.1	25.5	10
Gulu E	24.6	35.4	19.9	26.6	4
P 277	26.1	37.2	20.4	27.9	2
P 278	24.6	36.2	18.9	26.6	4
P 249	26.6	33.7	19.2	26.5	6
Engeny	22.3	28.6	19.3	23.4	14
Mean	25.5	33.4	19.4	26.1	
Range of sites mean	10-30	17-40	8-36		
No. of entries	36	36	25		
No. of sites	9	3	4		

Improved finger millet varieties

Despite considerable difficulties experienced as a result of the collapse of the former East African Community in 1977, the loss of breeding materials due to breakdown of cold storage, the departure of a number of expatriates; constant political unrest, lack of fuel transport and finance, poor management of the district variety trial centres to mention just a few; the finger millet breeding programme has progressed fairly well. Four varieties—Gulu E, Engeny, Serere 1 and P 224 have been released to farmers.

Over twenty varieties including eight crosses have consistently outyielded these recommended varieties, and thus offer bright prospects for release, for the benefit of the farmers. Seed of improved varieties is provided to farmers directly on a limited scale, or through the newly established Adaptive Development Project. Seed is also supplied through the Uganda Seed Project which multiplies, certifies, stores and distributes through Cooperative Unions.

LITERATURE CITED

Hulse, J.E., E.M. Laing and O.E. Pearson. 1980. *Sorghum and the Millets: their composition and nutritive value*. Academic Press. New York 1980.

Jameson, J.D. 1970. *Agriculture in Uganda.* Second Edition, Oxford University Press, pp. 127-138.

Khizzah, B.W. 1985. Finger Millet Breeding in Uganda, *Proceedings of the Regional Workshop on Sorghum and Millets Improvement in Eastern Africa,* Soroti, Uganda 22-26 July, 1985 pp. 23-25.

Mukuru, S.Z., V. Makumbi, B.W. Khizzah, S.E. Odelle, E. Atadan, C. Alibet and C. Etoju. 1974. Finger Millet Breeding. East African Agriculture and Forestry Research Organisation E.A.C. Record of Research.

Mukuru, S.Z., V. Makumbi, B.W. Khizzah, S.E. Odelle, E. Atadan, C. Alibet and C. Etoju. 1975. Finger millet Breeding. East African Agriculture and Forestry Research Organisation. E.A.C. Record of Research.

Mukuru, S.Z., V. Makumbi, B.W. Khizzah, S.E. Odelle, E. Atadan, C. Alibet and C. Etoju. 1976. Finger Millet Breeding. East African Agriculture and Forestry Research Organisation. E.A.C. Record of Research.

Okiror, M.A., S.Z. Mukuru, and V.M. Zake. 1976. Influence of male sterility on yield and yield characters in finger millet, (*Eleusine coracana* (L.) (Gaertn.) East African Agriculture and Forestry Research Organisation. E.A.C. Record of Research.

Zake, V.M., B.W. Khizzah, S.E. Odelle, E. Atadan, C. Alibet and C. Etoju. 1984. Progress Report 1981-1983. Sorghum and Millets Research in Uganda, Uganda Agriculture and Forestry Research Organisation.

15

IMPORTANCE, GENETIC RESOURCES AND BREEDING OF SMALL MILLETS IN KENYA

C. Mburu

INTRODUCTION

In Kenya, the millets of importance are Sorghum (*Sorghum bicolor*), finger millet (*Eleusine coracana*) and pearl millet (*Pennisetum typhoides*). Foxtail millet (*Setaria italica*) and proso millet (*Panicum miliaceum*) are of less importance.

Historical and archeological studies suggest that the early civilization in Eastern Africa depended heavily on sorghum and millets. Explosive expansion of the 'Bantu' in east, central and southern Africa must have depended substantially on crops like sorghum and millets adapted to the hot, low lying plains of this vast part of the African continent. With colonial incursions into Africa, the role of these crops began to change as new crops were introduced by explorers and settlers.

Though maize was introduced to the east coast of Africa as early as the fifteenth century, it did not assume importance until after the First World War, when active colonization of this region began. Maize, being less prone to bird damage, replaced sorghum and millets in their own native home of origin, the northeastern quadrat of Africa, where the greatest variability of both cultivated and wild forms of these crops exist. Available evidence shows that it is from here these crops originated and subsequently have spread to other parts of the continent and the world. The highlands of Ethiopia which form a linkage between the Middle East and Africa, are favoured as the centre for the enoblement of indigenous crops. Apart from bird damage, requirement in millet cultivation, changes in religious practices and other activities also brought progressive decline in the production of millets.

Maize established first in the areas of the Rift valley and western Kenya, and became a widely cultivated field crop in vast farming areas. It rapidly became the favourite staple food in western Kenya. Wheat, rice and industrial crops also replaced sorghum and millets in some other parts of Kenya. However, most of the area previously under sorghum and millets was taken by maize. Nevertheless, maize has not entirely replaced sorghum and millets. Finger millet has been a close associate of sorghum in waterlogged and striga-infested areas of Lake Victoria in western Kenya, bordering Uganda and to a lesser extent in the marginal rainfall areas (600 mm) of the eastern province of Kenya. Finger millet, due to its small seed size and storability without storage pests, has played an important role traditionally as a reserve crop. The small seed size of finger millet is advantageous as seeds dry faster compared to other cereals thus making it favourable in the hot, wet and humid areas of the Lake Victoria basin.

Pearl millet, being suited to hot, dry weather of medium and low altitudes, is mainly grown in the eastern parts of Kenya. Owing to its vulnerability to bird damage, the crop is cultivated to a limited extent. In fact its decline in Kenya in terms of area is more pronounced than that of sorghum and finger millet. Three other millets which deserve mention are proso millet, foxtail millet and teff. They are crops of less significance and minor importance in Kenya.

PRODUCTION TRENDS

As mentioned already only, two millet crops are important in Kenya, finger millet and pearl millet. Pearl millet is important in the eastern side, while finger millet though grown in several parts, is predominently cultivated in the western parts around Lake Victoria extending up to Uganda in the west and in parts of the Rift Valley in the east. It is difficult to give precise estimates of the area under these crops as they are grown in patches and in mixed stands. In eastern Kenya including the coast the estimated area is 33,000 ha under finger millet and 40,000 ha under pearl and other millets. The western sector of the country, i.e., the Western Province, Nyanza Province and parts of Rift valley, the estimated area for finger millet alone is around 35,000 ha. Table 1 shows area and production of millets in different provinces of Kenya. The figures do not indicate any particular trend either increase or decrease in hectarage and production for the given years. However, compared to earlier years, there is a distinct reduction in area and production of millets due to replacement of area by other crops especially maize and wheat. Cereal crops projections in terms of area and production for the years 1982-85 are given in Table 2.

Realizing the importance of millets, the Government of Kenya started a project for the improvement of millets including finger millet. The objective is to develop suitable cultivars and agronomic packages for the following problem areas where finger millet is predominantly grown.

1) Waterlogged and striga-stricken areas around Lake Victoria.

2) The hot, dry, medium and low altitude areas of the Eastern Province, and

3) The coastal strip and cold highlands of the Rift Valley region.

TABLE 1
Millet area (ha) and production (tonnes) in different provinces of Kenya during 1980-85

Province	District	1980		1981		1982		1983		1984		1985	
		Area	Prod.	Area	Prod.	Area	Prod.	Area	Prod.	Area	Prod.	Area	Prod.
Western	Kakamega	2722	—	2622	—	1032	—	941	747	980	838	—	—
	Bungoma	2736	—	2802	—	2385	—	2400	2536	4185	2848	3117	2496
	Busia	11620	—	11455	—	13540	—	*108320	*96355	13120	8474	—	—
	Total	17078	—	16879	—	16957	—	*111661	*99638	18285	12160	3117	2496
Eastern	Machakos	2100	1756	7433	6020	4859	3478	3894	4026	15500	14400	2000	164
	Kitui	20000	40000	—	—	50240	44725	22895	16410	22010	17043	29030	24934
	Embu	8694	9699	—	—	13374	12107	13800	12040	20200	18510	—	—
	Meru	8130	7860	—	—	8800	9544	10276	9473	9404	8854	21568	20475
	Total	38924	58315	7433	6020	77273	69854	50865	41949	67114	58807	52598	45574
Rift Valley	Nakuru	200	240	92	184	75	90	—	—	—	—	—	—
	Baringo	1908	1218	1508	723	1464	703	1171	656	—	—	—	—
	Keiyo Marakwet	2500	1810	1158	1344	2099	1344	916	364	—	—	—	—
	Kericho	2812	1792	4335	3468	1800	1440	—	—	—	—	—	—
	W/Pokot	550	352	1750	1400	1800	914	—	—	450	2880	93	74
	Nandi	—	—	256	102	—	—	50	20	150	60	160	96
	T/Nzoia	—	—	—	—	—	—	35	68	36	29	—	—
	U/Gishu	—	—	—	—	—	—	—	—	37	44	—	—
	Nraok	—	—	150	132	206	131	—	—	—	—	—	—
	Total	7970	5412	9249	7353	7444	4622	2172	1108	673	3013	253	170

Table 1 (contd.)

Province	District	1980		1981		1982		1983		1984		1985	
		Area	Prod.	Area	Prod.	Area	Prod.	Area	Prod.	Area	Prod.	Area	Prod.
	Kisii	7496	5356	5791	3917	2386	1527	—	—	4330	2311	—	—
	S/Nyanza	7130	5704	7381	4137	8034	1542	6417	4872	2882	2278	1968	1102
	Total	14626	11060	13172	8054	10420	6669	6417	4872	7212	4589	1968	1102
Coast	Lamu	70	—	—	—	—	—	—	—	—	—	—	—
	T/Taveta	37	—	—	—	—	—	—	—	—	—	—	—
	Kwale	—	—	—	—	—	—	—	—	89	26	146	64
	Total	107	—	—	—	—	—	—	—	89	26	146	64
Central	Murang'a	—	—	—	—	—	—	25	390	31	62	25	16
	Kirinyaga	—	—	—	—	—	—	194	158	129	161	100	58
	Total	—	—	—	—	—	—	219	548	160	223	125	74
N/Eastern	Isiolo	—	—	—	—	0.4	0.18	0.25	0.16	0.81	0.6	—	—
	Mandera	—	—	600	328	247.0	130.0	1700.0	822.0	950.0	665.0	1390	2220
	Garissa	—	—	2.8	1.5	3.0	6.4	3.74	—	5.1	2.5	2.5	1.2
	Wajir	—	—	17.2	9.3	—	—	12.0	—	2770*	—	709.0	369
	Total	—	—	620.0	338.8	250.4	136.58	1715.99	822.99	3665.91	668.1	2101.5	2590.2

* Indicates exaggerated figures

Source: Ministry of Agriculture and Livestock Development, Kenya, Crop Production Division, Food Crops Branch.

TABLE 2

Cereal crops projections 1982-85 (Area and Production figures in '000)

1981 (Estimated)			Total production	Annual growth rate (%)	Projected Production				
Crop	Area (ha)	Yield/ha			1982	1983	1984	1985	Units
Maize	1500.00	18 bags	27000	3.5	27845.0	28820	29828.0	30872.0	Bags (90 kg)
Wheat	134.30	20 "	2686	1.0	2726.3	2767.2	2808.7	2850.8	"
Triticale	10.0	22 "	220	7.6	440.0	473.4	588.4	547.4	"
Rice	12.2	25 "	305	7.6	328.2	353.1	379.9	408.8	Bags (70 kg)
Sorghum/millet	154.4	8 "	1315.2	4.7	1377.0	1441.7	1509.5	1580.4	Bags (80 kg)

Source: Ministry of Agriculture and Livestock Development, Kenya, Crop Production Division, Food Crops Branch.

GERMPLASM RESOURCES

Millets have been in cultivation in Kenya for a very long time. The country as mentioned above lies within the enclave of the origin and domestication of millets, which extends from the northeastern quadrat of Africa to the central and eastern parts of the continent. Naturally, the extent of variability present in local germplasm of finger millet and pearl millet is vast. In the West Agricultural Research Station (WARS) which is one of the main research stations for millet research, 1136 world collections of finger millet have been assembled and evaluated which included 125 local collections made during 1981 from Kenya.

Several germplasm collecting missions have been earlier undertaken by several foreigners, Van Arkel, Danton, Latham and Wood. No authentic records are available on year of collection, areas explored and nature of materials collected by these expeditionists. Millet collections totalled 11, 322, 7 and 120 respectively from these four missions.

In 1973, International Board for Plant Genetic Resources, funded a collection mission in collaboration with the Ministry of Agriculture, Government of Kenya and collected 602 sorghum, 263 finger millet and 48 pearl millet types. These were from western Nyanza, eastern and coastal provinces of Kenya.

In another collection mission to all the millet growing areas of the country, the team from Katumani Research Station, Eastern Kenya collected 185 pearl millet, 425 finger millet, 178 foxtail millet, 13 proso millet and three barnyard millet accessions. This station also introduced accessions from EAAFRO (Serere, Uganda), ICRISAT (India), USA, Ethiopia, India, CIMMYT (Mexico) and Botswana which included 1000 finger millet, 470 pearl millet, 625 foxtail millet, 193 proso millet, 29 barnyard millet and eight little millet. Part of these collections are still present in cold stores at Kitala in western Kenya, at Katumani in eastern Kenya and Maguga in Nairobi.

Finger millet collections from Kenya are found to show substantial variation for duration of maturity, plant height, tillering potential, panicle shape and size, grain colour, size, texture and threshability. *Eleusine indica* is often found in association with cultivated forms and the shattering types 'Shibras' invariably accompanied the local land races in their area of cultivation. From the above it is evident that there is a lot more to be collected and conserved in Kenya as regards millets particularly finger millet.

16

FINGER MILLET RESEARCH IN THE SOUTHERN HIGHLANDS OF TANZANIA

R.O.F. Mwambene

INTRODUCTION

Finger millet (*Eleusine coracana*) is grown in all regions of southern highlands of Tanzania, but is mainly concentrated in Sumbawanga and Nakani (Rukwa Region), Mbozi, Ileji and Mbeya (Mbeys Region) and to a lesser extent in Iringa and Ruvuma regions. It is grown for food, for home brewing to make light beer (Pombe), and as a cash crop. Its use for home-made light beer is now more important as most people have adopted maize as their staple food. While sorghum and millets are considered 'famine crops', among millets, finger millet is the most valuable cereal because of its good storability without any store pests and high nutritive value of grains.

Finger millet is grown under a shifting cultivation farming system, in which new land is cleared every year, burnt and ploughed using the hand hoe or oxen. Finger millet is then broadcast on a firm seed bed. Intercropping of finger millet is very common along with maize, sorghum, cassava, pumpkins, sunflower, sesamum, etc. After the harvest of finger millet, the field is used for raising crops such as common beans, maize, sweet potato or groundnut for one or two years then left fallow for six to seven years. Finger millet cannot be grown continuously on the same field due to heavy weed infestation and soil fertility depletion.

Although sorghum and millets are given high priority in the semi-arid areas of central and western Tanzania, with the concentration of research work at Ilonga Agricultural Research Institute (TARO, ILONGA), finger millet has retained its importance in the southern highlands of Tanzania, especially in the Ufipa and Mbozi Plateau. Sorghum and millet improvement programme at Uyole, concentrated on finger millet since 1970-71, when the institute was established. The report in this paper covers the highlights of the work carried out from 1970-71 to 1978-79 when the programme was shelved. The main objectives of the programme were:

1) Sorghum and millets breeding aimed at augmenting germplasm and breeding varieties suitable to different ecological zones.

2) Sorghum and millets agronomy aimed at putting together packages of inputs and practices which the farmers can adopt to increase productivity.

While trying to relate these objectives in its totality for sorghum and millets improvement and solving production problems in the country, it was found necessary to run trials and nurseries at Uyole and its sub-stations representing different ecological zones of the southern highlands. Consequently the trials and nurseries have been operating at different sub-stations and regional centres in collaboration with the national sorghum and millets co-ordinator stationed at TARO, Ilonga. Uyole station mainly concentrated research on high altitude sorghum and finger millet.

VARIETAL DEVELOPMENT

Traditional finger millet growers and consumers in the southern highlands of Tanzania continue to grow local finger millet types, and prefer local cultivars for food and beer making than improved varieties. When the breeding programme was initiated much effort was directed to make new introductions from Serere, Uganda and to evaluate them in a wide range of environments. In these trials, locals somehow were not included particularly in the initial years of testing. The data pertaining to grain yields of the varieties included in the east African finger millet trial at three major locations of the southern highlands of Tanzania are given in Table 1.

GENETIC RESOURCES

A field survey carried out in Ufipa plateau by a team of biologists and social scientists from Uyole, found finger millet to rank a high priority with the local farmers. This triggered a local finger millet collection and evaluation project by the plant breeding and genetics department. Preliminary evaluation of local finger millet collections at Uyole has shown that the southern highlands are rich in finger millet germplasm and the variability existing is large in each locality visited. The high altitude types of finger millet are quite productive, yielding up to 5 tonnes/ha. Having observed a lot of variability among local land races, a research programme was initiated at Uyole in 1976. This programme aimed at establishing local finger millet collections, describing their botanical characteristics, studying chemical composition of seeds and evaluating their yield potential and other agronomic characteristics. In two seasons, quite large collections were assembled and some promising varieties were identified (Table 2).

TABLE 1
Mean grain yield (kg/ha) in three major finger millet production areas of the southern highlands of Tanzania

	Locations			
	Uyole (Mbeya) 1800 m	Mbimba (Mbozi) 1500 m	Nkundi (Sumbawanga) 1810 m	Mean
Mbeya local	4,894	4,061	3,774	4,233
Sumbawanga	3,644	3,125	3,000	3,256
Engenyi	2,383	2,744	2,184	2,437
Serere local	2,328	2,500	2,151	2,326
P 283	2,283	2,467	2,130	2,293
P 224	1,939	2,847	2,195	2,327
Gulu E	1,856	2,153	1,762	1,924
Rombo Local	1,839	1,494	1,558	1,630
1/19	1,833	1,978	1,657	1,727
Eding	1,672	3,161	2,025	2,286
Mean	2,467	2,653	2,241	

TABLE 2
Yield of local finger millet varieties, tested at Uyole

		Grain yield (q/ha)			
Variety name	Place collected	1976-77	77-78	78-79	Mean
Chikwelekwele	Sumbawanga	44.5	44.7	44.4	44.53
Makukulu	"	—	39.4	33.3	36.35
Mambwe	"	50.0	51.9	36.1	46.00
Nameka	"	46.0	44.7	38.9	43.20
Katila	"	48.0	30.8	41.7	40.17
Mawutila	"	42.5	35.1	30.6	36.07
Amakazi	"	50.0	35.0	38.9	41.30
Amimakalala	"	51.5	33.9	38.9	41.43
Kalala	"	60.0	42.2	41.7	47.97
Chiza	"	50.0	34.6	33.3	39.33
Chiminuka	"	68.0	34.4	36.1	46.17
Machenchete	"	57.0	41.4	30.6	43.00
Kawulunge	"	51.0	33.6	36.1	40.23
Intiswe	Mbeya/Mbozi	47.0	48.1	19.4	38.17
Mbeya local	"	60.3	36.1	27.8	41.40
Tukuyu local	Tukuyu/Rungit	—	26.1	27.8	26.95
Mwangunu	Kyela	—	39.6	41.7	40.65
Ntunkane	"	45.5	36.1	30.6	37.40
Usangu local	Usangu Plains	41.5	33.2	38.9	37.87
Sumbawanga local I	Sumbawanga	.42.5	41.7	36.1	40.10
" local II	"	55.5	37.2	38.9	43.87

Table 3 presents data on chemical composition of some of the local finger millet and sorghum varieties grown in Tanzania. The data show a lower crude protein content in finger millet than in wheat. While wheat has high amount of pepsin soluble protein, most varieties of finger millet and sorghum show a low level. White seeded types one each in finger millet (Muwutila) and sorghum (White) relatively had low content of tannic acid, which is known to reduce solubility and digestibility of protein. The lysine content is 10-15 per cent lower in finger millet and 30-35 per cent lower in sorghum than in wheat. However, compared to wheat, finger millet has a very high calcium and sorghum has high iron content.

TABLE 3
Comparative chemical composition of finger millet, sorghum and wheat

Species and varieties	Fat %	Crude protein (6.25 × N) %	Pepsin HCl soluble protein	Ca %	P %	Fe mg/kg	Tannic acid %	Amino acids (% of protein)			
								Lysine	Histine	NH_3	Arginine
Finger millet											
Chikwelekwele (Brown)	1.1	7.3	5.7	0.35	0.25	25	0.41	2.83	2.33	3.10	4.29
Makukulu (brown)	1.1	7.4	5.8	0.31	0.26	28	0.40	2.87	2.33	3.18	4.46
Amakazi (brown)	1.1	7.4	5.8	0.32	0.27	25	0.44	2.87	2.40	3.17	4.37
Mawutila (white)	1.1	8.5	7.0	0.36	0.30	28	0.25	2.54	2.36	3.20	4.05
Tukuyu local (black)	1.1	8.0	5.9	0.45	0.33	37	0.42	2.68	2.41	3.15	4.24
Sorghum (high altitude varieties)											
Deep red types	1.0	10.4	5.1	0.022	0.33	59	1.23	1.99	1.92	2.86	3.91
White/brown types	1.0	10.7	6.8	0.030	0.38	140	1.33	2.19	2.04	3.05	3.87
White types	1.0	12.3	12.3	0.026	0.37	320	1.01	2.05	1.93	3.11	3.77
Wheat											
Tai	1.0	10.8	10.1	0.043	0.38	38	0.16	3.08	2.34	3.49	5.07

17

IMPORTANCE, GENETIC RESOURCES AND BREEDING OF SMALL MILLETS IN ZIMBABWE, WITH EMPHASIS ON FINGER MILLET

F.R. Muza

PRODUCTION TRENDS IN SMALL MILLETS

The important small millets in the world today are finger millet, barnyard millet, common millet, kodo millet and foxtail millet. In Zimbabwe the only important small millet is finger millet and is a traditional food crop. The other small millets are of less importance and some communal farmers collect these grains for food.

The areas of domestication of finger millet are the Highlands of east Africa stretching from Ethiopia to Uganda (Hilu and de Wet, 1976). Over 50 per cent of the world production of finger millet comes mainly from central and southern Africa. In Zimbabwe, based on Agritex data, finger millet occupies the third largest area after maize and pearl millet covering about 200,000 hectares. Finger millet requires more rainfall than pearl millet, another important food crop, and is therefore grown in natural regions II (750-1,000 mm per annum) and III (650-800 mm per annum) (Fig. 1). In these regions, the finger millet crop faces competition from others, especially maize, because of the favourable rains. As a result finger millet is usually allocated to land of low fertility, thereby reducing its productivity. In these regions the crop is mainly grown in the communal areas, and low yield of 0.5 t/ha is common. Some commercial and semi-commercial farmers have also started growing finger millet, getting higher yields of over 2 t/ha.

In 1984, finger millet became a controlled crop commanding a good selling price of $300/tonne, thus comparing well with other grain crops (Table 2). This has resulted in significant increase in finger millet production (Table 1).

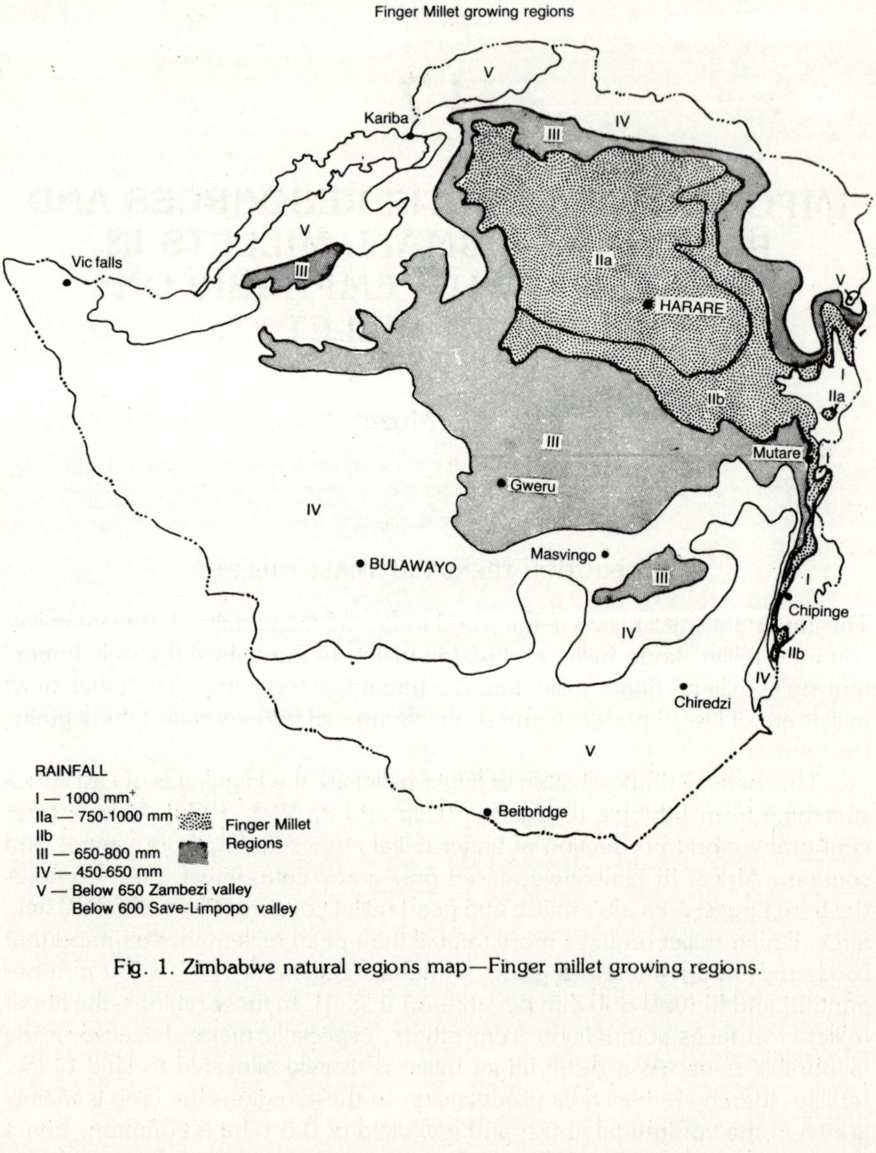

Fig. 1. Zimbabwe natural regions map—Finger millet growing regions.

TABLE 1
Production trends: finger millet

Period	Area (ha)	Production ('000 tonnes)
1950s	100,000	45-50
1960s	110,000	50
1970s	150,000	70-75
1980-81	181,000	140-150
1986	200,000	-150

TABLE 2
Producers's price of agricultural commodities—1985-86

Crop	Grade	Price ($/tonne)
White maize	A	180
Yellow maize	A	153
Groundnut	A	750
Sorghum	A	180
Wheat	A	300
Sunflower	A	340
Soybean	B	340
Edible beans	B	450
Pearl millet	A	250
Finger millet	A	300
Seed cotton	A	75c/kg

Source: Agritex data

GERMPLASM RESOURCES

Muza and Mushonga (1985) indicated that the use of primitive cultivars was a major factor for low yields in finger millet in Zimbabwe. Evolution of superior varieties will be witnessed if a wide genetic pool is available to work with. There has been a genetic erosion since finger millet became a cultivated crop in this region. This has been realized, and efforts are being made to preserve the germplasm of finger millet. As better varieties come in, the genes found in the primitive cultivars and land-races must be conserved.

In 1982, a germplasm collection trip was launched in Zimbabwe by the IBPGR, the Crop Breeding Institute (CBI) of the Department of Research and Specialist Services as well as ICRISAT scientists. A total of 286 cultivated and six wild and weedy accessions of finger millet were collected (Appa Rao and Mengesha, 1982). The samples differed in the form and shape of inflorescence—open with straight spikes, open with top-curved spikes or com pact with in-curved spikes (Appa Rao and Mushonga, 1985). These samples have been evaluated and characterized. In 1985, another collection trip was organized by IBPGR/CBI and a further 182 accessions were collected (Appa Rao, Mushonga and Muza, 1986).

At Harare station in Zimbabwe, a long-term cold storage room has been constructed with IBPGR/FAO assistance for storage of seeds. There is more need for further exchange of material. The collections from Zimbabwe have also been sent to ICRISAT, IBPGR and Fort Collins in the U.S.A. There is free import and export of genetic material in Zimbabwe. Also, more collections still are to be made from the southern parts of the country, which have not been covered by the earlier two collection expeditions.

As for the other small millets, we look forward to International Institutes like IBPGR and ICRISAT to identify the promising sources of germplasm and to facilitate their introduction to countries like Zimbabwe. Although the potential of the other small millets in Zimbabwe is not so obvious, we think it is worthwhile and necessary to have the germplasm.

BREEDING AND VARIETAL IMPROVEMENT

Apart from the recent introductions of improved lines, not much breeding and varietal improvement work has been done in finger millet in Zimbabwe.

During 1968-69, Mushonga effected 64 single head selections of finger millet and they were grown in single progeny rows in a completely randomized block design. He observed that pure lines differ significantly in yield, threshing percentage, grain weight per head, seed size, maturity period and plant height. High yielding lines possessed high grain weight per head, high threshing percentage, high number of seeds per head and early maturity.

The work on varietal improvement was again resumed in 1980. Samples made in germplasm collection trips were evaluated in observation nurseries. From the 1982 collections, 30 genotypes were selected based on grain yield, maturity and panicle shape. These have been evaluated at different sites in Zimbabwe during 1984-85. Some lines have yielded well at some stations (Table 3).

More selections have been made from the accessions collected in 1985. For the 1986-87 season, the following breeding activities are planned.

1) Finger millet variety trial—30 entries to be evaluated at four sites in Zimbabwe. This trial will also include the INDAF varieties from India.

2) Finger millet Regional Cooperative Trial (FMRCT) with 16 entries.

3) Finger millet Regional Introduction Trial (FMRIT) with 800 entries.

4) Further evaluation of 1985 collections.

5) Crossing programme by adapting hot-water emasculation technique.

We look forward to directions from International Organizations for improve-

TABLE 3
Finger millet trial 1984-85

Site	Gwebi	Makoholi	Mtopos	Variety mean	Rank
	Alt. 1488 m. Rainfall 750-1000 mm	Alt. Rainfall 1208 m 650-800 mm	Alt. Rainfall 1338 m 450-650 mm		
Variety		Grain yield (t/ha)			
TGR-367	4.4	3.0	2.4	3.30	1
" 54	3.8	2.8	1.8	2.79	2
" 164	4.4	2.2	1.6	2.72	3
" 316A	3.5	2.9	1.5	2.63	4
" 47	4.1	2.5	1.2	2.62	5
" 402	4.3	2.6	1.0	2.62	5
" 327	3.3	2.3	2.0	2.56	7
" 295	3.0	2.8	1.7	2.52	8
" 126	4.7	2.0	0.8	2.50	9
" 209B	3.6	2.3	1.5	2.48	10
" 389	3.9	2.0	1.5	2.47	11
" 316B	3.9	2.3	1.1	2.47	11
" 72	3.6	2.3	1.4	2.44	13
" 294	4.1	1.7	1.3	2.36	14
" 362	4.0	1.9	1.1	2.35	15
" 35	3.4	2.2	1.3	2.28	16
" 303	3.1	2.4	1.3	2.27	17
" 84	2.7	2.4	1.7	2.25	18
" 108	2.9	2.4	1.4	2.25	18
" 10	3.3	2.1	1.3	2.21	20
" 127	3.6	1.8	1.2	2.20	21
" 113	2.8	2.3	1.5	2.19	22
" 304	2.5	2.5	1.5	2.17	23
" 316	2.9	2.1	1.3	2.11	24
" 73	2.6	2.4	0.9	1.96	26
" 83	3.7	1.5	0.6	1.92	26
" 42	2.9	2.3	0.6	1.91	27
" 36	2.3	1.9	1.5	1.88	28
" 22	2.5	1.9	0.6	1.69	29
" 34	3.4	1.1	0.3	1.59	30

ment of the crop. If finger millet is pushed further into marginal areas due to competition from crops like maize, then varieties with water-stress resistance will be needed.

LITERATURE CITED

Appa Rao, S. and M.H. Mengesha. 1982. Germplasm Collection Mission to Zimbabwe. Genetic Resources Progress Report 41, ICRISAT, Hyderabad, India, pp. 1-16.

Appa Rao, S. and J.N. Mushonga 1985. Traditional Food Crops of Zimbabwe: I Finger millet. *Zimbabwe Agricultural Journal.*

Appa Rao, S., J.N. Mushonga and F.R. Muza. 1986. Collecting Plant Genetic Resources from Zimbabwe. Genetic Resources Progress Report, ICRISAT, Patancheru, India.

Hilu, K.W. and J.M.J. De Wet. 1976. Domestication of *Eleusine coracana. Econ. Bot.* **30:** 199-208.

Muza, F.R. and J.N. Mushonga. 1985. Progress Report No. 2. Zimbabwe Sorghum and Millets Improvement Project. International Development Research Centre (IDRC).

18

PRODUCTION TRENDS, GERMPLASM RESOURCES, BREEDING AND VARIETAL IMPROVEMENT OF SMALL MILLETS, WITH SPECIAL EMPHASIS ON TEFF IN ETHIOPIA

Seyfu Ketema

INTRODUCTION

In this paper eight millets are discussed. According to the information so far available five of them, *Setaria italica, Panicum sumatrense, Echinochloa colona, Paspalum scrobiculatum* and *Digitaria exilis* are not found in Ethiopia while the other three, *Eragrostis tef, Eleusine coracana* and *Panicum miliaceum* are known to exist in Ethiopia (Wolde Michael, 1977).

Teff (*Eragrostis tef*) is not only the dominant millet but also the most important cereal crop in Ethiopia. It is cultivated on over one million hectares annually while the other cereals including maize and wheat occupy less than a million hectares. Finger millet is a relatively minor crop, cultivated in less than 300,000 hectares annually (Table 1). There is no statistical data on the area and production of proso millet (*Panicum miliaceum*).

Both finger millet and proso millet play an insignificant role in the present day agriculture of Ethiopia and not much information is available as no extensive research is being carried out on these crops. Therefore, this paper essentially describes the work carried out on teff.

PRODUCTION TRENDS

The latest statistical figures presented in Table 1 show that both teff and finger millet area and production have not fluctuated drastically in the five years period 1979-83. The recently formulated food policy of Ethiopia places emphasis on

TABLE 1
National estimate of area under major crops in Ethiopia* during 1979-83
(Area: '000 ha)

Crop	1979-80	1980-81	1981-82	1982-83	1983-84
Cereals	4022.77	4711.70	4629.33	5029.23	4715.59
Teff	1513.33	1361.95	1331.45	1399.83	1317.95
Barley	909.83	830.94	810.35	908.00	796.31
Wheat	486.67	536.19	684.91	714.01	625.59
Maize	870.78	735.48	652.47	819.67	820.90
Sorghum	1326.29	979.07	844.26	905.65	913.61
Millet	215.87	232.87	226.49	225.15	215.30
Oats	—	35.20	79.40	56.92	25.93

Source: Central Statistics Office, Ethiopia
*Excluding Eritrea and Tigray.

maize, wheat and sorghum. How this food policy is going to affect the production of teff and other cereals in Ethiopia still remains to be seen.

Several developing countries have to face the challenge of feeding their growing population as well as increasing their export earnings to finance developmental activities in other sectors of their economy. In order to meet this challenge among many other things they will have to push their agriculture into marginal areas. The marginal agricultural lands in Ethiopia include the vertisols which are waterlogged and lack drainage, the areas with moisture stress and infertile lands. All these can be put back to production through appropriate management practices by restoring fertility, providing drainage and irrigation. These are expensive operations and beyond the immediate reach of many countries. An alternative is to develop crops that can be produced under such adverse conditions. One such crop in Ethiopia is teff. Its popularity among farmers is because of the following reasons.

1) Teff withstands waterlogged and anoxic conditions better than maize, wheat and sorghum. Farmers grow teff on vertisols that have drainage problems, where maize, wheat or sorghum cannot be grown.

2) Teff withstands moisture stress better than maize or sorghum. In many areas, maize and sorghum are planted around April. If these crops wilt due to moisture stress, the farmer will replough the land by the end of July or beginning of August and resow with teff. Therefore, teff is a rescue crop surviving and growing on the remaining moisture in the season.

3) Teff grain is not attacked by weevils. This means reduced post harvest loss in storage. In moisture stress areas where more than one sowing is a common practice and in years of total failure of rains the farmers have to store seed for long periods. Under such situations teff seed is an ideal one since it has no store pests.

4) It can be grown in areas where frost is a problem.

5) Teff straw is preferred by cattle over any other cereal straw and is an important source of feed during the dry season.

6) Teff commands a higher price in the market than all other cereal grains.

These advantages have probably made teff as an indispensable crop in Ethiopia.

GERMPLASM RESOURCES

The Plant Genetic Resources Centre of Ethiopia (PGRC/E) has the main responsibility for collecting and conserving germplasm of many crops. Their collections include 97 accessions of finger millet and 2175 teff accessions (Melaku, 1982). No collections of proso millet have been made, and only a few collections of finger millet are available at PGRC/E. However, this does not imply that there are no germplasm diversity of these crops in Ethiopia. It only implies that no organized collection of these crops has been so far made.

According to Vavilov, Ethiopia is the centre of origin for teff (Melak Hail, 1966; Tadese, 1975). Although the exact date of its domestication is not known there is no doubt that teff is an ancient crop of Ethiopia (Costanza, 1974; Tadese, 1975). It is believed that teff domestication first took place somewhere in the northern highlands of Ethiopia. Unger in 1866 discovered teff seeds in the Pyramids of "Dassur' built in 3349 B.C. and deduced that teff was grown in Egypt before the eighth century B.C. Tadese (1975) believes that teff could have been taken to Egypt via the Blue Nile by merchants or messengers.

Teff was introduced to other parts of the world by the Royal Botanic Gardens, Kew, which imported seed from Ethiopia in 1866 and distributed them to India, Australia, USA and South Africa. According to Tadase (1975), Burt Davy in 1916 introduced teff to California (USA), Malawi, Zaire, India, Sri Lanka, Australia, New Zealand and Argentina; Skyes in 1911, introduced it to Zimbabwe, Mozambique, Kenya, Uganda, Tanzania and Horuitz in 1940 to Palestine.

Teff is cultivated from sea level to 2,800 metres above sea level; on varying soils from waterlogged to well drained soils and in areas that have less than 300 mm to more than 1,000 mm, seasonal rainfall. No doubt this is a reflection of the wealth of diversity and the adaptability that teff possesses. Presently 2,175 accessions of teff have been collected and are maintained in Ethiopia. Nevertheless, these collections have not been systematically made and not yet fully characterized and evaluated. So, a proper strategy is still to be developed on systematic collection and conservation work.

Important variations in teff germplasm noted so far include variability for maturity period from 60 to 120 days; for plant height from 45 to 150 cm, and for culm thickness from 1.2 to 3.1 cm. A set of cultivars have been characterized and documented by Tadese (1975). Endeshaw (1978) believes that there are still many minor variations within these cultivars.

As in other crops, the recent drought in Ethiopia has caused a severe gene erosion of teff germplasm. I think the international community has the obligation to assist in the collection and conservation of teff germplasm, whose main source of diversity is found only in Ethiopia. This has to be done for the benefit of the present and future generations of mankind.

BREEDING AND VARIETAL IMPROVEMENT

Teff is a sexually propagated self-pollinated annual (Tareke, 1975; Tadese, 1975). It is an allotetraploid with 2n = 40 chromosomes (Tareke, 1975; Endeshaw, 1978; Tareke, 1981).

At present, the grain yield of teff is about 0.8 tonnes per hectare for the landraces and 1.7 to 2.2 t/ha for the improved varieties on cultivators holdings and 2.2 to 2.8 t/ha in research farms. Improvement work on teff through plant breeding started only in the late 1950s (Asrat, 1965; Tareke, 1974). This clearly shows that the improvement work on teff is relatively recent.

Since teff is the staple food of Ethiopia, extensive research on it is conducted only in that country. Unlike maize or wheat, which are consumed world wide the efforts of thousands of scientists are not available for the task of improving teff, and not much basic or applied research information is available on this crop.

The first efforts to improve teff are made through pure line selection from the land races. The limitation of this approach was soon realized and in order to create variability through gene recombination, intraspecific hybridization was attempted. This was not successful as the floral biology of teff was not well understood.

As an alternative to hybridization, mutation breeding work was started in 1972 to create new variability. The major objective of this approach was to create short stemmed lodging resistant, high yielding varieties. Although this objective is not yet fully realized, the following results have been achieved:

1) The effect of physical mutagens—gamma and x-rays and the chemical mutagens—ethylmethyl sulphonate, and sodium azide have been studied. The results show that gamma ray treatments of 250 krads and above are lethal. 150 krads of gamma ray treatment is considered as appropriate but varieties show different responses even at this dose.

2) For x-ray treatments, doses of 100-130 krads and for EMS treatments 2.5-4.7 per cent concentrations appeared optimum.

Teff flowers open during the early morning hours between 6.45 a.m. and 7.45 a.m. and they have a brief pollination time. Tareke (1975) reported first successful, intraspecific crosses. This opened a new era in breeding research in teff. The hybridization technique is a cumbersome and delicate one, as it involves hand emasculation followed by hand pollination. Seyfu (1983) has recommended a procedure of hand emasculation and hybridization. A skilled operator might dispense with hand emasculation by applying donor pollen before the anthers in the flower of the seed parent have dehisced. But selfing remains a high risk in this procedure.

The major objectives of the teff breeding programme are to develop lodging resistant varieties with high and stable yield adapted to various agroecological zones. To date many improved varieties have been developed (Table 2).

No one exactly knows how and why teff was domesticated in Ethiopia. However, the following facts may throw some light. It is speculated that the word teff was derived from the Arabic 'tahf', a name given to a similar wild plant used by semites in South Arabia in times of food scarcity (Rouk *et al.*, 1963; Costanza, 1974; Tadese, 1975; Endeshaw, 1978). According to Endeshaw (1978) seeds of *E. pilosa* which is believed to be the ancestor of teff was collected as food by people in many parts of Africa other than Ethiopia in times of famine. This suggests that the domestication of teff might have started in times of food scarcity.

This speculation of teff's domestication and the fact that there are several land races cultivated in low rainfall areas at the present moment indicate that teff has the potential to develop as an useful crop for several moisture stress areas in many parts of the world.

LITERATURE CITED

Asrat, F. 1965. Cereal and oil seed research. Branch Experiment Station, Debre-Zeit, Agricultural Research Centre, Alemaya Agricultural University, Dire Dawa, Ethiopia. 60 pp.

Costanza, S.H. 1974. Literature and Numerical Taxonomy of Teff (*Eragrostis tef*). M.Sc. Thesis, B.S. Cornell University, Illinois, USA, 110 pp.

Endeshaw, B. 1978. Biochemical and Morphological Studies of the relationships of *Eragrostis tef* and some other *Eragrostis*. M.Sc. Thesis, University of Birmingham, U.K. p. 41.

Melak Hail, M. 1966. Chemical composition of teff (*Eragrostis tef*) compared with that of wheat, barley and grain sorghum. *Economic Botany*, **20:** 268-273.

Melaku, W. 1982. Brief description of the plant genetic resources centre, Ethiopia. pp. 24-30. In: Berhane, G. (ed.) Proceedings of the Regional Workshop on Sorghum Improvement in Eastern Africa, Nazareth and Debre-Zeit Ethiopia, Oct. 17-21, 1982.

Seyfu, K. 1983. Studies on Lodging, floral biology and breeding techniques in teff (*Eragrostis tef* (Zucc) Trotter). Ph.D. Thesis, University of London, U.K. 122 pp.

Tadese, E. 1975. Teff (*Eragrostis tef*) Cultivars: Morphology and Classification: Part II. Debre-Zeit Junior College and Agricultural Experiment Station Bulletin Number 66, Alemaya University of Agriculture, Dire Dawa, Ethiopia, 73 pp.

Tareke, B. 1974. Studies on the problems associated with breeding teff (*Eragrostis tef* (Zucc.) Trotter). M.Sc. Thesis; Washington State University, U.S.A., 89 pp.

Tareke, B. 1975. A break-through in teff breeding technique FAO—Information Bulletin on Cereal Improvement and Production, **12:** 11-13.

Tareke, B. 1981. Inheritance of lemma colour, seed colour and panicle form among four cultivars of *Eragrostis tef* (Zucc.) Trotter. Ph.D. Thesis, University of Nebraska: Lincoln, Nehraska, U.S.A. 84 pp.

Wolde Michael, K. 1977. *A Glossary of Ethiopian Plant Names*. 2nd ed., Addis Adaba, Ethiopia.

TABLE 2

Improved varieties of Teff, their area of adaptation and some of their agronomic characters

Variety	Altitude (m)	Rainfall in the growing season (mm)	Days to maturity	Yield (q/ha)		Year of release
				In experimental station	In farmers field	
DZ-01-354	1600-2400	300-700	85-130	18-28	17-22	1970
DZ-01-99	1400-2400	300-700	85-130	18-28	17-22	1970
DZ-01-787	1800-2500	400-700	90-130	18-29	17-22	1978
DZ-01-196	1899-2400	300-700	80-113	16-24	14-16	1970
DZ-Cr-82	1700-2000	300-700	112-119	15-20	17-21	1981
DZ-Cr-37	1800-2000	134-500	82-90	15-27		1984
DZ-Cr-44	1800-2400	400-600	125-140	21-30	17-22	1981

19

IMPROVEMENT OF FINGER MILLET (*Eleusine coracana*) IN ETHIOPIA

Yilma Kebede and Abebe Menkir

INTRODUCTION

In Ethiopia, despite fluctuations between years and regions finger millet comprises about 5 per cent of the total cultivated area under cereals, and makes up a similar percentage of the total cereal production in the country (Table 1). The cultivation of finger millet is concentrated in the mid and lower altitude regions of Eritrea, Tigrai, Gojjam, Gonder and Wellega. In these regions, finger millet constitutes 10 per cent to 20 per cent of the total cereal production. From an area of just over 220,000 ha less than 200,000 tons of finger millet is produced, resulting in a national average of below 1 ton/ha.

TABLE 1

Estimated production of major cereal crops in Ethiopia

'000 tons

	1976/77	77/78	78/79	79/80	84/85
Teff	994.5	1022.4	1083.8	1144.0	873.8
Barley	894.5	689.9	696.8	772.0	749.2
Wheat	605.2	428.9	448.8	469.0	561.3
Maize	947.8	929.1	981.6	977.2	923.9
Sorghum	755.7	708.0	679.9	689.0	489.2
Millet	172.4	207.2	189.6	193.0	187.2

Source: Land use planning and regulatory department (1976-80 data)
1982 data book on land use and agriculture in Ethiopia Vol. I, Ministry of Agriculture, Addis Ababa
PMGE, ONCCP, CSO (1984/85 data).
Agricultural Sample Survey 1984/85 (1977 EC) Results of area and production 1985.

Finger millet is popular for making local beer and distilled spirit (Areki). The grain is also used for bread (injera), although other grains are more preferred for this purpose. In the major producing regions, it does relatively well in drier years and becomes a famine crop for farmers (Asrat, 1965).

The most important advantage of this crop is that the seed can be stored for a longer period without the use of insecticides. It is also considered to be free from the major pests and diseases, and unlike sorghum is not favoured by birds (Cloutier, 1984).

MILLET IMPROVEMENT—HISTORICAL SKETCH

The earliest reported work on finger millet was from Debre Zeit experiment station, where preliminary results indicated that as a group, white seeded types had higher yields compared to red and dark seed types (Asrat, 1965).

Later on, the Ethiopian Sorghum Improvement Programme considered the possibility of introducing millets as alternate crops for moisture stress areas of the country. In 1979, sixty millet accessions, ten each of finger, proso, little, kodo, foxtail and barnyard millets were received from ICRISAT and evaluated at the research station at Melkassa (Nazareth). Many of a these millets were not as early as expected. Only proso and foxtail appeared promising.

Another consignment of 247 foxtail millet and 35 barnyard millet accessions was received and evaluated in 1980. A trial consisting of some foxtail selections was carried out in subsequent years but the entries did not perform as expected.

FURTHER WORK

Recognizing the importance of finger millet as a potential dryland crop in the country, a programme was initiated in 1985, aimed at identifying high yielding, lodging and disease resistant lines for the major millet areas.

Finger millet is considered indigenous to Ethiopia (Huffnagel, 1961) and occupies diverse agro-ecological situations. There is a vast range of genetic variability in indigenous Ethiopian germplasm. Taking into account that assembly of the genetic resources and identification of important traits is essential for attaining breeding objectives, the Plant Genetic Resources Centre/Ethiopia has assembled over 700 accessions of finger millet both from indigenous and foreign locations. The majority of the indigenous collections are from Gojjam, Gonder and Wellega administrative districts. Characterization data indicated the range of variation in seedling vigour, days to maturity (102-163), number of fingers (6-11) and ears (16-72) length of fingers (5-14 cm) and plant height (40-142 cm) as well as seed color (light to dark).

So far, our research efforts have focused on isolating promising genotypes from indigenous collections grown at Melkassa and further evaluating them in possible areas of production. In conjunction with the Plant Genetic Resource Centre/Ethiopia, a total of 524 finger millet accessions were evaluated for various agronomic and morphologic characteristics in 1985 and 1986 (Table 2). Just under 50 per cent of the collections were indigenous. The rest were introduction from Zimbabwe, Burundi and Nepal. Out of these accessions, 44 early and 69 medium-late maturing types were selected and included in advanced observation nurseries (Table 3). Nurseries consisting of early maturing millets were grown and evaluated at low elevation locations (Melkassa, Miesso and Kobo). Medium-late maturing types were tested at intermediate (Pawe) and high (Adet) elevation sites. Based on agronomic excellence and adaptation as well as tolerance to lodging and head blast, some promising entries have been identified and included in variety trials.

TABLE 2

Selection from 1985 and 1986 grow outs of finger millet accessions planted at Melkassa

	Total evaluated	Source	Accessions advanced	Per cent selected
1985	186	IND.	40	22
	25	INTRO.	22	88
1986	63	IND.	36	57
	250	INTRO.	15	6

IND — Indigenous
INTRO — Introduction

TABLE 3

Classification of 1985 and 1986 finger millet selections based on maturity and location for testing

	Days of maturity	1985	1986	Test location
Early	<110	29	15	MK, MI, KB
Interm.	110-130	17	36	PW, AD
Late	>130	16		

MK — Melkassa, MI — Miesso, PW — Pawe, AD — Adet, KB — Kobo

In a finger millet preliminary variety trial grown at Melkassa and Kobo, yield levels averaged over locations varied from 1.5 to 3.4 tonnes/ha. About 50 per cent of the accessions produced yields of 2.5 tonnes/ha or better (Table 4). More variety and observation trials have been planned for low (Kobo, Miesso, Melkassa), intermediate (Pawe) and high (Adet) elevation testing sites during the coming season.

TABLE 4

Grain yield and days to flowering of selected finger millet entries (based on yield and earliness) from a preliminary variety trial grown at Melkassa (MK) and Kobo (KB), 1986

	Grain yield			Flowering		
Accession No.	MK	KB	Mean	MK	KB	Mean
	tons/ha			days		
203253	1.8	4.9	3.4	79	74	76
203261	2.0	4.4	3.2	81	78	80
100008X	2.5	3.3	2.9	75	73	74
100007X	1.9	2.8	2.4	74	73	74
100058	1.4	3.3	2.4	81	76	78
76T1 = 23 +	1.8	2.0	1.9	63	68	66

— X indigenous
— + early sorghum

CONCLUSION

Some progress has been made in the area of finger millet improvement in Ethiopia. Nevertheless, there is a lot to be done in identifying production constraints, refining recommendation domains, defining the requirements of the small farmers, and establishing several testing sites representing the major production zones. We are yet to initiate work on developing appropriate management systems and crop protection measures. Furthermore, work on nutritional characteristic of finger millet germplasm as well as for processing and utilization of the crop needs due attention.

LITERATURE CITED

Asrat Feleke. 1965. A progress report on cereal and oil seed research, 1955-1963. Branch Experiment Station, Debare Zeit, Ethiopia.

Cloutier, P. 1984. Agriculture-strategies for reclamation and development. Land use Planning and Regulatory Department of the Ministry of Agriculture and FAO, Addis Ababa.

Huffnagel, H.P. 1961. Agriculture in Ethiopia FAO, Rome.

IV

PHYSIOLOGY, CROPPING SYSTEMS, PRODUCTION TECHNOLOGY AND PESTS AND DISEASES IN ASIA

20

PHYSIOLOGICAL APPROACHES FOR IMPROVING PRODUCTIVITY OF FINGER MILLET UNDER RAINFED CONDITIONS

M. Udaya Kumar, V.R. Sashidhar and T.G. Prasad

INTRODUCTION

Finger millet, a C_4 plant, is an important grain crop in the southern states of India. It has a high production potential reaching up to 40-50 quintals per hectare under optimum conditions. However, the yield levels achieved are far below its actual potential because finger millet is predominantly grown under rainfed conditions (Fig. 1). Drought stress severely limits the yield of finger millet although it is reputedly one of the most resistant crops to drought. Even short periods of drought, during some stages of growth markedly reduce the yield.

Although drought causes more yield losses than the combined effect of all other abiotic stress factors, the progress made in enhancing productivity in a water limiting environment has been unfortunately insignificant. The chief reason for this slow progress in 'drought stress' research is the complexity of the problem, since, the magnitude, duration and the time of occurrence of drought is unpredictable and often compounded by the variations in temperature and relative humidity.

However, during this decade significant progress has been made in understanding the nature of stress injury and the adaptive mechanisms associated with growth and survival. Field evaluation programmes have been refined and several quick screening techniques developed for rapid screening of a large number of germplasm material to identify specific characters associated with higher productivity under moisture stress. In this paper different physiological approaches to increase the productivity of finger millet under

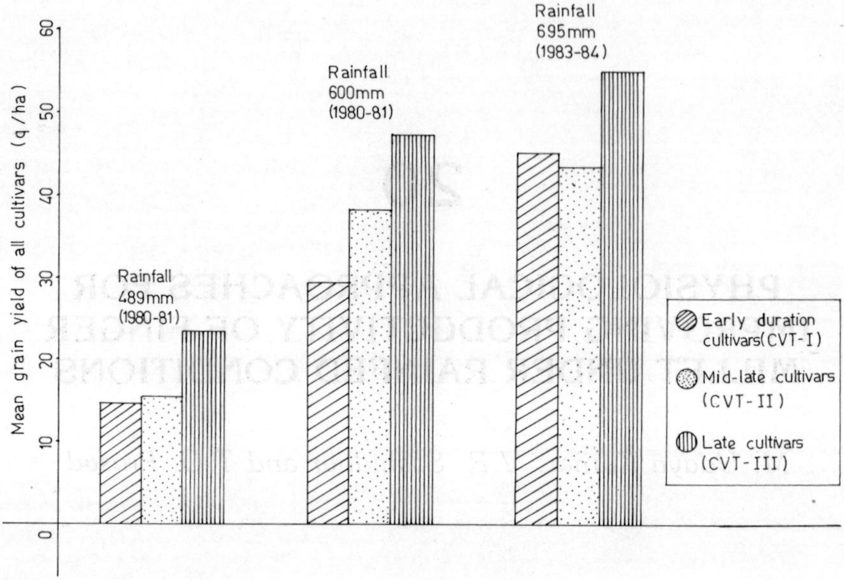

Fig. 1. Data showing the dependence of mean grain yield of three different duration groups of finger millet cultivars on the total rainfall received during the cropping season from June 1st to Nov. 30th (each rainfall value represents total rainfall from June to November).

moisture stress, are discussed. These approaches are based on our own results and also from studies conducted elsewhere.

In the first instance it is necessary to assess (1) the time, magnitude and duration of stress effect in a particular season, (2) the drought stress effect on growth as a constraint for productivity, (3) identifying the adaptive strategies of the plant for higher productivity under these conditions, and (4) the optimum growth period of the crop for maximum utilization of precipitation.

The rainfall distribution (Fig. 2) and water balance components of finger millet growing soils in Karnataka indicate that drought stress may occur either during early stages of growth or during mid-season i.e., during panicle development and anthesis. Analysis of the constraints in growth and productivity suggests that the following are the major factors.

i) Stress after sowing—Effect on seedling emergence and crop establishment
ii) Early season stress—Effect on early crop growth rate
iii) Mid-season stress —Effect on sink development and sink number.

GERMINATION AND SEEDLING SURVIVAL

Stress during germination and seedling establishment drastically affects crop stand and is often a major constraint especially in small millets with limited

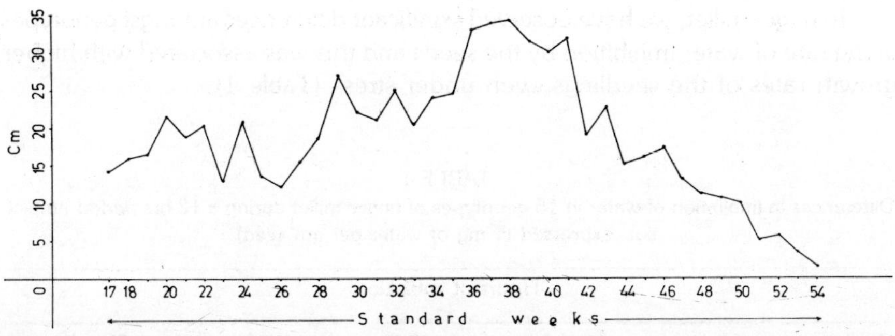

Fig. 2. Average rainfall (mean of 10 years from 1971-1980) during different months of main cropping season, Bangalore, India.

seed reserves. Germination and establishment is often affected under semi-arid conditions, where the soil surface is wetted inadequately and the rate of evaporation is high. In such circumstances, the seedling must compete with the process of atmospheric drying, for the rapidly diminishing moisture of the surface layers. Often, these layers dry out too rapidly for the seed to germinate or for the germinated seedling to extend its roots down into the deeper layers where available moisture can be found. Consequently the seedling may fail to survive even though the overall ecological conditions may be favourable for a mature plant. This problem is further compounded by the formation of crust, the extent and severity of which predominantly depends on soil characteristics. Significant species variation and also variation amongst genotypes within a species do exist in relative germination rates under these situations.

Apart from the pre-sowing moisture conservation measures generally adopted and the intrinsic water holding capacities of soils, seed characteristics associated with seedling vigour may determine the final crop stand. One of them is the intrinsic ability of the seedling itself to maintain higher growth rates which is a good reflection of its vigour. This would directly encourage survival by faster emergence before severe depletion of soil moisture occurred or indirectly by better osmotic adjustment by accumulation 'osmotically active solutes'. High seedling growth rates under stress are also favoured by higher rate of imbibition and the metabolic activity of the seeds. The latter facilitates uptake of water especially under low soil water potentials. The survival of the germinated seed during the stress period and its regrowth on stress alleviation is another important factor which determines seedling establishment. Hydration and dehydration tolerance of the germinated seeds by osmotic adjustment, better hormonal regulation and utilization of seed reserves is important.

The seed characteristics associated with better establishment may be:

i) High rate of imbibition
ii) Higher metabolic activity leading to high growth rates
iii) Hydration—Dehydration tolerance of the germinated seeds.

In finger millet, we have observed significant differences amongst genotypes in the rate of water imbibition by the seeds and this was associated with higher growth rates of the seedlings even under stress (Table 1).

TABLE 1

Differences in imbibition of water in 15 genotypes of finger millet during a 12 hrs period (imbibition expressed in mg of water per gm seed)

	Hours of imbibition				
Genotypes	2 hrs	4 hrs	8 hrs	12 hrs	Classification
GPU 1	281	311	362	412	High
ELC 7	259	301	374	286	High
PPR 1753	315	432	453	425	High
ELC 4	306	336	422	429	High
PR 202	268	290	369	366	High
ELC 6	299	313	370	366	High
ELC 9	272	356	427	392	High
GPU 10	218	244	331	413	High
RAU 10	167	165	359	279	Low
PES 83-2	125	157	248	277	Low
HR 7302	082	288	230	253	Low
GPU 6	125	157	257	267	Low
HR 911	115	145	208	279	Low
GE 2917	184	254	291	311	Low
GE 2978	157	206	302	311	Low

Differences in solute potential of the seeds were also observed depending on the prevailing conditions during panicle development. In a study with ten finger millet genotypes, it was shown that seeds obtained from a rainfed crop which had experienced stress, had higher germinability and seedling vigour when osmotic stress was imposed during germination. It was concluded that this higher germinability was possibly related to the higher solute content in the seeds particularly sucrose (Tables 2, 3). Such induced variability could be exploited. Similar results have been obtained with simple seed hardening techniques (Sastry *et al.*, 1969; Rajasekhar *et al.*, 1970). Genotypic differences in emergence of seedling under crust situations have also been observed (Table 4).

Although significant genotypic variations have been observed in several of these characteristics associated with seedling establishment, a breeding programme to incorporate these parameters is difficult and time consuming. It is worthwhile to develop agronomical techniques which can induce tolerance or facilitate stress avoidance thereby enhancing seedling vigour.

TABLE 2

Germination of finger millet seeds from stressed and non-stressed plants in water and in simulated stress with polyethyleneglycol (PEG) of −8 and −12 bars

Variety & treatment	Experiment - I		Experiment - II		
	H_2O 96 hrs	PEG −8 bars 96 hrs	Water 96 hrs	PEG − 8 bars 96 hrs	PEG − 12 bars 96 hrs
PES 176 T_1	100	80	100	60	50
T_2	100	4	100	15	20
PR 202 T_1	100	100	100	75	65
T_2	100	50	100	60	0
LSD for variety	—	13.9	—	15.5	7.5
LSD for treatment	—	25.5	—	20.5	21.5

T_1 — Rainfed stress with 100 kg N/ha
T_2 — Control (irrigated) with 100 kg N/ha

TABLE 3

Solute potential of finger millet seeds (− bars) from stressed and non-stressed plants grown under field conditions

	'A' Solute potential (Unit weight of seeds 500 mg)		'B' Solute potential (Unit number of seeds 200)	
Variety	T_1	T_2	T_1	T_2
PES 176	−11.2	−5.8	− 9.8	−7.4
PR 202	− 9.7	−5.0	−12.3	−6.7
IE 1022	− 9.5	−4.8	−10.5	−7.1
HR 911	−10.4	−7.8	− 9.2	−6.4
HR 374	− 9.2	−4.6	−11.7	−5.6
Indaf 5	− 8.8	−6.6	−11.2	−5.4
Indaf 9	− 7.1	−5.5	− 8.3	−6.9
DRKPES 1	− 8.4	−7.2	−12.3	−4.7
U 10	− 8.6	−3.8	− 9.1	−8.3
TNAU 294	− 9.1	−5.9	−10.2	−6.9

'A' — LSD for variety 1.25; LSD for treatment 1.75
'B' — LSD for variety 1.80; LSD for treatment 1.65
T_1 — Rainfed stress with 100 kg/N/ha
T_2 — Control (Irrigated) with 100 kg N/ha

TABLE 4

Genotypic differences in germination per cent of finger millet seeds in a crusted laterite soil

Variety	Control		Moisture stress		
	5 DAS	7 DAS	5 DAS	7 DAS	3 days after rewatering
Indaf 8	70.5	87.5	6.5	7.0	8.0
Hullebele	88.0	89.5	20.5	29.0	30.0
JNR 852	62.5	69.0	10.0	13.5	18.5
Indaf 5	48.0	58.5	19.5	37.0	39.0
ROH 2	66.0	69.0	20.5	29.5	30.0
PES 172	84.5	89.0	68.0	69.0	82.0
PR 202	78.5	82.0	38.0	37.0	45.0

MOISTURE STRESS AS THE LIMITING FACTOR FOR GROWTH AND PRODUCTIVITY

Depending on the stress situation, the mechanisms adapted by plants to drought stress are different. There are specific escape, avoidance and tolerance mechanisms. Their importance and relevance have been extensively reviewed (Hsiao, 1973; Levitt, 1980; Paleg and Aspinall, 1981). These mechanisms either favour survival under stress situation or help in maintaining good productivity under stress situations (Fig. 3).

In our concept, it is more important for a drought resistant crop to have characteristics associated with maximizing productivity under stress situations rather than to ensure mere survival.

The adaptive strategies for high productivity under rainfed conditions, in order of priority could be as shown in Fig. 4.

The best strategy for rainfed conditions is increasing the water harvesting and its utilization efficiency. The total productivity of any crop depends on the evapotranspiration (ET) water use efficiency (WUE) and the harvest index (HI).

Agronomical approach to enhance the moisture conservation and a few strategies to minimise water loss like mulching and practising optimum date of sowing have given rich dividends for enhancing productivity of rainfed crops.

Apart from these practices two other physiological processes associated with water harvesting and water conservation are: (1) root factors and (2) the plant processes associated with transpiration quotient (TQ).

ROOT FACTORS

An important feature of a drought resistant plant could be its deep root system. The relevance of root volume, spread and depth, relative energy allocation to roots, and the vertical conductances of the root system has been reviewed by Passioura (1981, 1981a). An extensive root system seems to be relevant only under a specific agroclimatic situation and it is soil and crop specific.

(1) Strategies for higher productivity with drought avoidance mechanisms:
 (a) Maintenance of water uptake
  Root characters
 (b) Water utilization efficiency
  Characters associated with low
  transpiration quotient (TQ)
 (c) Higher partitioning efficiency
  i) High HI
  ii) Remobilization of reserve carbohydrates
(2) Strategies for less reduction in productivity with drought escape mechanisms:
 (a) Developmental plasticity
  i) Postponement of flowering
  ii) Plasticity of tillering
(3) Strategies for less reduction in productivity with drought tolerant mechanisms.
 (a) High crop growth rates under stress and on alleviation of stress
  Characters at the organ level:
  i) Higher growth rate with low tissue water potential
  ii) Higher leaf expansion on alleviation of stress
  iii) Higher partitioning on stress alleviation.
  Characters at cellular level:
  i) Osmoregulation
  ii) Chloroplast integrity
  iii) Hormonal factors
  iv) Membrane integrity
(4) Strategies resulting in possible reduction in productivity with drought avoidance characters:
 i) Increase in stomatal and cuticular resistance
 ii) Reduction in radiation load
 iii) Reduction in evaporative surface.

Fig. 3. Adaptive mechanisms and their relationship with productivity under intermittent drought stress

(1) Higher water harvesting and utilization efficiency.
 (a) Water conservation mechanism (Agronomical).
 (b) Water harvesting by roots
 (c) Efficient water utilization (TQ).
(2) Developmental plasticity under stress
(3) Survival and growth under stress and higher crop growth rate after alleviation from stress.
(4) Partitioning and effective remobilization of reserves.

Fig. 4. Important adaptive strategies in finger millet for higher productivity under rainfed conditions

As soil water potential in the surface layers decreases, water retained in the deeper layers makes a larger contribution to ET. Often, in many shallow rooted crops like the small millets, when most of the moisture from the upper layers is exhausted, the plant is unable to extract water to satisfy the ET demand even though soil water available in the deeper layer is still high (Fig. 5). Under these conditions a deeper root system may have an advantage.

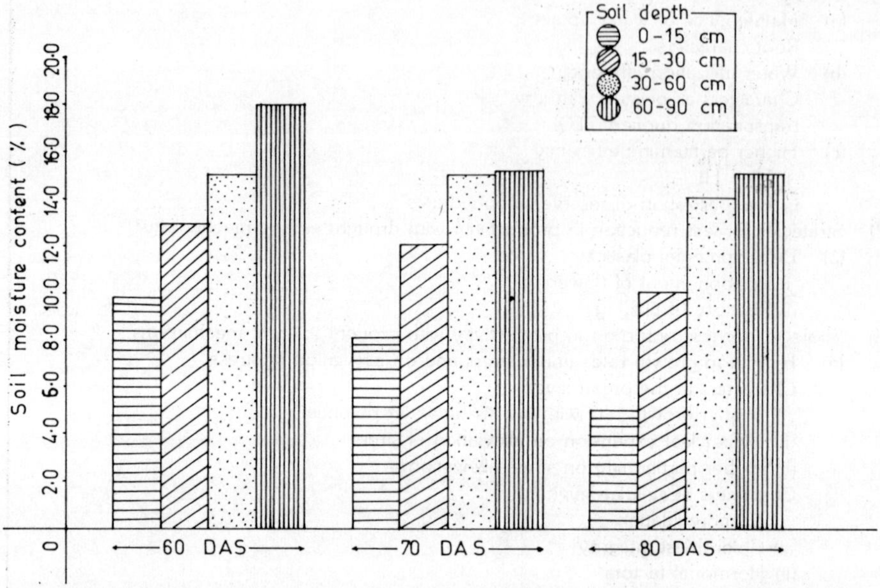

Fig. 5. Soil moisture depletion and the moisture content at various depths in a finger millet plot during a rainfree period in the laterite soils at the GKVK experiment station, Bangalore (Data from Annual Report of the Dryland Agri. Project 1974)

The following aspects need to be investigated further:

i) Development of suitable techniques for rapid and accurate measurement of root depth, spread and activity.
ii) Duration of the functional root system during the crop growth period.
iii) The characteristics associated with hydraulic conductivity (longitudinal resistance to flow) of root system.
iv) The relative allocation of carbohydrates to root systems and its significance.

Genotypic differences in root density especially in deeper soil layers, are well documented in some crop species like rice (Yoshida and Hasegawa, 1982). Since, finger millet is predominantly grown in soils with adequate moisture in the deep soil layers, it is necessary to develop a suitable programme to identify genotypes with higher water extraction capabilities.

PLANT PROCESSES ASSOCIATED WITH HIGH WATER USE EFFICIENCY

The physiological and biochemical factors associated with a low transpiration quotient (TQ) (high water use efficiency) are other important adaptive strategies of the plants for higher productivity under stress conditions.

Although the total evapotranspiration always shows a relationship with biomass production and productivity (Yield = Total ET × WUE × HI), the differences in productivity amongst genotypes at a given level of evapotranspiration are mainly attributed to variation in TQ or WUE. Hence, it is an important character under field conditions for efficient water utilization. The high WUE achieved by some species in semi-arid and arid conditions is often attributed to increased assimilation rate per unit water transpired (Bierhuzin, 1976).

Stomatal and mesophyll characteristics are basically responsible for the variation in TQ. However, under field conditions canopy characteristics have to be considered in terms of the relative rates of water loss and assimilation. The canopy conductance, a product of the stomatal conductance and Leaf Area Index (LAI) (Squire and Black, 1981), is an important determinant of productivity under field conditions. In this context, the total number of stomata is a more useful parameter as it takes into account the variability in both stomatal frequency and leaf area. A higher number of stomata per plant would increase the total transpirational water loss concurrently.

For crops grown under rainfed conditions, genotypes with low canopy transpiration rates are desirable (Jones, 1977). This can be achieved by identifying types with low conductances (Jones, 1979) or alternatively with genetically low leaf area (LA) or low number of stomata per plant but without sacrificing the ability to produce higher dry matter (DM), since the latter is highly correlated with grain yield in finger millet (Figs. 6, 7) (Sastry *et al.*, 1982).

Field experiments were conducted at our centre during summer and monsoon seasons to investigate the possibility of identifying genotypes with genetically low leaf area or leaf area duration (LAD), yet having good grain yield and biomass. In such genotypes the canopy water loss is likely to be relatively lower than in genotypes with larger leaf area (Figs. 8, 9). Since, earlier studies (Sastry *et al.*, 1982) had shown a high positive and significant correlation between dry matter production and grain yield in finger millet, the biological yield was used as the primary selection criteria to screen genotypes.

In a field experiment conducted with 100 medium duration genotypes a significant variation in total biomass, LA, HI, NAR and yield were observed. Among these genotypes there were distinct genotypes some with high leaf area, high dry matter and high harvest index and others with low leaf area, high dry matter and high harvest index (Table 5). Successive field experiments conducted in selected high and low leaf area types with high biomass and high harvest index have shown the possibility of identifying genotypes with low and high stomatal number (Tables 6 and 7). It is logical to assume that total canopy water loss would be low in genotypes with low leaf area or low stomatal number. Since the small millet crops, such as finger millet, are predominantly rainfed crops, this character assumes greater importance. If the total biological yield

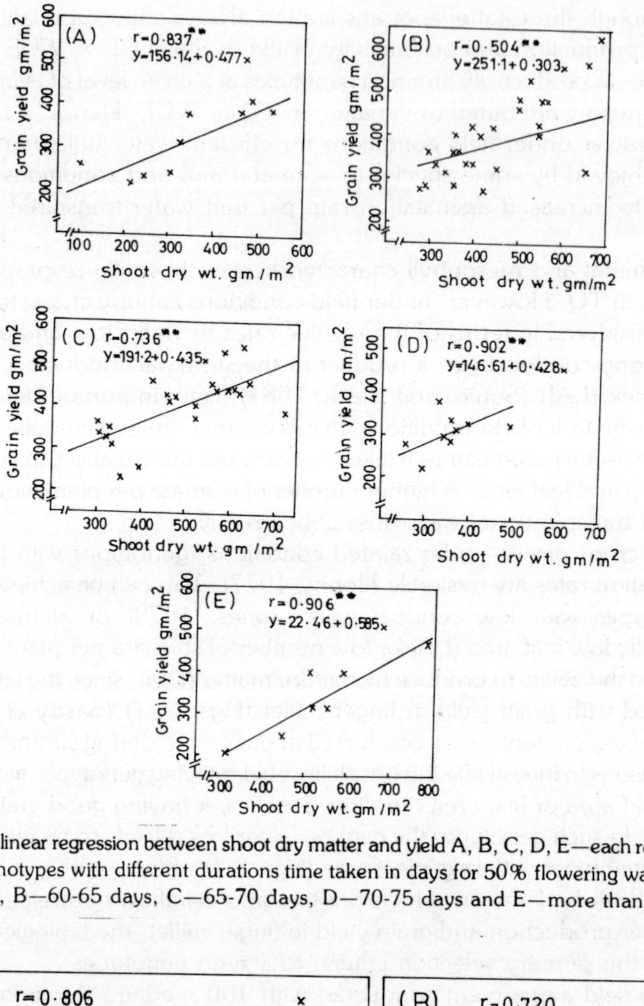

Fig. 6. Simple linear regression between shoot dry matter and yield A, B, C, D, E—each representing a group of genotypes with different durations time taken in days for 50% flowering was: A—less than 60 days, B—60-65 days, C—65-70 days, D—70-75 days and E—more than 75 days.

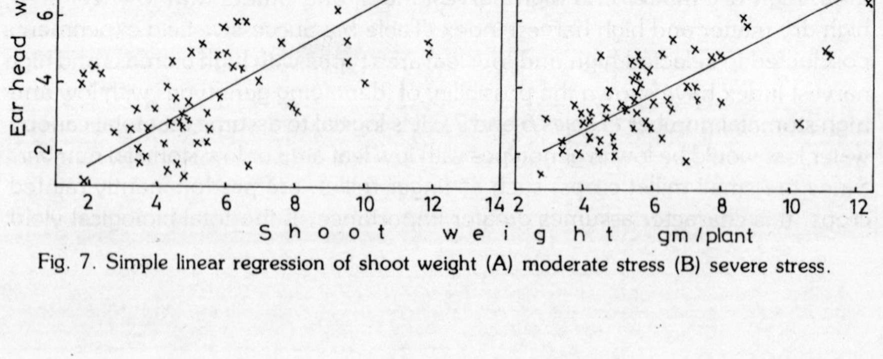

Fig. 7. Simple linear regression of shoot weight (A) moderate stress (B) severe stress.

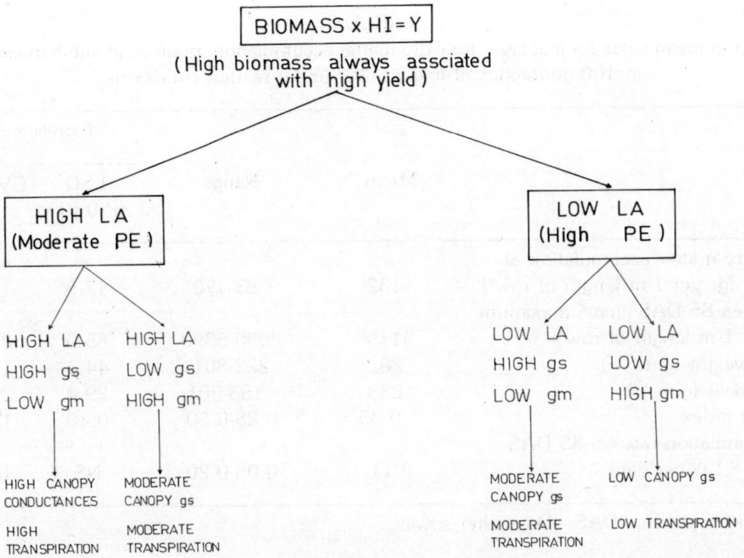

Fig. 8. Approach for identifying types with low canopy transpiration and high productivity.

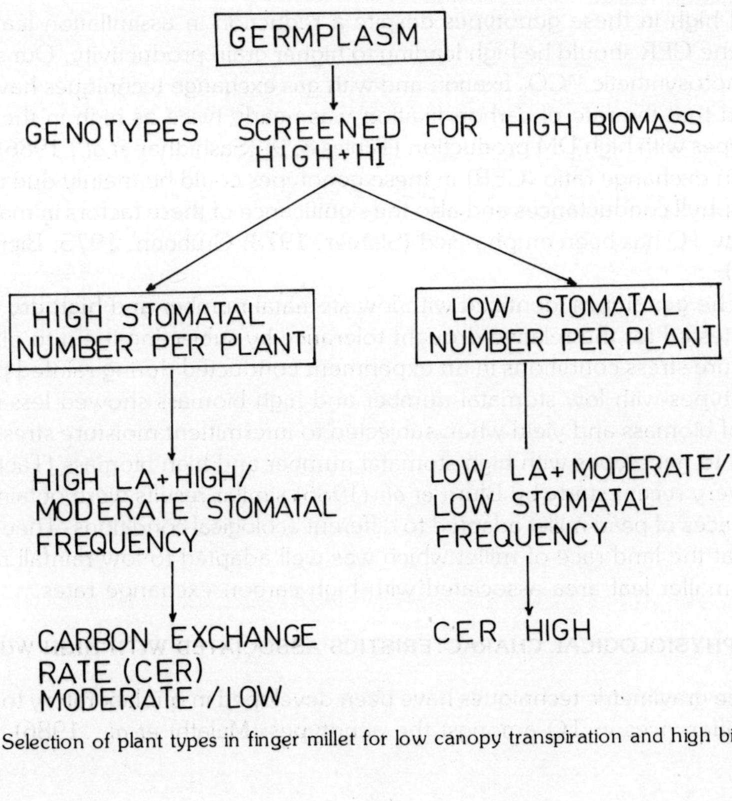

Fig. 9. Selection of plant types in finger millet for low canopy transpiration and high biomass.

TABLE 5
Range and mean value for leaf area, total dry matter accumulation, grain yield and harvest index in 100 genotypes of finger millet under rainfed conditions

	Mean	Range	Significance	
			LSD (P 0.01)	CV (%)
Total dry matter accumulation at harvest (g. per 1-m length of row[a])	132	83-170	17.5	15.9
Leaf area 85 DAS[b] (cm^2 maximum LA per 1-m length of row[a])	3135	2090-5390	85.0	23.0
Straw weight ($g.m.^{-2}$)	282	222-801	44.4	17.2
Grain yield ($g.m^{-2}$)	233	153-301	29.4	20.3
Harvest index	0.35	0.25-0.50	0.10	15.7
Net assimilation rate 65-85 DAS ($g.dm^{-2}$ LA per day)	0.11	0.05-0.20	NS	10.0

[a]Ten plants per m. [b]DAS—Days after sowing.

is still high in these genotypes despite a reduction in assimilation leaf area, then the CER should be high leading to higher grain productivity. Our studies on photosynthetic $^{14}CO_2$ fixation and with gas exchange techniques have confirmed that the rate of carbon fixation was nearly twice as high in these low LA types with high DM production (Tables 8, 9) (Sashidhar *et al.*, 1986). High carbon exchange ratio (CER) in these genotypes could be mainly due to high mesophyll conductances and also the significance of these factors in maintaining low TQ has been emphasised (Slatyer, 1973; Grubben, 1975; Bierhuzin, 1976).

The genotypes identified with low stomatal number and high dry matter were tested for the relative drought tolerance by subjecting them to different moisture stress conditions in an experiment conducted during rainfed period. Genotypes with low stomatal number and high biomass showed less reduction of biomass and yield when subjected to intermittent moisture stress compared to genotypes with high stomatal number and high biomass (Table 10). In a very recent study by Blum *et al.* (1986) similar results were obtained for land races of pearl millet adapted to different ecological conditions. They showed that the land race of millet which was well adapted to low rainfall regions had smaller leaf area associated with high carbon exchange rates.

PHYSIOLOGICAL CHARACTERISTICS ASSOCIATED WITH HIGH WUE

Precise gravimetric techniques have been developed in our laboratory to assess the differences in TQ amongst the genotypes (Malathi *et al.*, 1986).

TABLE 6

Variation in stomatal frequency (number per mm^2) in leaves at different canopy positions and the total number of stomata per plant in genotypes of finger millet with high and low leaf area

	Range and mean frequency in relation to leaf position					
	Top (L_1)	Middle (L_4)	Bottom (L_8)	Mean frequency	Range & mean LA 85 DAS	Mean stomatal No. per plant (X 10^5)
High LA:						
Adaxial surface	101-156 (125)	85-158 (117)	96-137 (111)	117	3190-5390	89.6
Abaxial surface	78-151 (118)	87-112 (98)	89-112 (100)	105	(1020)	
Mean				111.4		
Low LA:						
Adaxial surface	110-229 (148)	112-190 (126)	114-221 (139)	137	2090-2810	56.0
Abaxial surface	98-172 (118)	89-154 (108)	94-147 (116)	113	(2240)	
Mean				125.2		

TABLE 7

Genotypic variation in growth and yield attributes and stomatal number in high and low leaf area types of finger millet

Genotypes	Stomatal frequency	LAI	LAD (days)	NAR (g. dm^2 LA per day)	Stomatal number per plant (X 10^5)	DM at harvest (g. per 1-m length of row)[a]	Grain yield (g. per 1-m length of row)	HI
Category I: High LA, high DM and high HI:								
GE 1097	243	2.22	204.3	0.13	23.2	354	85.2	0.55
GE 476	257	4.79	245.6	0.11	27.8	406	112.5	0.35
GE 966	248	4.56	239.2	0.15	25.4	371	91.3	0.39
JNR 852	277	4.85	220.8	0.13	24.8	367	112.8	0.36
GE 821*	200	2.95	181.9	0.08	13.2	369	99.2	0.38
Mean	236	4.61	227.4	0.13	25.3	374	100.4	0.45
Category II: Low LA, high DM high HI:								
GE 187	257	1.79	108.3	0.21	10.4	333	87.4	0.42
HR 23A*	245	4.62	239.9	0.14	25.6	397	114.1	0.54
GE 325B	232	2.43	115.4	0.18	12.6	475	102.9	0.32
ROH 2	245	2.74	152.7	0.20	15.2	498	115.0	0.38
GE 94	200	2.03	152.9	0.15	13.6	364	81.9	0.54
Mean	236	2.50	132.3	0.18	12.9	417	96.8	0.42
L.S.D. ($P<0.05$)	32	1.01	62.2	NS	4.6	8	NS	NS

LAI—leaf area index; LAD—Leaf area duration; NAR—Net assimilation rate; DM—Dry matter; HI—Harvest index; [a]Ten plants per m. GE 821, HR 23A, classified as low and high LA types respectively in previous experiment, not consistant; eliminated for experiment 3, not included for mean.

TABLE 8

Photosynthetic $^{14}CO_2$ fixed by the leaves at different canopy positions in selected genotypes with high and low leaf area associated with high dry matter production

Genotypes	Fixation rate (cpm mg^{-1} dry wt.) Top	Middle	Bottom	Average specific activity (cpm mg^{-1} dry wt.)	Activity per plant (cpm)
High LA, high DM, high HI types:					
GE 476	20	18	10	16	175377
GE 1097	30	33	9	24	191810
JNR 852	17	17	14	16	172027
Low LA, high DM and high HI types:					
GE 187	38	44	25	36	166172
GE 325B	41	44	36	11	159852
Mean	—	—	—	48	178012
LSD (P 0.05)	—	—	—	8.2	NS

Each value is an average from 6 plants; LA — Leaf area; DM — Dry matter; HI — Harvest index.

TABLE 9

Photosynthetic rate in selected genotypes with high and low leaf area associated with high dry matter production

	Photosynthetic rate (mg. CO_2 dm^{-2} h^{-1})	
	At anthesis	10 days after anthesis
High LA, high DM, high HI types:		
GE 476	23.4	26.3
GE 1097	20.4	23.6
JNR 852	22.3	24.4
Mean	22.03	24.77
Low LA, high DM, high HI types:		
GE 187	36.5	37.2
GE 325B	33.4	34.7
Mean	34.9	35.9
LSD (P 0.05)	4.2	5.2

LA — Leaf area; DM — Dry matter; HI — Harvest index

TABLE 10

Growth and yield parameters in low and high stomatal number types under two moisture regimes

	Genotypes	Total dry matter (g/plant) 80 DAS		Leaf area (cm^2) (80 DAS)		Leaf area duration (LAD) days		Grain yield (g/m^2)	
		Control	Stress	Control	Stress	Control	Stress	Control	Stress
High stomatal number & high dry matter	GE 161	19.63	16.9	968	730	138.4	112.8	416.3	378.5
	GE 282	25.66	19.2	747	541	122.7	87.7	356.5	354.7
	JNR 852	27.06	17.2	918	812	156.2	110.8	566.3	442.8
	Mean	24.12	17.9	844	601	139.10	103.70	544.4	392.0
		(25.7)		(28.7)		(25.5)		(11.6)	
Low stomatal number & high dry matter	GE 325	26.22	20.18	543	535	100.8	87.2	362.5	352.10
	HR 23A	23.84	21.82	634	532	113.6	96.4	457.5	440.10
	Mean	25.03	21.00	588	533	107.20	91.80	410.0	396.1
		(15.9)		(10.0)		(14.30)		(3.4)	
LSD ($P<0.05$)	Treatments	3.84		129.3		—		NS	
	Varieties	NS		204.4		—		91.5	

(a) Control—Non-stressed block was given three protective irrigations to alleviate moisture stress at critical stages.

(b) Rainfed stress block. Values in the parentheses indicate per cent reduction due to stress.

By determining the cumulative water transpired and the dry matter accumulated during the crop growth period, genotypes differing distinctly in TQ were identified (Table 11). Apart from TQ, variation in cumulative water used (CWU) and other physiological characters like leaf area, transpiration rate per unit leaf area, stomatal behaviour and net assimilation rate (NAR) were observed. The major factors contributing for low TQ were high NAR and low transpiration rate per unit leaf area.

Based on these studies it was possible to identify genotypes with varying levels of cumulative water use and transpiration quotient.

1. High CWU and high TQ
2. Low CWU and high TQ
3. High CWU and low TQ
4. Low CWU and low TQ.

Our studies have shown that the genotypes belonging to the latter two groups are better for rainfed conditions. Genotypes with high CWU and low

TABLE 11

Transpiration quotient (TQ) in selected genotypes of finger millet differing in LA but having high biomass

Genotypes	Cumulative water used (CWU) (ml)	TDM (g)	Leaf area (cm^2)	Transpiration quotient (TQ) CWU/TDM
Selected genotypes with low LA/high DM				
GE 94	6675	36.5	767	182
GE 325	5820	23.5	583	247
ROH 2	7147	23.4	926	305
GE 3209	7452	29.2	505	255
GE 2327	5749	23.0	969	249
GE 2338	5759	35.1	1039	186
GE 2444	7892	35.1	1088	224
GE 2008	8368	41.3	1537	202
Mean	6857	30.3	926	231
Selected genotypes with high LA/high DM				
GE 1097	6458	27.7	1137	233
GE 2976	6499	33.1	1832	196
GE 3255	6666	32.2	1154	213
JNR 852	8107	41.1	1542	197
GE 476	6544	27.7	1292	236
GE 3243	7207	35.9	2110	200
GE 2969	6152	26.6	1556	231
Mean	6804	31.9	1517	225

TQ types have an advantage when moisture available in the deeper layers can be harvested. In this type high CWU is possibly associated with greater stomatal conductances resulting in a high water demand. However, if moisture is severely limited it is imperative to select types with a low CWU associated with a low transpiration quotient. Again in these types the low CWU may presumably be a consequence of lower canopy conductances.

In general, the small leaf area types showed low TQ associated with low CWU. Genotypes with moderate or larger leaf area associated with high NAR and high transpiration showed high CWU with low TQ (Fig. 10).

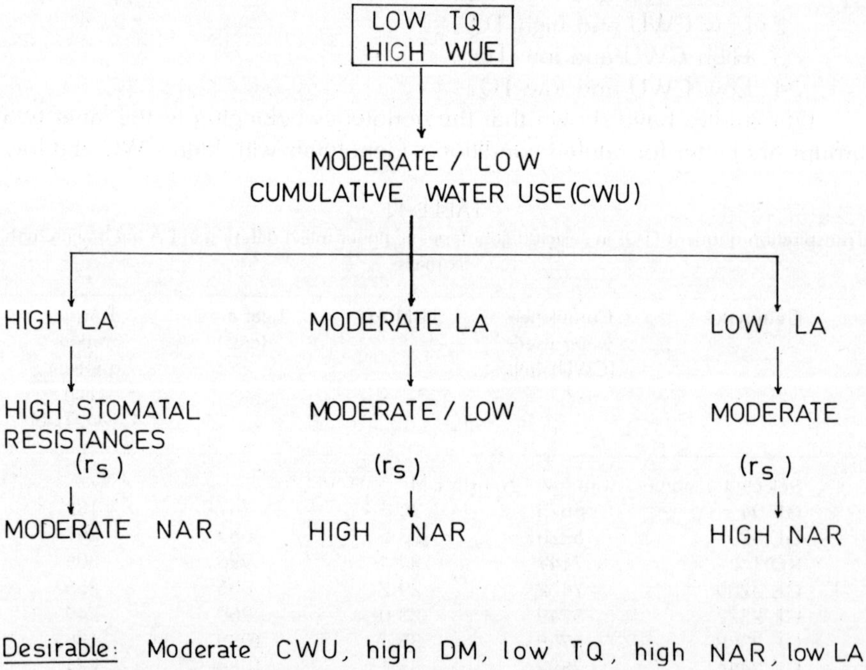

Fig. 10. Characteristics associated with low transpiration quotient (TQ).

The desirable characters seem to be low TQ with moderate or high CWU with high dry matter. The morphological and physiological characters associated with such types could be:

i) Low leaf area (small leaf size/number)
ii) High dry matter
iii) High NAR/photosynthetic rate preferably by virtue of high mesophyll conductances
iv) Low canopy transpiration
v) High partitioning efficiency.

PHOTOSYNTHETIC RATES AND ITS IMPORTANCE TO PRODUCTIVITY UNDER RAINFED CONDITIONS

Biomass production is predominantly dependent on canopy photosynthesis. Though both leaf area and photosynthesis contribute to biomass production, increasing the photosynthetic efficiency is advantageous especially for crops grown under rainfed conditions. Genotypes with high photosynthetic rate may still produce high biomass with small leaf area. Such types will have a lower transpiration leaf area, and could be expected to have a low transpiration quotient and high water use efficiency. The concept of low leaf area and high photosynthetic efficiency and high biomass types of small millets having an advantage under rainfed conditions has been established recently (Gurumurthy, 1982; Sastry *et al.*, 1982; Sashidhar *et al.*, 1984, 1986).

The genotypic differences in photosynthetic efficiency (PE) are often arrived at by measuring PE in a single leaf. A precise measurement of mean canopy photosynthetic rate is difficult.

The mean photosynthetic rate over crop growth period can be calculated by leaf area duration/dry matter (LAD/DM) ratios. Higher the value, lesser the photosynthetic rate and vice versa. This concept of LAD/DM can be extended and widely adopted as a preliminary screening technique for determining the canopy photosynthetic rate in different genotypes. Differences in PE of different leaves of canopy, diurnal fluctuations in PE and problems associated with plant architecture like mutual shading are taken care of by determining LAD/DM ratios. However, differences if any, in dark respiration will slightly alter this value. The genotypes selected for high biomass and low leaf area were shown to have high PE in finger millet (Sashidhar *et al.*, 1984) and mungbean (Devendra, 1986).

Significant genotypic variation exists in the photosynthetic rate in finger millet genotypes and several plant characteristics were shown to be associated with high PE and high translocation efficiency of photosynthates. The leaf vein frequency, the ratio of veinal width to leaf width, the mean veinal width and mean inter veinal width showed significant relationship with PE and also translocation of photosynthates (Table 12). Some of these characteristics have been shown to have high heritability value and genetic advance (Perumal, 1982) and could be used in the breeding programme to improve photosynthetic traits.

DEVELOPMENTAL PLASTICITY

In spite of the inbuilt mechanism for low transpiration quotient in finger millet, the crop experiences severe moisture stress in many locations during early stages of growth even with a good degree of soil moisture conservation practices. Stress induced plasticity in postponing the flowering and development of new tillers on stress alleviation are often suggested as adaptive mechanisms under such situations.

TABLE 12
Range, mean in leaf characteristics of the first leaf at flowering and its relationship to photosynthetic efficiency in 42 finger millet genotypes

	Range	Mean	CD	Correlation coefficient 'r' value
Leaf width (μ)	4700-7300	6169	578	−0.41
Total vein number/leaf	24.0-32.3	28.2	2.1	0.59
Total major vein number	6.0-7.3	6.5	0.9	0.03
Total minor vein number	17.7-26.0	21.6	2.0	0.58
Minor/major vein number ratio	2.5-4.13	3.28	0.5	0.50
Leaf vein frequency	3.88-5.61	4.82	0.5	0.78**
Ratio of veinal width/leaf width	21.14-33.2	28.1	2.3	0.76**
Mean veinal width (μ)	48-69	58	2.2	0.76**
Mean inter veinal width (μ)	123-203	151	15	−0.82**
Phloem width (μ)	397-697	555	84	0.50
Leaf thickness at base (μ)	200-356	333	59	0.40
Specific leaf weight (mg/cm^2)	3.83-4.32	4.22	0.2	0.53
Total veinal width (μ)	1364-2072	1737	205	0.52

Range in PE 12238-22172
(^{14}C fixation gm^{-1} leaf dry weight)
Mean 19001
CD 5% 2786

Medium duration cultivars have better plasticity both in terms of postponment of flowering during stress and production of new tillers on stress alleviation as compared to early cultivars. However, the information available on variations in developmental plasticity in finger millet is scanty.

PLASTICITY IN TILLERING

Mid-season drought stress effect on overall productivity has been shown to be less in tillering genotypes with an ability for tiller development on alleviation of stress (Alagarswamy, 1981). However, in finger millet, the specific advantage of tillering types under stress situations has not been elucidated so far. In many genotypes of finger millet, the productivity of successive tillers reduces drastically (Tables 13, 14) and the late formed tillers and nodal tillers formed after stress alleviation contribute very little to grain weight.

TABLE 13
Number of ears and their grain weight in different tillers in finger millet

				Primary and secondary		Side shoot ears	
	Total ears/ plant	Grain weight g/plant	Main ear weight (g)	Number of ears	Grain wt./ ear (g)	Number of ears	Grain weight/ ear (g)
Variety—I	6.91	22.66	5.51	3.19	3.95	2.72	1.70
Variety—II	6.96	15.80	3.43	3.08	2.69	2.88	1.51

TABLE 14
Relative contribution of basal and nodal tillers to the total tillers in finger millet (Rainfed kharif crop, 1981)#

		Basal tillers			Nodal tillers			
Genotypes	Total tillers/plant	A Unproductive tillers	B§ Productive nonharvestable	C Productive harvestable	A	B	C	Total harvestable tillers
HR 23 A	4.85	0.30	0.13	3.80	0.16	0.30	0.16	3.96
	(100)*	(6.20)	(2.70)	(78.4)	(3.3)	(6.20)	(3.30)	(81.7)
JNR 852	5.98	0.20	0.16	4.80	0.23	0.46	0.13	4.93
	(100)	(3.30)	(2.7)	(80.3)	(3.80)	(7.7)	(2.20)	(82.5)

* — Values in parenthesis indicate per cent of total tillers
§ — Tillers with ears but not mature at the time of harvest of main ear
# — Rainfed crop suffered stress at panicle initiation (PI) and anthesis.

In finger millet, a relationship exists between productivity and mean ear weight, but not ear number per plant (Fig. 11). Thus in recently developed genotypes, the higher yield potential is the result of enhanced mean ear weight. The plasticity in tillering as an adaptive mechanism under rainfed situations needs to be thoroughly investigated.

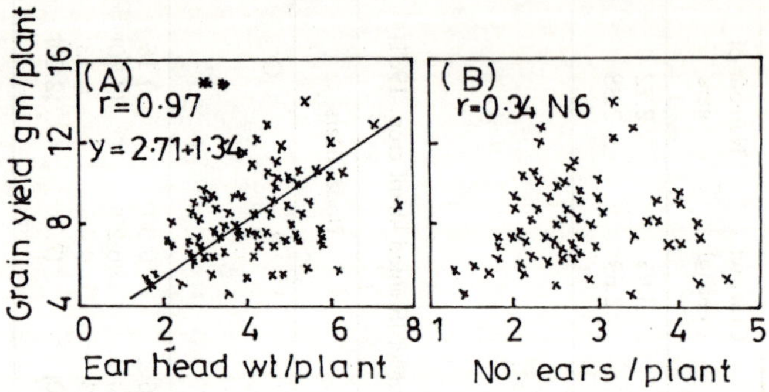

Fig. 11. Simple linear regression between mean ear weight, number of ears/plant and yield (A) average ear head yield (B) number of ears per plant/yield.

HIGH CROP GROWTH RATE ON ALLEVIATION OF STRESS

The differences in productivity under rainfed situations are often attributed to differences in crop growth rates (CGR) on alleviation of stress. The functional leaf area at the end of stress period and resumption of leaf growth and its activity on alleviation of stress determines the CGR.

The leaf expansion even at low tissue water potential (under stress) is generally marginal and very little genetic variation exists in this character. In finger millet, our study has shown that as leaf water potential decreased, there was a rapid cessation of leaf elongation indicating that it is a very sensitive character to moisture stress. The threshold water potential at which leaf elongation was reduced by 50 per cent in maize was shown to be −4 bars (Acevedo *et al.*, 1971; Boyer, 1970). However, in finger millet, there is a resumption in leaf elongation on stress alleviation and in some genotypes the leaf elongation rates exceed that of the control (Figs. 12 and 13) (Vishwanath, 1977). Similar results were obtained by Ludlow and Ng (1976) in *Panicum maximum*.

The resumption of leaf expansion and the NAR depends on the intrinsic dehydration tolerance mechanisms like osmo-regulation, maintenance of membrane integrity, reduced photo-inhibition and photo-oxidation properties, and hormonal aspects.

Genotypic variations in CGR during stress are very limited and if they do exist, no systematic approaches were made to identify such types. However, marked genotypic variation in CGR on stress alleviation exists and it is essential to identify the biochemical and physiological parameters associated with them for further exploitation of genetic variability.

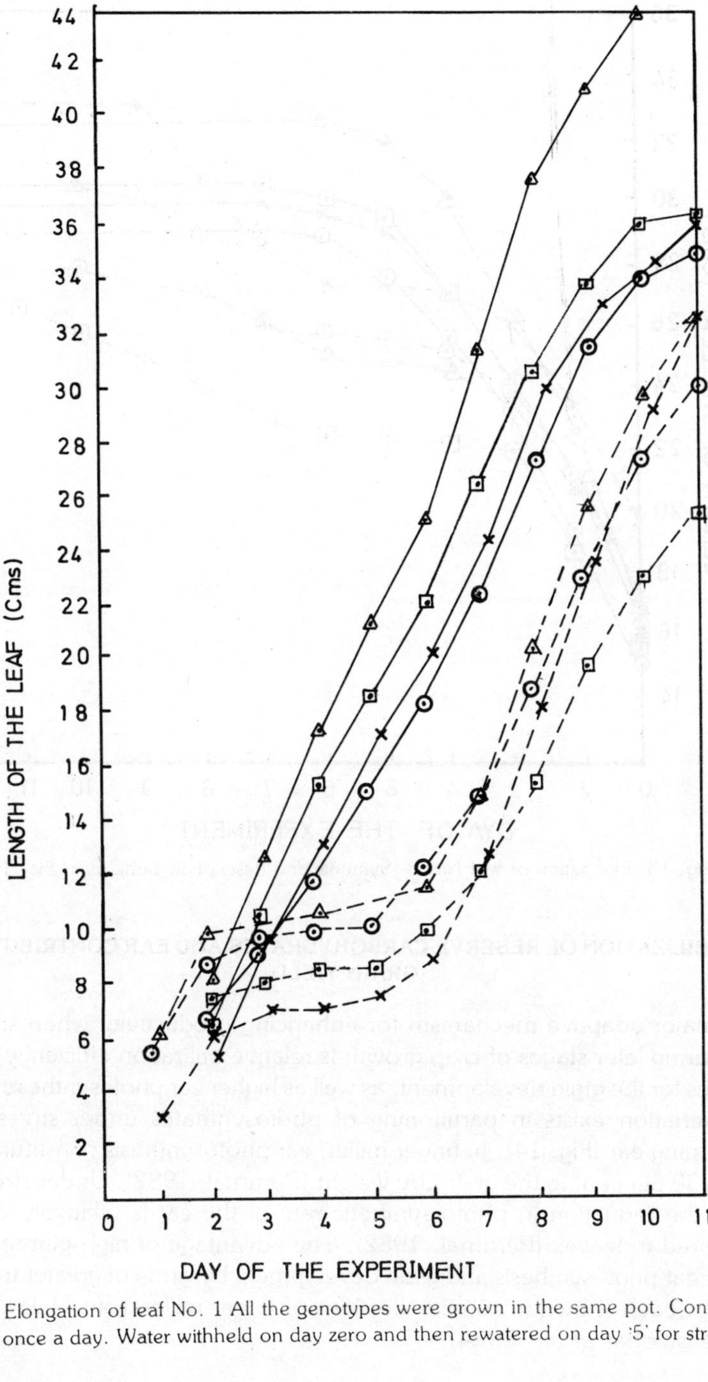

Fig. 12. Elongation of leaf No. 1 All the genotypes were grown in the same pot. Control were watered once a day. Water withheld on day zero and then rewatered on day '5' for stress pots.

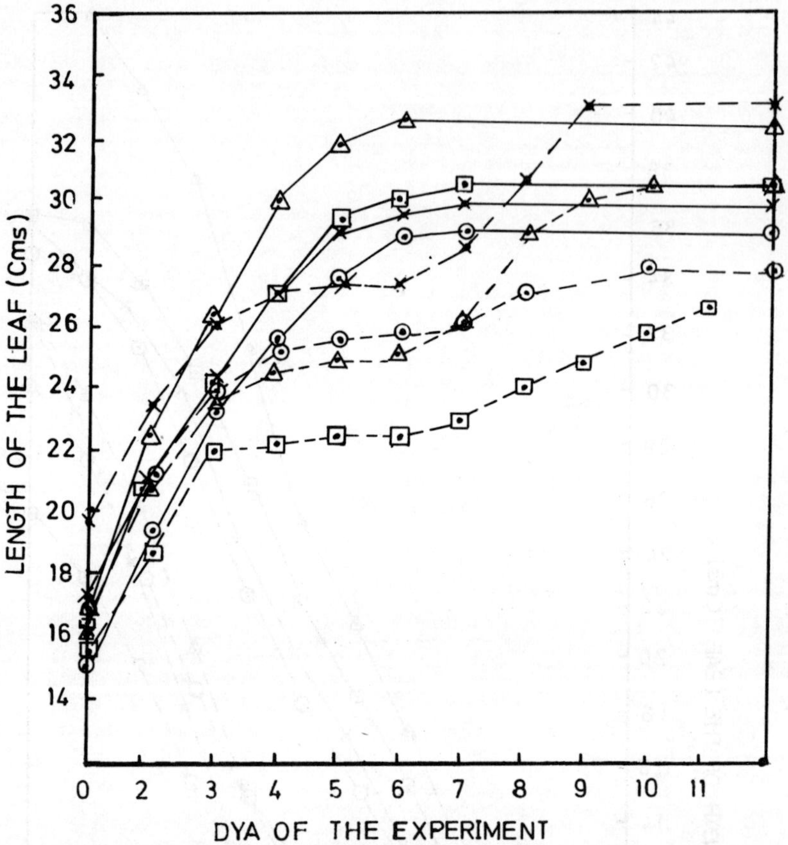

Fig. 13. Elongation of leaf No. 2 (Symbols and description same as in Fig. 12).

REMOBILIZATION OF RESERVE CARBOHYDRATES AND EAR CONTRIBUTION TO GRAIN YIELD

One major adaptive mechanism for enhancing productivity when stress occurs during later stages of crop growth is relative utilization efficiency of stem reserves for the grain development, as well as higher ear photosynthesis. Significant variation exists in partitioning of photosynthates under stress to the developing ear (Fig. 14). In finger millet, ear photosynthesis constitutes nearly 5 to 30 per cent to the grain dry weight (Perumal, 1982). Under stress condition the reduction in photo-synthetic rate of the ear is relatively very less compared to leaves (Perumal, 1982). The advantage of high glume size for higher ear photosynthesis and grain development by virtue of greater.translocation has been shown in some collections of finger millet from Malawi (Table 15) (Sashidhar *et al.*, 1984).

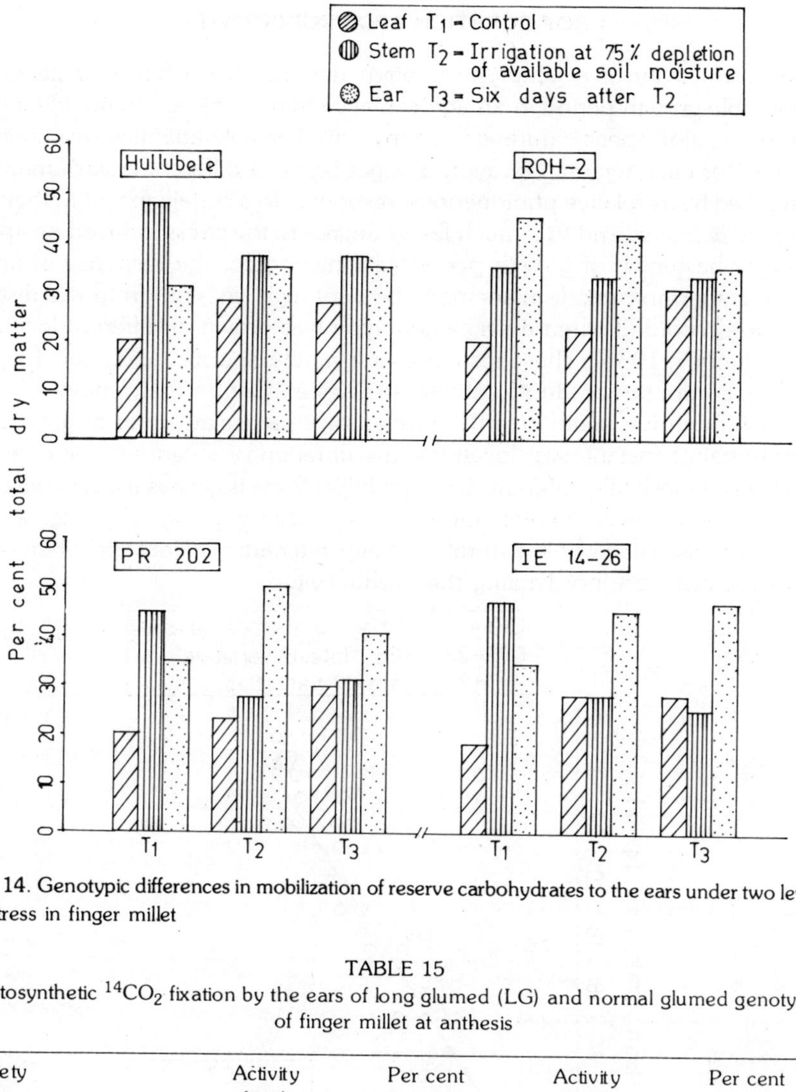

Fig. 14. Genotypic differences in mobilization of reserve carbohydrates to the ears under two levels of stress in finger millet

TABLE 15

Photosynthetic $^{14}CO_2$ fixation by the ears of long glumed (LG) and normal glumed genotypes of finger millet at anthesis

Variety	Activity fixed (CPM/g)	Per cent increase over check (Indaf-5)	Activity fixed (CPM/organ)	Per cent increase over check (Indaf-5)
EG 2970 (LG)	16276	113	27069	218
GE 3301 (LG)	18614	143	39596	365
GE 2973 (LG)	15098	128	52958	622
PES 176	11131	—	10394	—
INDAF-5	7638	—	8519	—
CD at 5% P	3225	—	5225	—

CROP DURATION AND PRODUCTIVITY

The productivity of a genotype is often dependent on full exploitation of favourable growth period in an agroclimatic region. This led to the identification of location specific duration groups with desirable adaptive mechanisms suitable for each region. Duration of a genotype in different environments is controlled by its relative photoperiodic response, to a certain extent to thermoperiodic response and to a much lesser degree to the stress induced postponment or hastening of growth periods. In this regard, the response of finger millet to thermo-periodism seems to be very high, as seen from the distinct variation recorded when a single genotype was grown at different locations (Anonymous, 1986). These responses are varietal specific, and often the problem is compounded due to abiotic stresses existing in that region.

Although the mean growth duration of a particular group of genotypes may remain constant over locations, the duration of a genotype at different locations is markedly different. In finger millet there is always a direct relationship between growth duration and biomass productivity (Fig. 15). Such a relationship is also seen within a duration group in a particular location when other abiotic stresses are not limiting the productivity.

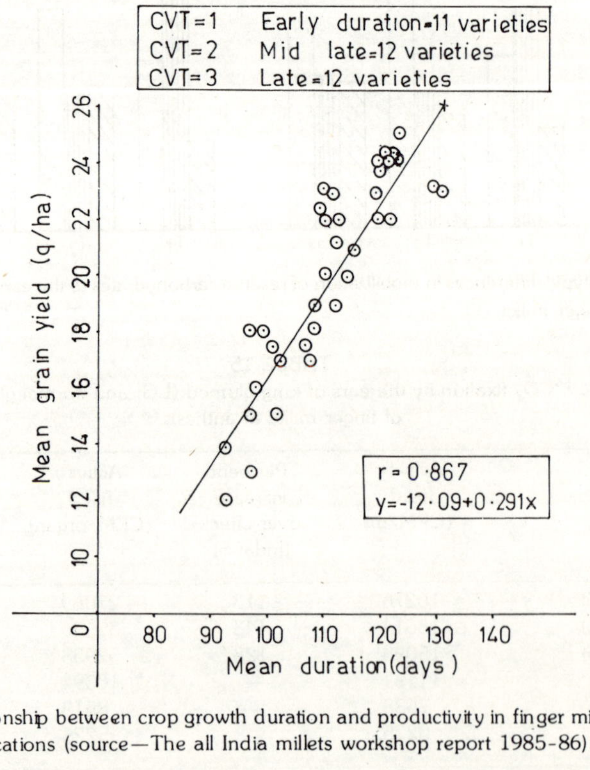

Fig. 15. Relationship between crop growth duration and productivity in finger millet genotypes, mean of 24 locations (source—The all India millets workshop report 1985-86).

One of the approaches to exploit the entire crop growth period is to identify suitable quantitative photosensitive types. In this region the rainfall in the beginning of the season is erratic, unpredictable and often gets delayed. With a photoinsensitive type flexibility is not possible because the early sown crop matures early without exploiting the complete growth period and the late sown crop would fill the grains under moisture stress and cold conditions. Also many high yielding genotypes evolved are photoinsensitive and maturity period depends on the sowing date. Therefore, it is necessary to use different genotypes which suit the different sowing dates. Quantitative photosensitive genotypes are best suited and adjust to the flexibility in sowing dates (Table 16). However, one should be cautious with photosensitive genotypes since they possess limited range of adaptations and often possess low partitioning factors.

TABLE 16
Photoperiod response of quantitative photosensitive varieties when sown in May and July

	Days to anthesis	
	May sowing	July sowing
GE 3011*	112	86
GE 3499*	122	88
GE 3480*	122	84
GE 2790**	88	78
PR 202**	80	72
JNR 7-1**	80	72
ELC 4 **	88	74

* Photosensitive
** Photointensive or weakly photosensitive

The quantitative short day plants have an advantage particularly when precipitation in an agroclimatic zone is bimodal. Such a situation occurs in some regions in Karnataka where premonsoon showers are adequate enough to support plant establishment and growth. However, premonsoon showers are followed by a rainfree period of 4 to 6 weeks before the monsoon sets in. An already established crop (if it survives the stress period) in this situation would have higher crop growth rate (CGR) on stress alleviation (monsoon period) and therefore these genotypes would be more productive. Under such situations, the desirable characteristics of the genotypes would be:

i) Establishment of the crop during premonsoon period.
ii) Tolerance to moisture stress for a period of 4 to 6 weeks.
iii) Higher crop growth rate on stress alleviation.
iv) Photosensitive nature of the genotype.

REFERENCES

Anonymous 1986. Annual Report of the All India millet Improvement Project, Finger millet, ICAR.

Acevedo, E., T.C. Hsiao and D.W. Henderson. 1971. Immediate and subsequent growth responses of maize leaves to changes in water status. *Pl. Physiol.* Lancaster **48:** 631-6.

Alagarswamy, G. 1981. Some aspects of production physiology of pearl millet. Lecture delivered at the summer institute on production physiology of dryland crops held at the Andhra Pradesh Agricultural University, Hyderabad pp. 9-12.

Anonymous 1981. Annual Progress Report of the All-India Co-ordinated Millets Improvement Project, Progress report on finger millet, ICAR.

Bierhuzien, J.F. 1976. Irrigation and water use efficiency In: O.L. Lange, L. Kappen and E.D. Schulze (ed.) *Water and Plant Life*, Springer, Berlin, pp. 421-431.

Blum, A. 1982. Evidence for genetic variability in drought resistance and its implications in plant breeding. In: *Drought resistance in crops with emphasis on rice*, pages 53-71. IRRI, Manila, Philippines.

Blum, A. and C.Y. Sullivan. 1986. Comparative drought resistance of land races of sorghum and millet from dry and humid regions. *Annals of Botany* **57:** 835-846.

Boyer, J.S. 1970. Leaf elongation and metabolic rates of corn, soybean and sunflower at various leaf water potentials. *Pl. Physiol.*, **46:** 233-5.

Devendra, R. 1986. Identification of morphological and physiological parameters and their stability in different seasons in green gram (*Vigna radiata*) and blackgram (*Vigna mungo*). Ph.D. thesis submitted to the Univ. of Agri. Sci., Bangalore, India.

Gaastra, P. 1956. Photosynthesis of crop plants as influenced by light, carbon dioxide, temperature, and stomatal diffusion resistance. *Mededelingen Land-bouwhoge School, Wageningen* **59:** 1-68.

Grubben, G.J.H. 1975. La culture de L' amaranthe, legume-feuilles tropicales avec reference special au sub-Dahomey. *Mededilingen Landbouwhoge School, Wageningen*, **75-6:** 1-223.

Gurumurthy, B.R. 1982. Desirable growth characteristics in relation to productivity in finger millet (*Eleusine coracana* Gaertn.): A physiological analysis M.Sc. (Agri) dissertation. Univ. of Agri. Sci., Bangalore.

Hsiao, T.C. 1973. Plant responses to water stress. *Ann. Rev. Plant Physiol.*, **24:** 519-570.

Jones, H.G. 1977. Transpiration in barley leaves with differing stomatal frequencies. *J. of Expt. Bot.*, **28:** 162-168.

Jones, H.G. 1979. Stomatal behaviour and breeding for drought resistance. In: *Stress physiology in crop plants* (eds. H. Mussel and R.S. Staples), pp. 407-428, New York, Wiley.

Levitt, J. 1980. Water, radiation, salt and other stresses. In: *responses of plants to environmental stresses*, Vol. II. Academic Press, New York.

Ludlow, M.M. and T.T. Ng. 1976. Leaf elongation rate in *Panicum maximum* var. *trichollum* following removal of water stress. *Aust. J. Plant Physiol.*, **4:** 263-272.

Malathi Chari, K.S. Arun, T.G. Prasad and M. Udaya Kumar 1986. Genotypic differences in water use efficiency amongst soybean genotypes. *Indian J. Plant Physiol.*, (In Press).

Paleg, L.G. and D. Aspinall. 1981. *Physiology and biochemistry of drought resistance*. Academic Press. Sydney.

Passioura, J.B. 1981. Water collection by roots. Pages 39-53. In: L. Paleg and D. Aspinall, eds. *The physiology and biochemistry of drought resistance in plants*, Academic Press, Sydney.

Passioura, J.B. 1981a. The role of root system characteristics in the drought resistance of crop plants. Pages 71-83. In: *International Symposium on principles and methods of crop improvement for drought*, IRRI, Philippines.

Perumal, K.R. 1982. Genotypic variation in photosynthetic efficiency and translocation and its relation to leaf character, growth and productivity in finger millet (*Eleusine coracana* Gaertn.) Ph.D. thesis.

Rajasekhar, B.G., S. Rama Rao, K.D. Krishna Sastry and K.N. Mallanna. 1970. Effect of seed hardening on yield of ragi (*Eleusine coracana*) at different levels of fertility. *Mysore J. Agri. Sci.*, **4:** 271-7.

Sastry, K.S.K., S. Rama Rao and K.M. Appaiah. 1969. Effect of pre-treatment of seeds of *Eleusine coracana* to tolerance to high concentrations of sodium chloride and mannitol solutions. *Mysore J. Agri. Sci.*, **3:** 47-54.

Sashidhar, V.R., T.G. Prasad, A. Seetharam, M. Udaya Kumar, and K.S. Krishna Sastry. 1983. Ear photosynthesis in long and normal glumed genotypes of finger millet (*Eleusine coracana* Gaertn.) *Indian J. Plant Physiol.*, **4:** 359-362.

Sashidhar, V.R., T.G. Prasad, S.J. Patil, M. Udaya Kumar and K.S. Krishna Sastry. 1984. The balance between leaf area and photosynthetic activity in determining productivity of fox-tail millet (*Setaria italica*) under rainfed conditions. *Experimental Agriculture* (Great Britain) **21:** 241-247.

Sashidhar, V.R., B.R. Gurumurthy, T.G. Prasad, M. Udaya Kumar, A. Seetharam and K.S. Krishna Sastry. 1986. Genotypic variation in carbon exchange rate, functional leaf area, and productivity in finger millet (*Eleusine coracana*): An approach to identify desirable plant types for higher water use efficiency under rainfed conditions. *Field Crops Research,* **13:** 33-146.

Sastry, K.S.K., M. Udaya Kumar and H.R. Viswanath. 1982. Desirable plant characteristics in genotypes of finger millet (*Eleusine coracana* Gaertn) for rainfed conditions. *Proc. Indian Nat. Sci. Acad.,* **48:** 264-270.

Slatyer, R.O. 1973. The effect of internal water status on plant growth, development and yield. In: *Plant responses to climatic factors* (ed. R.O. Slatyer), Proc. Uppsala Symp., UNESCO, Paris, pp. 177-188.

Squire, G.R. and C.R. Black. 1981. Stomatal behaviour in the field. In: *Stomatal physiology* (eds. P.G. Jarvis and T.A. Mansfield), pp. 223-234. Society for Experimental Botany, Seminar series.

Viswanatha, H.R. 1977. Evaluation of finger millet (*Eleusine coracana* Gaertn.) genotypes for relative drought tolerance based on physiological parameters and field performance. M.Sc. (Agri) thesis submitted to the Univ. of Agri. Sci., Bangalore, India.

Yoshida, S. and S. Hasegawa. 1982. The rice root system. Its development and function. In: *Drought resistance in crops with emphasis on rice.* Pages 97-114. International Rice Research Institute, Manila, Philippines.

21

CROPPING SYSTEMS AND PRODUCTION TECHNOLOGY FOR SMALL MILLETS IN INDIA

B.R. Hegde and B.K. Linge Gowda

INTRODUCTION

Small millets occupied an area of 6.3 m ha and produced 4.7 million tons of grain in India during the year 1983-84. Finger millet (ragi) alone occupied 41 per cent of the area under small millets and contributed about 64 per cent to the total production. The area under small millets is slowly declining with these lands diverted to other crops of higher money value. Whereas, the area under finger millet has marginally increased from 2.21 m ha in 1949-50 to 2.60 m ha in 1984-85, the area under small millets has decreased from 5.42 m ha to 3.63 m ha during the same period (Fig. 1).

Millets have a wide adaptability and can be grown in extremes of soil and climatic conditions. Many of the millets are short duration crops and are able to produce good green fodder within a span of 40-45 days. Few crops respond as markedly to modest applications of fertilizers as ragi and other millets do. They produce as much as 24 to 27 kg grain/ha/day in class III lands where other cereals yield only about 20 kg/ha/day (Gautam, 1977).

Millet grains are highly nutritious and in many respects are superior to rice and wheat. For example, dehusked proso millet has a protein content of 12.5 per cent whereas wheat has about 11.8 per cent. Finger millet is known for its high Ca content of 344 mg/100 g edible grain compared to wheat with only about 41 mg/100 g grain. At present the industrial use of these millets is limited except finger millet which is used for malting and brewing.

In spite of several favourable attributes in the millets, their further expansion will be severely limited unless efforts are substantially increased to improve their production technology. The present productivity is hardly 1160 kg for finger millet and 460 kg/ha for other small millets. The national target is to

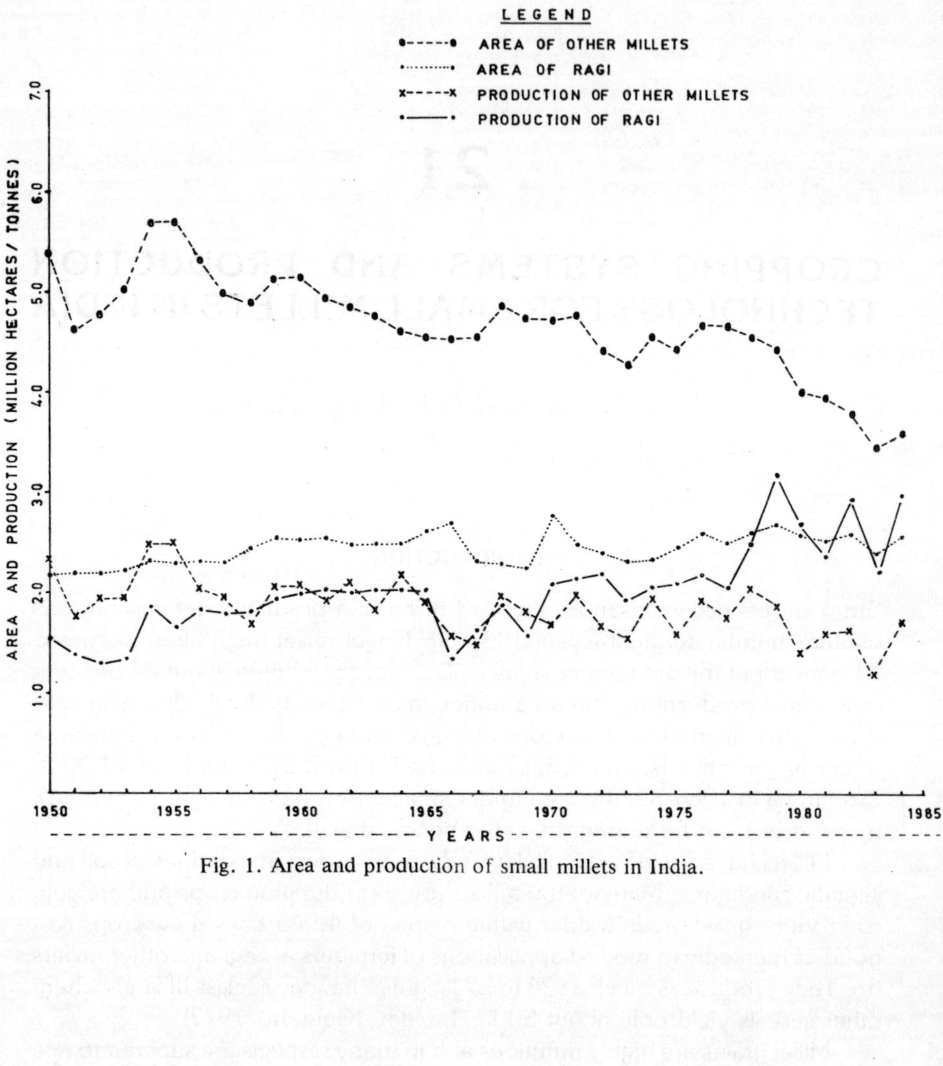

Fig. 1. Area and production of small millets in India.

increase the production of finger millet to 5.25 million tons keeping the area constant at about 2.5 million ha and maintain the production of other small millets at about 1.5 million tons even when the area is reduced to about 2 million ha, by 2000 AD. This calls for raising the productivity of finger millet to 2100 kg/ha and that of other millets to 750 kg/ha.

There are several constraints limiting the higher productivity in small millets.

1) Excepting finger millet, millets are grown on poor shallow and marginal soils mainly under rainfed conditions. Some of these millets are still grown in the hilly areas under shifting cultivation which is one of the most primitive ways of crop production.

2) Seeds are often broadcast. This is a major bottle neck in taking inter-cultivation operations and effective weed control.

3) The mixed cropping practices adapted by the farmers are mostly suited to sustenance agriculture and many of them are not remunerative.

4) Small millets are often cultivated under unmanured and unfertilized conditions.

5) Non-adoption of improved varieties and timely agricultural operations like tillage, sowing, weeding and interculturing has resulted in non-remunerative returns.

6) Millets are consumed mostly in the regions of their production. There is neither market avenue nor industrial use for these grains.

As a result of continuous efforts in the State Agricultural Universities and All India Coordinated Research Project on Millet Improvement (AICMIP), suitable varieties and management practices have been evolved for different agroclimatic regions in the country. A small millet improvement project financed by the IDRC (International Development Research Centre, Canada) was started in 1979 in five centres which has contributed substantially to the production technology of small millets. An effort has been made in this article to compile the available information on the production technology for cultivation of these millets.

FINGER MILLET (*Eleusine coracana* GAERTN.)

Among millets, ragi or finger millet has an unique place and is the only millet which has been able to touch an average productivity level of more than a tonne per hectare. The crop has a wide range of seasonal adaptation and is grown in varying soil and temperature conditions. It can be grown throughout the year if moisture is adequate and if temperatures are above 15°C. It has adapted to conditions prevailing from sea level to an altitude of 3000 m. Though finger millet is a warm season crop, recent developments in breeding have brought out varieties which can tolerate cold.

Finger millet is cultivated in different soil types but is mainly grown on red and laterite soils. Alluvial and black soils are also suitable provided the soils are well drained. As growing conditions differ from state to state, the choice of appropriate variety for a place depends on the agro climatic situations of the region. In the northern states, particularly in higher elevations, early maturing varieties (90-100 days) are required; and medium to late duration varieties (110-120 days) are preferred in the plains and southern states. As finger millet has a remarkable capacity for recovery, it is considered an excellent dryland crop. The minimum water requirement for successful growing of the crop is 400 mm, but can be grown in higher rainfall areas. In India, it is cultivated mostly in rainy season under dryland conditions with only a small area under irrigation. Out of the total area of 2.60 million ha under the crop in the country, only 0.35 million ha receives irrigation.

An excellent review of the world literature on the *Eleusines* is provided by Rachie and Peters (1977). A recent review of the production technology for rainfed finger millet is provided by Hegde and Seetharam (1985).

Tillage

The land preparation for finger millet varies from a little cultivation as in the slash and burn system to 5-6 ploughings with wooden ploughs before harrowing. Finger millet is generally cultivated in red and laterite soils which become hard on drying and form crust. To increase the infiltration and conserve moisture, tillage every year is essential in such soils.

Deep ploughing with iron mould board plough soon after the harvest of the previous crop or in May-June followed by cultivation with peg tooth cultivator is found advantageous. This practice increased the grain yield by about 200-300 kg per ha over the farmer's practice of 2-3 wooden ploughings.

The seeds of finger millet being very small in size need to be sown at a shallow depth of 2-3 cm. In very loosely prepared soils, the seeds are likely to sink to deeper layers if rains occur. Therefore, soil compaction just below the seeding zone is recommended. The sequence of tillage is ploughing, passing a cultivator and then a tine tooth cultivator to smoothen the surface and crush the clods.

To keep the weeds under control and the surface loose, 2-3 intercultivations are essential. If the crust forms on the surface within four days after sowing, a cylindrical spiked crust breaker can be rolled on the field. The farmers use a thorny brush harrow for crust breaking in absence of the availability of the crust breaker.

Time of sowing and method of establishment

As a dryland crop, finger millet is sown in July in the southern States and in May-June in northern States. This is recommended in view of about 120 days required for maturity of medium duration varieties. Shorter duration varieties can be sown later but their yield potential is usually low. When long duration varieties are sown late in *kharif*, they come to heading 2-3 weeks in advance. In general, the yield levels are reduced when sown late except in stable varieties like PR 202.

Varieties suitable for different seasons are identified and are recommended for each region.

TABLE 1
Varieties of finger millet recommended for different regions and seasons

State	Season	Variety	Month of sowing
Andhra Pradesh	Early Kharif	AKP 2	May
	Kharif	VZM-1, Sharada, Godavari and Kalyani	July
	Rabi	Simhadri and Kalyani	August-September
Bihar	Kharif	BR 2, BR 407, RAU 8, HR 374	June
	Summer	BR 2, BR 407	February
Gujarat	Kharif	Gujarat Nagli-1	June-July
Himachal Pradesh (Low altitudes)	Kharif	VL 204, Local	June
Himachal Pradesh (High altitudes)	Kharif	Locals	May
Karnataka	Kar	HR 374, Indaf 9	April-May
	Kharif	Indaf-1, Indaf-3, Indaf-8, HR 911, HR 919	June-July
	Late Kharif	PR 202, Indaf-5, Indaf-9, HR 374	August
	Rabi	Indaf-7, Indaf-9	September-October
	Summer	Indaf-5, Indaf-9, HR 911	January-February
Maharashtra	Kharif	B 11, A 16, E 11, 28-1, 50-1	June
Madhya Pradesh	Kharif	JNR 852, IE 28, EC 4840, JNR 1008, JNR 981	June-July
Orissa	Kharif	Dibyasingha, B-4-10-56, Bhudei Local	June
	Rabi	Dibyasingha, B-4-10-56 Bhudei Local	January
Tamil Nadu	Nargalipattam	Co 7, Co 10, K 5, K 6, Indaf-5	May-June
	Chittaraipattam	PR 202, Co 11, K 6, K 7, Co 12,	April-May
		K 5, K 6, Co 11, Co 12, K 7	May-June
	Adipattam	PR 202, K 5 and K 7	June-July
	Purattasipattam	K 6, Co 11	September-October
Uttar Pradesh (Plains)	Kharif	Nirmal, T-36b, PES-176	May-June
Uttar Pradesh (Hills)	Kharif	VL 204, Local	June

Source: Seetharam (1986).

Finger millet is established either by drilling or broadcasting the seeds in the prepared land under dryland conditions; but under irrigated conditions, it is invariably transplanted. Transplanted finger millet has less weed problem and usually yields higher. Transplanting is also advantageous under multiple cropping systems as the land is occupied by the crop for a shorter time.

A medium duration variety of finger millet when sown in the second or third week of July always yielded higher than when sown in August (Ashok *et al.*, 1979). When planting is delayed, establishing the crop by transplanting can help in maintaining the yield level.

TABLE 2

Yield of finger millet as influenced by method of establishment in September

Method of establishment	Yield of finger millet (kg/ha)		
	3 years average	2 years average	
	Indaf-8	Indaf-8	Indaf-5
Transplanted	2650	2565	2593
Drilled	1330	761	711

Source: Anonymous (1984).

For transplanting, three week old seedlings raised in beds are used. Studies conducted in several locations of the Millet Improvement Project indicated that in a medium duration variety, 25-30 day old seedlings are ideal for transplanting.

According to the peasants' practice, seedlings could be kept in the nursery for a week for every month of the crop's total life period. Thus, a variety of 120 days duration could be left in the nursery up to four weeks.

Spacing and plant population

Finger millet is often broadcast with heavy seed rate (about 25 kg/ha) and thinned out later by cross cultivation, both lengthwise and breadthwise with the help of tine hoes. However, this results in non-uniform stand and too high a population. An optimum population has been observed to be 400 to 800 thousand plants per ha. Yields decline with higher populations.

When a row spacing of 25 cm is adopted, a plant to plant spacing of 10 cm gives about 400 thousand plants per ha. In a study conducted by Murthy and Hegde (1981), it was observed that the number of tillers in Indaf-5 variety reduced from 3.8 per plant at 270 thousand plants to about 2.1 at 800 thousand plants per ha, but there was no significant difference in yield. It appears that with reduced number of tillers, the contribution from individual earheads will be more and the total yield remains the same as the contribution of the main shoot and the primary tiller is about 71 per cent of the yield in individual plants (Krishnamurthy, 1973).

Normally 25-30 cm row spacing is recommended for dryland finger millet keeping in view the convenience for intercultivation and weeding. However, studies at Bangalore, have shown that wider rows up to 45 cm gave similar yields as 25 cm row spacing. In four years out of 12, which were low rainfall years, the yields were higher due to wider spacing. But in three wet years, the yields were more in closer spacing. Wider row spacing, apart from having an advantage in intercultivation for a longer period, also helps in accommodating intercrops like soybean.

Manures and fertilizers

Finger millet responds well to N application all over the country. On an average the response varied from 6 to 23 kg per kg N (Gautam *et al.*, 1982). Studies at several centres of the Millet Improvement Project also indicated that the response to N was of the order of 23.1 kg per kg N at low level of 20 kg N which came down to 19.9 kg per kg N at 60 kg N per ha.

Among the forms, Calcium ammonium nitrate (CAN) had advantage over other sources for slightly acidic soils of the south (Ananthanarayana *et al.* 1971). Neem cake blended with ammonium sulphate and urea gave better results than their straight forms at equivalent N levels (Subbaiah *et al.*, 1982).

The response of finger millet to phosphate application has been favourable. The studies conducted at Bangalore gave a response of 16.3 kg grain per kg P_2O_5 at a low level of 30 kg per ha. This figure reduced to 14.7 at 60 kg P_2O_5 per ha (Anon., 1984).

Response of the crop to potash application was generally lower and was between 6 and 9 kg grain per kg of K_2O applied at levels up to 60 kg K_2O per ha.

The importance of fertilizers in achieving higher yields is well documented. In the absence of adequate fertility in soil, even the high yielding varieties gave yields on par with the locals.

TABLE 3

Variety and fertilizer interaction in finger millet at Bangalore

Variety	Grain yield (kg/ha) (Average of two years)	
	Low fertility	Recommended fertilizers
Local (Hullubele)	1391	1859
Improved (Indaf-5)	1260	2161

Source: Anonymous (1984).

Fertilizer recommendations vary from State to State. The general recommendation in Karnataka is 50, 37.5, 25 kg N, P_2O_5 and K_2O for dryland and 100:50:50 N, P_2O_5 and K_2O for irrigated conditions. All P_2O_5 and K_2O are to be applied at sowing whereas, nitrogen is applied in split doses. In a study at Bangalore, 50 kg N/ha gave a yield of 2430 kg grain when applied at planting compared to 2650 kg/ha when the same dose was applied in two splits and 2760 kg/ha when applied in three splits. However, it is essential that all the N is applied before flowering (Havanagi and Hegde, 1983).

Seed-cum-fertilizer drills are available for sowing simultaneously both seeds and fertilizers side by side or in the same row. This usually resulted in about 30 per cent increase in the grain yield at the same level of fertilizers. Even mixing of seeds and low nitrogen granular fertilizers like DAP was found beneficial. In one of the studies at Bangalore, application of 100 kg DAP per hectare through seed-cum-fertilizer drill resulted in a yield of 2537 kg/ha, whereas the same dose mixed with seeds and drilled gave 2464 kg/ha. When the fertilizer was broadcast and seeds were drilled the yield was only 2056 kg/ha. A slight reduction in population occurred when seeds were mixed with fertilizers, but this did not affect the yield adversely (Anon., 1986).

Application of farm yard manure (FYM) has been observed to help in reducing the requirement of fertilizer nitrogen to obtain the same yield levels. On an average, the yield levels at 10 t FYM/ha were on par with fertilizers alone at 50 per cent of the recommended level (Havanagi and Hegde, 1983). In another study it was observed that a combination of organic manures like FYM, poultry manure or sericulture waste with urea always resulted in a higher yield than any inorganic source independently (Anon., 1983).

Among different forms of biofertilizers, *Azospirillum brasilense* has been tested with finger millet in seven locations over a period of six years. In the first three years of the study, *Azospirillum* was used only as a seed treatment and in the subsequent years the seed treatment was combined with FYM.

Based on the studies right from 1980, it is concluded that the seed treatment with Azospirillum has an effect equivalent to the application of 20 kg N/ha (Table 4).

Weed management

Finger millet is grown predominantly as a mixed crop under rainfed conditions during Kharif. Added to this, the practice of broadcasting the seeds commonly followed by the farmers poses difficulties in weed control by intercultivation. Most commonly observed species of weeds with finger millet also belong to the grass family and are difficult to distinguish in the early stages. However, the situations under irrigated conditions where the seedlings are transplanted is entirely different. In a transplanted crop, the land is thoroughly prepared and the seedlings grow faster soon after establishment. This helps in suppressing the weeds.

TABLE 4
Influence of Azospirillum on the grain yield of finger millet

Azospirillum for seed treatment (average of two years)		Azospirillum with FYM (Average of 3 years)	
Treatments	Yield (kg/ha)	Treatments	Yield (kg/ha)
T_1: Control	1653	T_1: Control	1753
T_2: Azospirillum (Seed treatment)	1839	T_2: Azospirillum (Seed treatment)	1969
T_3: Half the recommended nitrogen only	2226	T_3: Azospirillum with FYM in furrows	2102
T_4: T_3 + Azospirillum (seed treatment)	2255	T_4: T_2 + Half the recommended N	2492
T_5: Full dose of recommended N	2440	T_5: T_3 + Half the recommended N	2608
T_6: T_5 + Azospirillum (Seed treatment)	2776	T_6: 40 kg N only	2894

Source: Annual reports of AICMIP for the years 1980-1985.

Effective weed control chemicals for finger millet and their method of application under different agroclimatic situations are yet to be perfected. Therefore, competition with weeds for moisture, nutrients and light in the initial stages is inevitable. However, when the competition is extended beyond 3-4 weeks, substantial reductions in grain yield have been reported (Hegde *et al.*, 1983). In general, the yield reduction was of the order of 48-50 per cent (Patro and Das, 1972; Sundaresh *et al.*, 1975).

Pre-emergence application of nitrofen and neburon at 0.5 to 1.0 kg a.i./ha have been recommended in sole crop of finger millet (Shankaran *et al.*, 1974; Nanjappa and Hosmani, 1986). As a post-emergence application, both in drilled and transplanted finger millet 2, 4-D at 0.5 to 1.0 kg a.i./ha has given effective control of weeds (Lingegowda *et al.*, 1974).

Both in transplanted and directly drilled stands of finger millet, 2-3 intercultivations and one or two hand weedings are needed. Even when chemicals are used for weed control, intercultivation becomes essential to provide a favourable root zone environment.

Water management

Finger millet is one of the hardiest among millets. It requires between 40 and 65 cm water when irrigated at an optimum level of moisture depletion of 50-60 per cent available soil moisture at the surface 30 cm depth of soil (Patil *et al.*, 1969 and Kaliappa *et al.*, 1974). Tillering and pre-flowering stages are the most critical stages with respect to moisture stress (Reddy, 1976). In general, finger millet needs to be irrigated at an interval of 8-12 days during summer and winter months. But during kharif, 3-4 well timed irrigations are enough to produce as much yield as a crop fully grown under irrigation.

Rainfed finger millet experiences intermittent dry and wet spells. For a successful production of any crop under such situations, efforts are essential to conserve as much rain water as the soil can hold and dispose off the surplus smoothly without causing any erosion. Deep ploughing and intercultivation open up the hard surface crust and improve the rate of infiltration and water holding capacity of soil. Corrugations and furrows at intervals are being followed in this crop which further enhance the opportunity for moisture storage.

Recent recommendations under dryland conditions are to harvest the surplus run-off water in farm ponds and provide one or two irrigations at times of prolonged breaks in rainfall. A study conducted over four years in the dryland project at Bangalore revealed the possibility of increasing the crop yields by 100 per cent and stabilizing the yield of finger millet by providing one or two protective irrigations under unfavourable rainfall situations.

Cropping systems

Under dryland conditions, finger millet is grown during the rainy season. In south India, rows of finger millet are commonly intercropped with other crops like fodder sorghum, field beans, niger, castor and pigeonpea. It is not uncommon to find mixtures of other millets like bajra or little millet in dryland finger millet fields. These intercrop components are sown through separate tubes in a bullock-drawn seed drill along with finger millet simultaneously at an interval of 2.5 to 3 m. Quite often, mustard is sprinkled all over the field.

Finger millet is grown in rotation with other dryland crops like groundnut, horsegram, sorghum, other millets, cotton, tobacco and sesamum. Under irrigated conditions, it is usually grown after paddy in areas where water is not sufficient for second crop of paddy. Finger millet is also grown after sugarcane, potatoes, onions, carrot, chillies, etc. In such cases, the crop is usually transplanted under irrigation as a sole crop.

Under rainfed conditions, growing other crops as mixtures with finger millet is a rule rather than an exception. Under a subsistence system of farming, this appears to be the most acceptable practice. However, with adoption of improved management practices, traditional mixed/intercropping systems are found to be non-remunerative.

As seen from Table 5, traditional intercropping may not be profitable under high fertility levels but it is advisable under low fertility conditions. Further, the same studies indicated that inclusion of fodder intercrops like maize or pearl millet in wider intervals of 7:1 row proportion is better than putting them closer.

Recent studies have shown that finger millet and soybean drilled in alternate rows at 22.5 cm apart can give as much as an entire crop of finger millet (about 2500 kg/ha) with 200-300 kilograms bonus yield of soybean.

TABLE 5

Yield of finger millet as influenced by intercropping and fertility management

Level of fertility	Grain yield (kg/ha)	
	No intercrop	Intercropped with fodder sorghum
Farmers level of fertilizers (low fertility)	1093 (Rs. 708)	1057 (Rs. 773)
Recommended level of fertilizers	1513 (Rs. 1251)	1258 (Rs. 1184)

Source: Hegde *et al.* (1980).

Note: Figures within the brackets indicate the net income of the whole cropping system including the intercrops.

The traditional system of growing pigeonpea with finger millet as in South India is less profitable due to suppression of pigeonpea crop in the early stages of growth. In order to avoid this suppression and give some lead to pigeon crop, efforts were made to sow pigeon pea in May in paired rows with a spacing of 3.3 m between pairs. Finger millet was sown in the interspace between two such pairs of pigeonpea in the month of July, i.e., after 1½ months of sowing pigeonpea. A furrow is opened between pigeonpea rows to achieve an effective inter-terrace management in the field. Thus the system is a combination of intercropping and inter-terrace management. This system is compared with other systems in Table 6.

TABLE 6

Crop yields as influenced by intercropping of pigeonpea in finger millet

Treatments	Crop yields (kg/ha)	
	finger millet	Pigeonpea
1. Both pigeonpea and finger millet sown in July	2130	275
2. Pigeonpea sown in May and finger millet in July	1274	1274
3. Entire finger millet sown in July	2279	—

Source: Dryland Project, Bangalore (Anon., 1984).

The system is already recommended to the farmers and is popularly known as 'paired row technique'. A pigeonpea variety of about 6 months duration and a medium duration finger millet variety of about 110 days are suitable for this type of intercropping.

KODO MILLET (*Paspalum scrobiculatum* L.)

Kodo millet is a long duration crop (110-130 days) compared to other small millets and grows well on shallow as well as deep soils. The seeds have an excellent storage life and can be stored even up to 100 years. Out of the total area of about 2.1 million ha under this crop in India, 1.3 million ha are in Madhya Pradesh.

Being a crop of tribal areas and marginal lands, it did not receive much attention till recently. With the establishment of Paspalum improvement centre at Dindori in Madhya Pradesh in 1979, systematic studies have been initiated on the production technology of the crop.

Seeding time and method of sowing

A population of 600-800 thousand plants per ha appears to be optimum for the crop. This is achieved by adopting a seed rate of about 15 kg/ha and a spacing of 22.5 cm between rows and 5 to 7.5 cm between plants. The observations at Dindori, Madhya Pradesh that a crop sown at 22.5 cm but unthinned within the row and a broadcast field yielded on par, prove that the crop responds to higher population (Table 7).

TABLE 7

Effect of sowing time and plant spacing on the yield of kodo millet

	Yield (kg/ha)						Mean	
Plant spacing	Unthinned		22.5×7.5 cm		22.5×5.0 cm			
Sowing time	Grain	Straw	Grain	Straw	Grain	Straw	Grain	Straw
Last week of June to first week of July	1840	4890	1650	3900	1730	4250	1740	4340
Second week of July	1500	4020	1500	3860	1220	3280	1400	3720
Last week of July to first week of August	550	2330	530	2370	440	2120	510	2270
Mean	1300	3450	1230	3380	1130	3220		

Source: Annual reports of the IDRC Centre, Dindori (MP) for the years 1981, 1982 and 1983.

Kodo millet responded very favourably to dry seeding where the rains are assured in June. Sowing 10 days before the onset of monsoon resulted in the highest grain and straw yield compared to any sowings taken after the onset of monsoon. However, this is not likely to be universal. At Dindori, the June rains are certain and the monsoon rains cease by the end of September. Under such situations, delayed sowing subjected the crop to stress at later stages. Further, early sowing had comparatively lesser incidence of dead hearts caused by shootfly damage.

Cropping systems

Kodo millet is mixed with cereals (finger millet and maize) and oilseeds (niger, groundnut and soybean) in the proportions of 1:1 and 2:1. A combination of kodo millet with groundnut in the row ratio of 1:1 was found better than a single crop of the millet. The combination yielded 360 kg grain of the millet and 680 kg groundnut pods per hectare as against 940 kg grain/ha of the millet under sole crop system.

Response to manures and fertilizers

The crop responds well to a moderate application of N and P_2O_5. Response to added potash has not been observed.

TABLE 8
Response of kodo millet to N and P fertilizers

Levels of N (kg/ha)	Grain yield (kg/ha) (Average of 3 years)	
	O P_2O_5	20 kg P_2O_5
0	1087	1240
20	1603	1673
40	1760	2306
60	1953	1963

Source: Annual report of the IDRC Centre at Dindori (M.P.) for 1980, 1981, 1982.

In studies conducted by Kaushik and Gautam (1981), the yield with N application at 60 kg/ha was 2.32 t/ha compared to 1.41 t/ha without N. The optimum dose was worked out to be 54 kg N/ha. In another study at Bangalore, Linge Gowda *et al.* (1977) observed highest response at 25 kg N with a grain yield of 1.63 t/ha compared to 0.85 t/ha without any nitrogen.

The crop responded favourably to the application of FYM with about 200 kg grain for every 5 tons of the manure. Even though the crop gave an additional yield of 150 kg/ha with *Azospirillum* seed treatment, the response was not statistically significant. It is concluded that there is a need to identify more efficient strains of *Azospirillum* specific to *Paspalum*.

Intercropping systems

During kharif season, millets are grown predominantly in hilly tracts either as pure or mixed crops with sesame, niger and pigeonpea. Shifting cultivation is also in vogue where niger and kodo mixtures are grown for 3-4 years then giving a rest, for 2-3 years. Systematic studies to identify an acceptable intercropping system are meagre.

A trial was conducted at Rewa, Madhya Pradesh involving three intercrops with kodo *viz.*, greengram, sesamem and soybean. As observed from this two year trial, the grain yield of kodo millet was higher when intercropped with greengram.

It appears that there is a good scope for intercropping with kodo millet, especially with long duration pulses like pigeonpea.

FOXTAIL MILLET (*Setaria italica* BEAUV.)

Foxtail millet is essentially a grain crop of about 100 days duration suited to conditions of low and moderate rainfall ranging from 500 to 700 mm. It can be grown in higher altitudes (up to 1830 m above MSL) and is an important foodgrain in the foothills of Himalayas. It is a crop grown almost throughout the year in different parts of the country.

In India, cultivation of foxtail millet is mainly confined to the lower Deccan Plateau including high lands of Andhra Pradesh, Karnataka and Tamil Nadu, which account for about 90 per cent of the area in the country.

Foxtail millet is mostly grown mixed with other crops like cotton, castor, pigeon pea, bajra, groundnut and finger millet. It is also grown as a pure crop, particularly in black cotton soils where it is followed by a rabi crop like coriander in favourable seasons or by safflower or horsegram in years of less rainfall. In the hilly regions of North India, foxtail millet is sown with other kharif crops and matures in about 2 months, providing food during scarcity periods. In Punjab, Himachal Pradesh and U.P. it is grown from June-July to September-October either as a border or as a mixed crop with several kharif crops.

Foxtail or Italian millet may well have unrealized potential and the Chinese have claimed exceptionally high yields sometimes exceeding 11,000 kg/ha (Rachie, 1975). However, in India the yield of rainfed pure crop varies from 400-800 kg of grain and 1000-2000 kg of straw per ha. Generally cooked like rice or made into porridge, it makes a food which is considered to be very nutritious. The grain must be pounded or otherwise husked before cooking to remove the tightly enclosed glume.

Time of sowing

Early sowing in the monsoon always produces higher yields than later sowings. The longer duration varieties gave higher fodder yield when sown early, as shown in Table 9.

The reduction in yield due to delayed planting was not compensated by increasing the population. Studies conducted at Dholi in Bihar and at Nandyal in Andhra Pradesh revealed that a population density between 440 (22.5 × 10 cm^2) and 890 thousand plants per ha (22.5 × 5 cm^2) did not vary the yields but the sowing dates had significant effects. Dry seeding in the month

TABLE 9
Effect of sowing time on the yield of foxtail millet

1980			1981		
	Yield (kg/ha)			Yield (kg/ha)	
Time of sowing	Grain	Straw	Time of sowing	Grain	Straw
July 22	2390	5420	July 8	1770	5030
July 31	1630	4810	July 18	1730	4170
August 8	1370	4810	August 12	1230	3930
August 19	960	3850	August 22	1300	3530
C.D. at 5%	220	580	C.D. at 5%	170	220

Source: Annual reports of the IDRC Millet Centre, Nandyal, Andhra Pradesh.

of June was as advantageous as sowing soon after the receipt of monsoons in July.

Even during summer under irrigation at Nandyal, the crop showed a positive response to early sowing in January. Sowing on 1st January gave 3330 kg grain/ha which reduced to 2630 kg/ha when sown on 31st January, whereas sowing on 20th of February gave only 1050 kg/ha. Tiller number per metre length reduced from 48.3 for January 1st planted to 19.0/m for February 20th sowing.

Response to fertilizers

The response of foxtail millet to the application of nitrogen was positive up to 40 kg N/ha at Dholi. Averaged over three varieties, the response per kg of N was 31 kg up to 20 kg N, at 40 kg N the response decreased to 23 kg grain per kg nitrogen. Further studies at Nandyal showed that the response could be of the order of 28 kg per kg N, up to a level of 60 kg N/ha.

Similar studies in Tamil Nadu during 1972 and 1973 showed that on an average, 30 kg N/ha increased the yield from 590 kg to 1005 kg/ha (Gautam, 1976). Response to P application was not as marked as in case of N. Studies conducted at Nandyal showed a response of about 6.5 kg grain per kg of P_2O_5 up to 20 kg P_2O_5/ha.

In similar studies conducted in Karnataka, a combination of 60 kg N + 20 kg P_2O_5 gave the highest yield of 1.32 t/ha compared to 0.68 t/ha when fertilizers were not applied (Hosmani *et al.*, 1975).

Response to biofertilizers

A two year trial conducted at Nandyal indicated that foxtail millet (Var. Arjuna) gave higher yield with *Azospirillum brasilense* especially in the presence of nitrogen. The influence was both on grain as well as straw through increase in the number of productive tillers, as shown in Table 10.

TABLE 10
Effect of *Azospirillum* and nitrogen on the yield of foxtail millet

Treatment	Grain yield (kg/ha)	
	1980	1981
Control	1750	600
Azospirillum only	1750	800
10 kg N/ha	2120	1100
20 kg N/ha	2400	1310
40 kg N/ha	2430	—
10 kg N + Azospirillum	2700	1430
20 kg N + Azospirillum	2750	1800
40 kg N + Azospirillum	3500	—
C.D. at 5%	690	320

Source: Annual report of the IDRC Millet Centre, Nandyal.

Similar studies in Karnataka showed that foxtail millet responded to the treatment with *Azotobacter chroococcum*. The yields with 0, 30 and 60 kg N/ha were 0.85, 1.10 and 1.25 t/ha respectively. The corresponding yields with *Azotobacter* treatment were 0.93, 1.15 and 1.31 t/ha (Raj *et al.*, 1979).

Organic manures in the form of FYM had a significant influence on the grain yield of Setaria at Nandyal. It appears that there was a significant interaction of FYM and fertilizer nitrogen as seen from Table 11. It is interesting to note that 60 kg N/ha along with 10 t FYM gave as high as 2850 kg grain/ha compared to 380 kg/ha under control (650 per cent increase).

TABLE 11
Effect of organic and inorganic fertilizers on foxtail millet

Levels of N (kg/ha)	Grain yield (kg/ha) Level of FYM		
	0	5 t/ha	10 t/ha
0	380	580	830
20	1000	1100	1260
40	1490	1830	2220
60	1240	2520	2850

Source: Annual report of the IDRC Millet Centre, Nandyal.

Cropping systems

Foxtail millet is commonly grown as a mixed crop either with cotton or pigeon pea. An experiment conducted for two years at Nandyal showed that

2 rows of Setaria and 1 row of cotton produced higher returns compared to entire crop of Setaria.

At Hagari, it was established that Setaria and cotton could be advantageously grown at 12:6 or 6:3 row proportions as strip crops for effective soil conservation (Rao *et al.*, 1975). Foxtail millet was observed to be a good soil binder with an antierosion value of 5.7 compared to 1.6 for cotton.

LITTLE MILLET (*Panicum miliare* LAM) (Renamed *P. sumatrense* ROTH)

Little millet or sama, is a quick growing short duration cereal which withstands both drought and water logging. As in other small millets, its cultivation is restricted to tribal areas and marginal lands. Frequently grown as a monocrop but often mixed with other cereals, millets, pulses or oilseeds. Little millet in early kharif followed by niger is a common sequence in Orissa.

The little millet improvement centre started functioning in 1979 and is located at Semiliguda in the State of Orissa.

Time of sowing and plant density

The seeds are often broadcast and sometimes drilled with the onset of monsoon in the month of June. Systematic studies from 1979 to 1983 at Semiliguda have shown that early sowing in June gave higher grain as well as straw yield compared to later sowings. The practice resulted in higher number of effective tillers and reduced the incidence of midge.

TABLE 12

Yield of little millet at Semiliguda as influenced by date of planting

	Grain yield (kg/ha)					
	Variety Koraput Local					
Sowing time	1979	1980	1981	1982	1983	Mean
June first fortnight	1070	740	670	890	1650	1004
June second fortnight	1000	240	470	630	990	666
July first fortnight	240	160	100	460	530	298
July second fortnight	210	100	—	—	—	155

Source: Annual Reports of the IDRC Millet Centre, Semiliguda.

When the crop was sown in first fortnight of June, a higher plant density of about 890,000 plants per ha, gave higher yield compared with either lower populations or very high populations which were not thinned. Planting later than first fortnight of June resulted in lower yields, but there was no significant interaction between sowing date and plant population. Highest yields under all dates of sowing were obtained at 890,000 plants per ha.

Response to fertilizers

A moderate application of 20 kg N and 20 kg P_2O_5/ha was optimum for little millet, as shown in Table 13.

TABLE 13
Grain and straw yield of little millet as influenced by fertilizer application (averaged over three varieties and two seasons, 1980 and 1981)

	Yield (kg/ha)			
	Grain		Straw	
Levels of N (kg/ha)	P_0	P_{20}	P_0	P_{20}
0	767	922	2785	3355
20	830	1242	3395	4585
40	958	1213	4075	5260
60	848	1143	3520	4860

Source: Annual Progress Report of IDRC Millet Centre, Semiliguda.

It is interesting to note that the trends repeated over number of years and trials in several other locations also resulted in the same level of fertilizers to be optimum and profitable. It was further observed that nitrogen should be applied in two equal splits for higher grain production.

Farm yard manure was beneficial and at 10 t/ha was able to replace about 20 kg N.

Biofertilizers

The influence of biofertilizers especially *Azospirillum* was studied on little millet at Semiliguda, Ranchi in Bihar, and Dindori in Madhya Pradesh. The results of the trials are presented in Table 14. Based on these results it was concluded that seed treatment with *Azospirillum* saved 4-10 kg N/ha. Applying *Azospirillum* along with FYM did not increase the yield significantly over seed treatment alone.

Cropping systems

Mixing of pigeonpea seeds with little millet is a common practice with the tribal people. The proportion of 1:1 to 1:3 between pigeonpea and little millet resulted in higher monetary returns than growing little millet as an entire crop. Mixing of little millet (75 per cent) and black gram (25 per cent) resulted in the maximum grain production of 1530 kg/ha compared to entire little millet with 730 kg/ha during 1983 at Ranchi.

TABLE 14
Effect of *Azospirillum brasilense* on the yield of little millet

Treatment	Grain yield (kg/ha)		
	Semiliguda (Average of 2 years)	Ranchi	Dindori
Control	700	1110	1140
Azospirillum only	800	1020	1710
10 kg N/ha	970	—	—
20 kg N/ha	1310	1110	1990
10 kg N + Azospirillum	1180	1110	1840
20 kg N + Azospirillum	1450	—	—

Intercropping of pigeonpea has been tried in different row ratios. A two year trial at Semiliguda with pigeonpea showed that the total monetary return with intercropping was more than an entire crop of little millet.

PROSO MILLET (*Panicum miliaceum* L.)

Proso millet is a quick growing short duration cereal with low moisture requirements. It can be grown throughout the year. Two quick crops of proso millet are taken during March-June under irrigation in Bihar. In South India, proso millet is raised during rabi in black soils on stored soil moisture. Farmers usually grow this crop on residual fertility especially on fields vacated by potato, peas, mustard and wheat. An intensive work in proso millet agronomy was taken up at Dholi in Bihar under the proso millet improvement programme.

Sowing time

Proso millet crop is usually planted by broadcasting. Under experimental conditions, however, line planting has been observed to do better. Row to row distance should be kept at 22 cm and plant to plant 7.5 cm (Choudhari and Rai, 1982). This would roughly need a seed rate of 10-12 kg/ha. For better germination the seeds should be soaked for 24 hours in water and should not be planted deeper than 4 cm. Planting dates vary from February to middle of April (Summer season). However, the optimum sowing time has been identified to be the middle of March as evidenced by the studies conducted at Dholi.

Suitable varieties

A three year study conducted at Dholi brought out that BR 7, an improved variety, performed better than the rest recommended in the region. The performance of BR 7 was uniformly superior to the other entries in all the years under trial.

TABLE 15
Performance of selected proso millet varieties

Varieties	Origin	Grain yield (kg/ha) (Average 1979, 1980, 1981)
MS 4872	Coimbatore	1810
MS 1914	Coimbatore	1770
PM 29	Jodhpur	1870
Shyamcheena	Dholi	1960
BR 7	Dholi	2210
Local	Dholi	1750

Source: Annual Progress Report of the IDRC Millet Centre, Dholi.

Fertilizer requirement

The crop responded favourably to the application of nitrogen at low levels. Under experimental conditions at Dholi, the yield increased from 1350 kg/ha without any nitrogen to 1840 kg/ha with nitrogen applied at 40 kg/ha. While there was no significant response to application of potash, the response to P application up to 40 kg P_2O_5/ha was significant. For better effects, all the P should be applied as basal, whereas, N should be applied in two equal splits at sowing and 35-40 days after sowing.

Irrigations

Since proso millet is basically a summer crop in Bihar, irrigation is crucial for obtaining high yields. Usually a total of three irrigations are required for February planted crop, four for March and five for April planted crops. Flowering and grain filling stages are critical and sufficient moisture should be there in the field at this time (Choudhari and Rai, 1982).

Biofertilizers

The response to the application of *Azospirillum* was favourable either through seed treatment or through FYM. However, the differences were not significant.

Intercropping

The possibility of sowing proso millet + pigeonpea as intercrops in September (post-monsoon season) followed by proso millet in summer (March sowing) was experimented. Another system extensively tried was with greengram in different row proportions. Though the systems brought higher returns than proso millet alone, the yield levels of both the component crops were appreciably lower in intercropping.

BARNYARD MILLET (*Echinochloa frumentacea* LINK.) (Renamed *E. colona*)

The cultivation of barnyard millet or Sawan dates back to ancient times. Barnyard millet is grown during kharif in shallow soils with low moisture holding capacity while rice is planted in deep soils with better moisture availability. Frequently rice and barnyard millet are found in the same field.

The grains of barnyard millet are used as a staple food and the dehulled grains are cooked just like rice. Its stover is an important source of fodder for animals during winter months. The crop is cultivated in the states of Madhya Pradesh, Maharashtra and Tamil Nadu in India. Barnyard millet is one of the most important millets in the hills of Uttar Pradesh occupying an area of about 200 thousand ha. The barnyard millet improvement centre is located at Almora, Centre for Hill Agriculture in Uttar Pradesh which started functioning on June 1, 1979.

Planting time

Barnyard millet is grown as a kharif crop under rainfed conditions. To provide good amount of moisture throughout the growth period, earliest opportunity is taken to sow the crop and even dry seeding is in practice.

Sowing from March up to the end of May resulted in similar yields. When sowings were delayed till the middle of June there was a slight reduction. However, with short duration varieties likeVL-8, the sowing could be delayed up to the middle of June.

Studies at Kanke in Bihar revealed that sowing with the onset of monsoon gave the highest yield. Similar studies at Dholi showed that the sowings could be delayed till the first week of July but further delay reduced the yield. At higher populations the reduction was small, indicating that there exists a possibility of maintaining the yield even when sown late.

Planting density

Barnyard millet has tillering ability, therefore the difference in yield over a narrow range of population is not significant. A trial conducted at Almora for three years from 1979 showed that there was no difference in yield due to variations in row spacing from 20 to 30 cm and inter-row spacing from 10 to 25 cm. Thus a population range of 133,000 to 500,000 plants per ha resulted in similar yields. On the basis of these results, an interrow spacing of 20-25 cm and an intra-row spacing of 10-15 cm is recommended for Almora region. Studies at Dholi for three years (1981 to 1983) indicated that a closer spacing of 22.5 × 5 cm (890,000 plants/ha) produced higher yields than wider spacings.

Method of seeding

Even though barnyard millet can be established by broadcasting, drilling the seeds in the recommended row spacing of 25 cm, or transplanting yielded more than any other method of establishment. This is shown in Table 16 using the variety VL-11.

TABLE 16
Influence of method of establishment on the yield of barnyard millet

Planting method	Yield (kg/ha) (Average of 1982 and 1983) No fertilizers	Recommended fertilizers	Mean
Broadcasting	1270	2100	1680
Line sowing	1780	2770	2270
Transplanting	1860	3160	2510

Source: Annual Progress Report of the IDRC Millet Centre, Almora.

It is evident from the foregoing table that line sowing should be encouraged in barnyard millet.

Fertilizer application

The crop responds to the application of fertilizers especially N and P. Systematic studies carried out at Almora, Dholi and Rewa since 1979, indicated that in general, a dose of 40 kg N and 20 kg P_2O_5 per ha is adequate for a rainfed crop of barnyard millet.

A good response was observed on the farmers field to the application of N up to 40 kg per ha. On an average of two years, the yield was 2830 kg/ha with 40 kg N to rainfed barnyard millet compared to 1390 kg/ha without any fertilizer nitrogen. Application of the recommended level of nitrogen in two equal splits, 50 per cent at sowing and 50 per cent at tillering stage, produced maximum grain yield.

Fertilizer application along with FYM is beneficial and FYM helped in reducing the nitrogen requirement of the crop.

Weed management

Keeping the field weed free from the initial stages of crop growth was found essential to achieve a higher yield. If the weeds are left in the field beyond 30 days, there was a drastic reduction in the yield as shown in Table 17.

Effective chemicals may have to be identified to keep the land free of weeds at earlier periods of crop growth.

Crop sequences and intercropping

The traditional practice of finger millet fallow-barnyard millet-wheat was identified to be less efficient as the land is left fallow in rabi after harvesting

TABLE 17
Effect of the time of first weeding on the yield of barnyard millet

Time of first weeding	Grain yield (kg/ha)	
	1981	1982
30 DAS	680	1600
45 DAS	720	770
60 DAS	410	620
75 DAS	40	400
No weeding till harvest	40	330

Source: Annual reports of the IDRC Millet Centre, Almora.

finger millet. Instead, a rotation of barnyard millet-pea-barnyard millet-wheat, or Echinochloa--Chick pea were found to be advantageous both for total yields and total returns (in place of chick pea, lentil or finger millet may be used). Not many systematic studies are available on the intercropping and mixed cropping with this crop.

OUTLOOK FOR SMALL MILLETS

Small millets have an untapped potential under adverse soil and climatic conditions. Even though the area under small millets is slowly decreasing, there will be still about 6 million hectares under small millets in India where other crops have less chance of adaptation.

Because of their comparative photoinsensitive nature, short growing season and low moisture demand, millets can be very well fitted into multiple cropping systems both under irrigation as well as dry farming conditions. During scarcity years they can provide nutritious grain as well as valuable fodder in a short span of time. Their long storability under ordinary conditions has made them 'famine reserves'. This aspect is perhaps the most important for Indian agriculture where the crop production suffers due to the vagaries of the monsoon. There are types to suit a wide range of rainfall situations which can be used for mid-season corrections when rains are delayed.

The review of the work done in India discussed in the previous pages, has brought out the potential of several millets in achieving higher yields. Adoption of simple agronomic practices like timely sowing of recommended varieties, fertilizer application at moderate levels, weeding and intercultivation could increase the yield level by 200 to 300 per cent. Not much has been achieved in identifying acceptable cropping systems to suit different agroclimatic situations. Further, research efforts are concentrated to only a few centres even though the millets are cultivated in almost all the states under contrasting climatic conditions. Research efforts need to be intensified in all the regions of their production to provide location specific recommendations.

TABLE 18
Varieties of small millets recommended for different states

State	Kodo millet	Foxtail millet	Little millet	Proso millet	Barnyard millet
Andhra Pradesh	PSC 10	Arjuna, SIA 326, Chitra	Co 1, Co 2	Varada	Local, Co 1
Bihar	—	Local, SIA 326	V 15, V 17	BR 7, Ramcheena, Shyam cheena	RAU 2, RAU 3, RAU 9
Karnataka	PSC 1, PSC 2, JNK 364	K 221-1, RS-118	Co 1, Co 2, PRC 3	Co 3	—
Maharashtra	—	Arjuna, SIC 3	—	Varada, No. 11	—
Madhya Pradesh	IPS 147-1, RPS 41, Keharpur, RPS 123, JNK 364, PSC 1, PSC 2, RPS 62-3	—	Dindori-1, Dindori-2, PRC 3	—	—
Orissa	—	—	Koraput Local, SS-81-1, Bodo Sawan, PRC 3.	—	—
Tamil Nadu	Co 2, K-1	Co 4, Co 5, K 2	Co 1, Co 2, K1	K 1, Co 2, Co 3	K 2, Co 1
Uttar Pradesh	PSC 1, PSC 2	—	VL 5, VL 7	BR 7, Ram Cheena, Shyam Cheena	VL 8, VL 11, Anurag

Millet grains are normally consumed in the areas of their production and very little finds its way to the market. Nevertheless, their place in industrial uses and livestock feeds is already well established. Estimations by the National Commission on Agriculture, indicate that the demand for small millet grains, excepting finger millet, is not likely to increase due to availability of other preferred cereals and the declining trend in area under these crops is likely to continue. It means that all out efforts are needed to increase the production in the existing area by adopting improved varieties and management practices.

SUMMARY

Intensive research was started on various aspects of millet production during the year 1969 under the auspices of the All India Coordinated Millet Improvement Project. Five special centres were established later in 1979 for the improvement of small millets, viz., kodo millet, foxtail millet, little millet, proso millet and barnyard millet under the assistance of International Development Research Centre (IDRC), Canada. Research efforts at these centres were concentrated mainly in the fields of breeding, agronomy, entomology and pathology.

As a result of intensive efforts, the national yield level of finger millet has been raised from 700 kg/ha in 1949-50 to 1160 kg/ha during 1983-84. In case of other small millets, national yield has increased from 421 kg/ha to 461 kg/ha during the same period. The quantum jump in case of finger millet has been made possible through the breeding of varieties to suit different agroclimatic conditions and seasons of sowing, fertilizer application at moderate levels, and changing over from traditional intercropping systems to improved systems. In other small millets, traditional practices are still widely adopted in the absence of more profitable and acceptable cropping systems. However a number of improved varieties have been released, as shown in Table 18.

Cultivation of millets in India is restricted mainly to the poorer sections of the society and optimum production technology is rarely adopted. Large numbers of farmers still practise broadcasting to establish the crop, even when improved seed drills are available. Adoption of this primitive practice comes in the way of intercultivation and effective weed control. Similarly, the traditional systems of mixed cropping are found suitable only for subsistence farming. For higher productivity either a sole crop or an improved intercropping system has to be adopted.

REFERENCES

Ananthanarayana, R., K. Krishnamurthy and L.A. Dixit. 1971. Effect of different nitrogenous fertilizers in rainfed ragi. *Ann. Arid Zone*, **10:** 43-47.

Anonymous. 1984. Highlights of dryland research in red soils of Karnataka (1971-83). Mimeographed report of the AICRP for Dryland Agriculture, Bangalore Centre.

Anonymous. 1986. The first fifteen years. Bulletin of the AICRP for Dryland Agriculture, Bangalore.

Ashok, E.G., V.C. Reddy and B.K.L. Gowda. 1979. Effect of sowing dates and spacing on grain yield of finger millet varieties. *Mysore J. Agric. Sci.*, **13**(2): 134-136.

Choudhari, L.B. and B. Rai. 1982. Production technology for proso millet. *Indian Farming* **32**(1): 13-15.

Gautam, R.C. 1976. Nitrogen requirement of *Setaria italica* under rainfed conditions. *Indian J. Agron.*, **21**(3): 297-298.

Gautam, R.C. 1977. Minor millets and package of practices for their cultivation. Indian Farmers Digest, **10:** 13-17.

Gautam, R.C., B.K.L. Gowda, R.R. Patil and S.K. Kaushik. 1982. Nitrogen requirement of finger millet at different agroclimatic centres. *Indian J. Agron.*, **27**(4): 344-346.

Havanagi, G.V. and B.R. Hegde. 1983. Response of pearl millet and finger millet crops to fertilizer management under rainfed conditions. *Fert. News.*, **28**(9): 62-68.

Hegde, B.R., N. Munikrishna Reddy and G.V. Havanagi. 1980. Intercropping with ragi in Karnataka. *Indian J. Genet.*, **40**A: 130-135.

Hegde, B.R. and A. Seetharam. 1985. Finger millet and other minor millets. In: *Efficient management of dryland crops.* Central Research Institute for dryland agriculture, Hyderabad, pp. 79-101.

Hegde, B.R., A.P. Vishwanath, N. Munikrishna Reddy, T. Sathyanarayana and G.V. Havanagi. 1983. Crop weed Competition in dryland ragi. *Mysore J. Agric. Sci.*, **17:** 315-319.

Hosmani, M.M., V.S. Gidnavar, P.V. Hugar and T.K.P. Setty. 1975. Response of Navane (Foxtail millet) to N and P application. *Indian J. Agron.*, **20**(2): 173-174.

Kaliappa, R., S. Venkatachalam, K.M. Nachappan, K.V. Selvaraj, S. Rajaram and H. Manoharan, 1974. Influence of soil moisture stress on yield of ragi. *Madras agric. J.*, **61**(9): 773-775.

Kaushik, S.K. and R.C. Gautam. 1981. Response of kodo millet varieties to nitrogen and phosphorus levels under rainfed conditions. *Indian J. Agron.*, **26**(2): 168-170.

Krishnamurthy, K. 1973. Partitional analysis of grain yield from different shoots as influenced by nitrogen in finger millet. *Agron. J.*, **65**(6): 856-858.

Linge Gowda, B.K., M.G. Rajappa and H.N. Sundaresh. 1974. Agronomic practices for rainfed ragi. *Curr. Res.*, **3:** 13-14.

Linge Gowda, B.K., Doddahanumaiah and B.C. Suryanarayana. 1977. Haraka responds very well to nitrogen fertilizers. *Curr. Res.*, **6**(2): 21-22.

Mann, H.S. and M.B.L. Saxena. 1977. Minor Millets for Rajasthan. *Indian Fmg.*, **27**(4): 17.

Murthy, S.N.N. and B.R. Hegde. 1981. Tillering in relation to intra-row competition in ragi under rainfed conditons. *Indian J. Agron.*, **26**(3): 337-338.

Nanjappa, H.V. and M.M. Hosmani. 1986. Weed control under cropping systems in drill sown rainfed finger millet. *Mysore J. agric. Sci.*, **20:** 9-18.

Patil, V.S. *et al.* 1969. Annual Progress Report, Major River Valley Project, Siruguppa.

Patro, C.K. and N.L. Das. 1972. A note on the performance of Stam, F-34, Ansar-529 and MCPA on control of weeds in ragi. *Indian J. Weed Sci.*, **4**: 69-72.

Rachie, K.O. 1975. *The millets: Importance, utilization and outlook.* International Crops Research Institute, Hyderabad, pp. 62.

Rachie, K.O. and L.V. Peters. 1977. *The Eleusines: A review of world literature*, ICRISAT, Hyderabad, pp. 179.

Raj, K.G., V.C. Patil, and G.D. Radder, 1979. Effect of nitrogen levels and *Azotobacter* inoculation in Setaria. *Research Bulletin of Marathawada Agricultural University*, **3**(12): 163-165.

Rao, G.M., M.V. Kulkarni, G.V. Havanagi, B.V. Venkat Rao and S.V. Patil. 1975. UAS Stn. Series, 1.

Reddy, V.C. 1977. Effect of soil moisture stress at different growth stages and yield of finger millet. M.Sc. Thesis submitted to UAS, Bangalore.

Shankaran, S., E. Subbaiah E. and A. Rajagopal. 1974. Studies on chemical weed control in relation to irrigation levels in finger millet. *Madras Agric. J.*, **61**(8): 433-438.

Seetharam, A. 1986. Package of practices for higher productivity for ragi and minor millets (In press).

Subbaiah, S., K.M. Ramamurthy, H.J. Francis, R. Suresh Kumar and G.V. Kothandaraman. 1982. Influence of nitrogenous fertilizers with and without neem cake blending on the yield of finger millet (*Eleusine coracana* G.). *J. Indian Soc. Soil Sci.*, **30**(1): 37-43.

Sundaresh, H.N., M.G. Rajappa, B.K.L. Gowda and K.S. Krishna Shastry. 1975. Critical stages of weed competition in ragi under rainfed conditions. *Mysore J. Agric. Sci.*, **9:** 582-585.

22

DISEASES OF SMALL MILLETS AND THEIR MANAGEMENT IN INDIA

S. Viswanath and A. Seetharam

INTRODUCTION

The principal members of the small millets group are finger millet or ragi (*Eleusine coracana*), Italian or foxtail millet (*Setaria italica*), common or proso millet (*Panicum miliaceum*), little millet (*Panicum miliare*) (renamed *P. sumatrense* kodo millet (*Paspalum scrobiculatum*) and barnyard millet (*Echinochloa frumentacea*) renamed *E. colona*). All these are cultivated in one or more states in India and occupy 4.5 per cent of the total cultivated area. Many diseases affect these millets and cause enormous losses which in the aggregate, may account for millions of rupees. The important diseases affecting each of these millets are described under each crop.

FINGER MILLET DISEASES

Ragi, one of the millets grown on a large area in India, is attacked by many diseases caused by fungal, bacterial and viral pathogens. They are blast, helminthosporiose, smut, downy mildew, foot rot, bacterial leaf spot, bacterial wilt, mottle streak and streak virus.

Blast

Of the several diseases that affect ragi, blast caused by *Pyricularia grisea* is the most serious one causing considerable grain loss in many ragi growing regions. The disease occurs almost every year during rainy season but the extent of crop loss depends on the severity and the time of onset of disease. Environmental conditions particularly rainfall, temperature and humidity are the most important pre-disposing factors on the severity of blast. According to McRae (1922) the grain loss due to blast could be over 56 per cent, while

Venkatarayan (1947) reported more than 80 per cent yield loss in old Mysore State. In order to get some insight on the disease intensity and extent of annual crop loss due to blast, a study was undertaken during 1980 at five locations and the data are presented in Table 1.

TABLE 1
Blast incidence and grain yield loss in finger millet at different locations in India

Locations	Blast incidence (per cent)		Grain yield loss (per cent)
	Neck blast	Finger blast	
Almora	40.0	46.2	23.4
Bangalore	8.1	29.3	36.4
Dholi	52.5	8.3	34.0
Jabalpur	70.1	22.4	30.9
Kovilpatti	35.2	15.3	13.6
Mean	41.2	24.2	27.8

The pathogen can infect the crop at all stages, from seedling to post-flowering phase. The symptoms at the seedling and tillering stage are the appearance of small brown, circular to elongated spots on leaves which eventually develop into large elongated or spindle shaped areas, the centre of the spots being greyish. The spots may coalesce to involve a large area of the leaf blade.

Rath and Mishra (1975) reported that neck infection causes significant loss in grain number, grain weight and significant increase in spikelet sterility. However, the crop loss will be greater when the disease appears on the neck and ears during flowering and grain development phase. At this stage the disease may appear on peduncle and/or on finger causing neck and finger blast respectively. Depending on the time and severity of infection the infected ears become completely chaffy or produce shrivelled grains.

The pathogen harbours in glumes, straw as well as on some of the graminaceous weeds. The seeds also have been found to carry the pathogen in the pericarp and endosperm. But embryo infection has not been observed. A temperature of 25-30°C, humidity of 90 per cent and above, cloudy days with intermittent rainfall, are favourable for the rapid spread of the disease. The disease prevails from the month of May to September, but it may extend even up to November, when the favourable conditions are prolonged (Anonymous, 1979). The maximum disease incidence has been observed in the crop sown in the month of August (Table 2).

The fungus spreads mainly by air borne conidia. The initial inoculum probably coming from weeds or some cereal plants acting as collateral hosts. The fungus may also persist in plant debris and to some extent in the shrivelled

TABLE 2
Seasonal incidence of neck and finger blast

Sowing date	Disease incidence (%) Neck blast	Finger blast
June 22nd	1.6	5.7
July 7th	4.3	8.3
July 22nd	11.4	23.5
August 7th	46.3	72.6
August 22nd	31.3	60.4
September 22nd	12.0	52.1
October 22nd	1.8	2.6
November 22nd	0.0	0.9
December 22nd	0.0	0.0
January 22nd	0.0	0.0
February 22nd	0.0	0.0
March 22nd	0.0	0.0
April 22nd	0.0	0.0
May 22nd	4.6	5.1

grains i.e. from the infected ears which give rise to the initial infection in the nursery from where it may spread to the main field.

Control of the disease by using chemicals has shown that spraying of mancozeb or carbendazin or kitazin once at the seedling stage and twice at the earhead stage reduces the disease incidence. This was found to be economical also.

In addition to the chemical control, evolution of varieties with inbuilt genetic resistance is the best means of combating disease problems in any crop. Such an approach is more relevant in a crop like, ragi which is predominantly grown by poor and marginal farmers who have little means of controlling diseases through chemicals. Almost all cultivars, both local and improved, presently under cultivation are susceptible to blast disease though the level of susceptibility reaction varies from cultivar to cultivar. So, breeding for blast resistance assumes greater importance in ragi. The success of such a programme depends on the identification of stable resistant sources and its subsequent utilization in breeding. As a first step in this direction, 1941 of the available 4500 germplasm, accessions were evaluated under field conditions with high disease prevalence. The results are given in Table 3 (Anonymous, 1984). Twenty eight entries showed resistant reaction for both neck and finger blast, and only 9 entries showed combined resistant reaction for leaf, neck and finger blast as shown in Table 4.

TABLE 3
Evaluation of germplasm entries to leaf, neck and finger blast

Disease reaction	Leaf blast		Neck blast		Finger blast		Neck and finger blast*	
	No. of entries	% entries	No. of entries	% entries	No. of entries	% entries	No. of entries	% entries
Resistant	18	0.93	281	14.48	30	1.55	28	1.45
Moderately resistant	109	5.62	420	21.64	372	19.17	268	13.81
Moderately susceptible	310	15.98	397	20.46	491	25.30	—	—
Susceptible	671	34.57	431	22.21	542	27.93	—	—
Highly susceptible	833	42.92	412	21.23	506	26.07	—	—
Total	1941		1941		1941			

*Entries showing moderate resistant reaction for both neck and finger blast only have been considered.

TABLE 4

Reaction of entries showing resistant reaction to both neck and finger blast

Sl. No.	Entry No.	Leaf blast (grade)	Neck blast (% incidence)	Finger blast (% incidence)
1	GE 75	6	0.00	0.25
2	156	5	0.00	0.00
3	158	7	0.00	0.00
4	243	7	0.00	0.59
5	281	4	0.00	0.43
6	396	6	0.00	0.57
7	406	6	0.00	0.00
8	568	2	0.00	0.44
9	639	5	0.00	0.59
10	669	4	0.00	0.00
11	705	3	0.00	0.00
12	834	5	0.00	0.79
13	844	5	0.00	0.17
14	942	5	0.00	0.44
15	965	5	0.00	1.00
16	1044	4	0.00	0.90
17	1055	6	0.97	0.90
18	1126	5	0.00	0.80
19	1293	4	0.00	0.00
20	1309	5	0.50	0.44
21	1348	6	0.00	0.70
22	1407	6	0.00	0.68
23	1409	4	0.00	0.43
24	1423	7	0.00	0.00
25	1546	4	0.00	0.00
26	1709	7	0.00	0.46
27	1855	3	0.00	0.00
28	1916	5	0.00	0.36

Breeding was initiated at GKVK centre, Bangalore, by utilizing IE 1012 as the resistant parent and PES 176, HPB 7-6, PR 202 and Indaf 8 as high yielding adopted parents. Preliminary evaluation of F_5 progenies for blast and yield data suggests in general that both neck and finger blast incidence was higher in early maturing progenies within the cross. It is noteworthy that five progenies each in PR 202 × IE 1012 and in Indaf 8 × IE 1012 crosses were outstanding with significant high yield over their parents coupled with high tolerance to both neck and finger infection.

SEEDLING BLIGHT OR LEAF BLIGHT

Butler (1918) stated that blight caused by *Drechslera nodulosum* is also one of the serious diseases of ragi. It has been recorded in India and other countries. In India, this disease is prevalent in Madhya Pradesh, Andhra Pradesh, Maharashtra, Uttar Pradesh, Bihar, Tamil Nadu and Karnataka.

The disease affects all the parts of the plants like base of the plant, culms, leaf sheath, leaf blade, neck and fingers. Germinating grains may be killed before the seedling emerges above the soil. After emergence, symptoms appear as minute oval, light brown lesions on the leaf blade. These develop into elongate lesions and turn dark brown. Several such lesions coalesce to form large patches on the leaf blade. In some cases the basal portion of the seedling gets affected resulting in seedling death. The pathogen affects the stem, sheath, neck and finger of the earhead causing chaffiness and discolouration of the seed. Consequently there is considerable reduction in the yield. According to Grewal and Pal (1965) seeds were found contaminated with *D. nodulosum*.

The primary infection is caused by the pathogen on the seed. The fungus remains viable on the stubbles and plant debris. Secondary spread is through air borne conidia. According to McRae (1932) the optimum temperature for infection is 30-32°C and can occur between 10 and 37°C. Several other cereals are infected by this organism including *Setaria italica, Eleusine indica, Echinochloa frumentacea, Panicum miliaceum, Sorghum vulgare* and *Zea mays* (Mitra, 1931; Mitra and Mehta, loc. cit.). High humidity with intermittent rains during the period of emergence of ear and before grain formation causes heavy ear infection and reduction yield.

As the disease is primarily seed borne, seed treatment with any organomercurial compounds will control the pre-emergence damping off seedling blight. The secondary infection in the field can be reduced by protective sprayings with suitable fungicides. In addition to chemical control of the disease, studies have been conducted to identify genetic resistance in the germplasm collection. According to Coleman (1920) considerable differences were observed in the reaction of different varieties in Karnataka. The varieties with green glumes exhibited greater infection than those with purple glumes.

Wilt or foot rot

Wilt or foot rot caused by *Sclerotium rolfsii* is mainly a soil borne disease. The disease is prevalent in all regions wherever the crop is cultivated. The affected plants appear pale green and stunted. The infection occurs at the base of the plants, involving leaf sheath and culms. These soften and become brown at the place of infection, eventually the plants wilt, lodge and dry up, white fan like mycelial growth is evident between the sheath or on the stem at the basal region. Later, minute, mustard seed like, tan coloured Sclerotia are formed on the surface.

The fungus is mainly soil borne and active during rainy season and it is not economical to control the disease through chemicals. However, cultural methods like deep ploughing before sowing, crop rotation with non-graminaceous crops and maintaining optimum soil conditions will help to avoid this disease.

Downy mildew or green ear disease

Downy mildew or green ear disease caused by *Sclerophthora macrospora* affects the plants and produces two types of symptoms. The occurrence of this disease was observed in old Mysore State in 1930 (Venkatarayan, 1947). In 1948, the severe outbreak of this disease was again recorded in Mysore State. The damage was so severe in some fields as to render the crop not worth harvesting. Since then, it is known to occur in Tamil Nadu and other states in India.

The affected plants become stunted with shortened internodes, leaves arising close together and the plants assume a bushy and bunched appearance. The leaves are pale green. The lemma, palea and sometimes glumes change into leafy structures. The proliferation takes place first in the basal spikelets and afterwards others become involved. Finally the whole ear presents a bush like appearance (Thirumalachar and Narasimhan, 1949).

Raghavendra and Safeeulla (loc. cit.) reported the fungus as internally as well as externally seed borne. The fungus has a wide host-range including *Eleusine indica,* maize, wheat, oat, *Eragrostis pectinacea* and *Digitaria marginata.* Physiologic specialization is well developed in the species, so that some races are confined to particular hosts.

As the pathogen is easily carried in fragments of proliferated parts mixed with seed, the seed treatment with organomercurial compounds will check the primary infection. Providing good drainage in the low lying field, proper crop rotation methods, roguing of the infected plants and elimination of wild grasses and related hosts will reduce the disease incidence.

Smut

Earlier, smut disease caused by *Melanopsichium eleusinis* was only of minor importance. After the introduction of new high yielding varieties, it is gaining importance in Karnataka. In Karnataka, two types of symptoms have been observed (i) occurrence of smut sori in the grains and also on the main rachis or peduncle of the inflorescence, (ii) the second type of symptom is the reduction and shrivelling of entire inflorescence, the floral organs converted into sac-like smut sori. The disease appears at the time of grain formation. Few grains in an ear are affected, but these are scattered throughout the ear. The sori are globose, greenish at first and later turn dirty black. Spores are formed in the cavities found in the sori, mixed with a gelatinous matrix but later the mass becomes pulvurent (Mundkur and Thirumalachar, 1946).

The pathogen is neither systemic nor externally seed borne. Floral infection is by the wind borne spores. At present there is no control measure which can be advised, but *in vitro* studies have shown that systemic compounds like Aliette and Vitavax inhibit fungus growth. Attempts have been made to identify resistant lines from the germplasm collection against this disease.

Virus diseases

Several viruses have been reported to infect ragi. Rao *et al.* reported a sap transmissible virus, while Jagannathan *et al.* reported one that is transmitted by aphid. A leaf hopper transmitted virus was reported from Karnataka, which is different from the above, having ragi has the lone host (Yaraguntaiah and Keshavamurthy, 1969). In 1973, Mariappan *et al.* reported, from Tamil Nadu, a streak virus which was transmitted by delphacid insect *Sogatella* sp. which appeared to be the same virus prevalent in Karnataka. However, another virus, producing streak symptoms transmitted by a leaf hopper vector *Cicadulina chinai* (Ghauri) was reported from Karnataka (Nagaraju *et al.*,1982).

Two viruses have been observed on finger millet in Karnataka, namely mottle streak and streak. The mottle streak virus causes chlorosis, broken streaking, striping, mottling and severe yellowing of leaves. The streak virus symptoms are continuous, straight streaking on the leaves parallel to the veins, stunting, shortening of internodal length and production of more tillers. Mottle streak virus is transmitted by leaf hoppers *Cicadulina bipunctella bipunctella* and *Cicadulina chinai*, whereas, streak virus is transmitted only by *Cicadulina chinai* (Nagaraju *et al.*, 1982).

Virus vector relationship of mottle streak virus and its vector *Cicadulina bipunctella bipunctella* was studied. The results revealed that the vector takes a minimum acquisition feeding period of 12 hours to acquire the virus on source plant and after that it has an incubation period of 8 days to become viruliferous. The minimum inoculation feeding period of 12 hours is necessary to transmit the virus to test plant. The symptom expression of the disease is observed 9 days to one month after inoculation.

Similarly the virus vector relationship of streak virus and its vector, *Cicadulina chinai* was reported by Nagaraju and Viswanath (1981). The minimum acquisition and inoculation feeding period have been found to be 6 hours and 30 minutes, respectively. The incubation period within the vector has been found to be one hour. The virus remains infective in the insect throughout its life span. It takes 7 days after inoculation for the symptom expression in the test plant.

Mottle streak virus incidence is more prevalent than streak virus in the field. However, the streak virus is more virulent resulting in the mortality of the plant.

The epidemiological studies of mottle streak virus indicate that the disease incidence is maximum in the March sown crop and the least in the July sown crop. Maximum leaf hopper population was observed in the November sown crop and the least in the July sown crop. The spread of the disease was not uniform throughout the year (Anonymous, 1979). Further, there was positive correlation between the incidence of the disease and jassid population (Table 5).

TABLE 5
Seasonal incidence of ragi viruses and its leaf hopper vector population at Bangalore

Sowing date	Mottle streak virus (%)	Streak virus (%)	Vector population (per sq. mtr)
June 22nd	2.90	1.21	2.77
July 7th	2.18	0.10	1.00
July 22nd	0.58	0.00	0.66
August 7th	1.97	0.19	3.71
August 22nd	2.97	0.17	1.83
September 22nd	0.98	0.14	1.62
October 22nd	2.02	0.07	2.66
November 22nd	3.09	0.07	5.91
December 22nd	0.86	0.22	2.07
January 22nd	1.15	0.02	1.34
February 22nd	3.04	0.07	2.45
March 22nd	7.26	0.07	2.61
April 22nd	6.84	0.06	2.21
May 22nd	1.89	0.02	1.87

The host range studies show that the mottle streak virus can infect 9 plant species, whereas streak virus has a wider host range in that it can infect crop plants like Bajra, sorghum, maize, wheat, barley and oats also, in addition to the 9 plant species which mottle streak can infect. However, four wild species of Eleusine and three grass species were found to be immune to this virus (Nagaraju *et al.*, 1982).

In the varietal screening trial to find out genotypes resistant to the mottle streak virus, some degree of resistance has been seen among the African collections in the germplasm, whereas, all accessions tested were found to be susceptible to streak virus.

The loss of grain due to mottle streak virus was found to be 53.9 per cent (Table 6). This loss could be minimized by spraying Rogor 15-20 days after planting. In the case of streak virus, the loss has gone up to 100 per cent when they were inoculated in early stage i.e. up to 30 days old seedlings (Nagaraju, *et al.*, 1981). The loss is as high as 60-70 per cent when they were inoculated even at the time of earhead emergence (Table 7).

TABLE 6
Effect of mottle streak virus infection on different characters of ragi plants

Characters	Infected	Healthy (control)
No. of tillers	1.90	2.32
Straw weight per plant (gm)	35.60	63.00
Earhead weight per plant (gm)	10.20	25.40
Grain yield per plant (gm)	7.40	19.60
Loss of straw weight over control (%)	43.49	
Loss of grain weight over control (%)	62.74	
1000 grain weight (gm)	3.01	3.4

FOXTAIL MILLET DISEASES

Foxtail millet is one of the small millets grown on a large area in India. In India, it is cultivated in Andhra Pradesh, Madhya Pradesh, Karnataka, Tamil Nadu and Maharashtra. Blast, rust, downy mildew and smut are the most important diseases affecting this crop.

Blast

Blast caused by *Pyricularia setariae* is one of the diseases causing considerable grain loss in many states. The disease occurs almost every year during the rainy season. Environmental conditions, particularly rainfall, temperature and humidity are the most important factors affecting the severity of this disease.

The spots are seen on the leaf blade. They are surrounded by a dark brown margin. The spots are small and scattered and measure 2-5 mm in diameter. When the disease appears in severe form, the leaves wither and dry up. Unlike the blast of rice or finger millet, the infection is not evident on the neck or ear.

According to Kulkarni (1969) studies on the pathogenicity and cultural, physiological and morphological characters of *Pyricularia setariae* isolates causing blast of *Setaria italica* showed differences indicating the existence of at least four physiological forms of the fungus.

Palaniswamy *et al.* (1970) reported that in inoculations, *Pyricularia setariae* infected *Setaria italica*. Young plants up to 40 days old were highly susceptible to the disease which was found to be seed borne and to some extent soil borne. The fungus lost its viability on stored-seeds after 75 days.

The isolate from *Setaria* readily infects finger millet, pearl millet, wheat and *Dactyloctaenium aegyptium*.

Control of this disease by using chemicals is not economical. However, studies have been conducted to identify genetic resistance within the germplasm collection. Singh *et al.* (1976) found the varieties SR 118, SR 102, ISc 709, 701, 703, 710, 201, JNSc 33, 56, RS 179 and ST 5307 was resistant.

TABLE 7
Effect of streak virus infection on different characters of ragi plants

Observations	Age of the seedlings at inoculation (in days)						
	10	20	30	40	50	60	Control
Mortality of infected plants (%)	100.00	100.00	50.00	46.00	40.00	0.0	
No. of tillers	3.40	2.88	2.67	2.69	2.68	2.70	2.70
Plant height at maturity (cm)			23.50	26.50	36.50	36.50	64.60
Earhead weight per plant (gm)			0.60	1.50	2.80	2.89	6.50
Straw weight per plant (gm)			1.94	2.19	3.17	6.29	16.85
Grain yield per plant (gm)			0.21	0.49	1.15	1.54	5.52
Loss of grain yield over control (%)	100.00	100.00	96.19	91.12	79.16	72.10	
1000 grain weight (gm)			0.48	1.10	2.18	2.27	2.98

Rust

Rust caused by the fungus *Uromyces setariae italica,* is known to be prevalent in all the states in India, wherever this crop is grown, but was of little economic importance hither to (Butler, 1918). But in recent years there were several reports of the rust assuming serious epiphytotic proportions and destroying the crop before the ears have formed.

Numerous minute, brown urediosori appear on both sides of the leaf. They are brown, oblong pustules often arranged in linear rows. Pustules are also produced on leaf sheaths, culms, and stems. The rust affects the crop in all stages of growth. The greatest extent of damage occurs when the infection commences before the flowering stage. When the infection takes place after the grains are set, no appreciable reduction in yield is noticed. Premature drying of foliage is caused by early and heavy infection and the plants may dry up before heading or even the ears that are formed may develop only few grains.

The life cycle of the fungus has two stages, the urediosori and the teliosori.. The uredia are the first to be seen. Urediospores germinate, producing one or more germ tubes capable of infecting the host. The telia are formed on the leaf blade, leaf sheath and stem and are larger in size than uredia. The teliospores are single celled, pedicellate, oblong, globose, yellowish brown, with a smooth, thick walls, which are much thicker at the apex than at the base.

The urediospores infect the host and produce uredia in seven to ten days. If the crop is grown throughout the year as in many parts of India, the fungus can perpetuate in its uredial stage, with the collateral hosts possibly playing a part in its perpetuation. The rust appears within 20 to 25 days of sowing and the intensity increases as the plant grows older. The telia appear at the time of maturity of the crop.

Most of the agronomically important varieties are susceptible to this disease. However, some of the entries in the germplasm collection have been found to be tolerant to this disease. Further, their use as parents in hybridization with local varieties may lead to the evolution of resistant varieties.

Smut

In India, smut caused by *Ustilago crameri* is prevalent in Karnataka, Andhra Pradesh, Tamil Nadu and Maharastra. Sundararaman (1921) found that in some parts of districts of the Southern States, nearly 75 per cent of the grains in an ear were affected. Inhalation of the spores during threshing operations may cause hay fever or asthama among the labourers (Fisher, 1953).

Symptoms

The sori are seen in the flowers and the basal parts of the palea. The fungus affects most of the grains in an ear but sometimes the terminal portion of the spike may escape. The sori are pale greyish in colour and measure 2 to 4 mm in diameter. When crop matures the sori rupture and produce dark powdery mass of spores.

The fungus is externally seed borne. Certain amount of soil borne infection has also been observed in some dry areas. Wang (1943) recorded that the dikaryotic hyphae penetrated the tissues of two day old seedlings by mechanical pressure. The invading hyphae are systemic, mainly concentrated towards the apical portions and at the time of flowering replace the ovaries, producing septate hyphae which transform into chlamydospores.

Control

Since the disease is mainly seed-borne, it can be controlled by treating the seed with organomercurials or steeping the seed for 10 to 30 minutes in 2 per cent copper sulphate solution or 0.5 per cent formalin for about 30 minutes.

DOWNY MILDEW OR GREEN EAR

Downy mildew, (*Sclerospora graminicola*) is prevalent in India, China, Japan, Russia, the southeastern countries of Europe and America. In India it is prevalent in Maharashtra, Tamil Nadu, Karnataka, Andhra Pradesh and Bihar. In certain years it has caused loss up to 50 per cent in the foxtail millet crop.

Symptoms

There are two types of infections reported; the primary infection, which starts with the seedling is systemic, while the secondary infection occurs on older plants is local. Primary infection causes chlorosis of the plant and the leaves turn whitish. The terminal spindle fails to unroll, becomes chloratic and later turns brown and gets shredded. Whitish bloom of sporangiophores and sporangia develop on the surface of the affected leaves under humid conditions. When the infection is mild, the plant may develop ears, but the floral parts are proliferated into green leafy structures, hence the name 'green ear'. In the spikelet the glumes, lemmae, paleae, stamens and pistils are turned into leafy structures of variable size. Sometimes only a portion of the ear may be affected, with the remainder producing normal grains.

Secondary infection causes formation of restricted chlorotic lesions on the younger leaves. On these the downy fungal growth may be seen under humid conditions.

Infection

The fungus is an obligate parasite. Primary infection is mainly from soil borne oospores or on the grains. The sporangia are produced on the host and spread throughout the field by wind, causing secondary infection when the favourable conditions prevail.

CONTROL

The disease is partly seed borne. Seed treatment with an organomercurial gives partial control by eliminating the surface borne oospores. Eradicating the diseased plants before the oospores are formed helps to reduce the inoculum potential in the field.

Minor diseases

In addition to the above mentioned diseases, *Ephelis,* bacterial blight and viral streak have been observed on this crop, but they are of little economic importance hither to.

LITTLE MILLET DISEASES

In India little millet is cultivated in many states. It is a hardy, short duration crop and can withstand both water-logging and drought. It is generally grown on poor lands. The rust is the most common disease of this crop.

Rust (*Uromyces linearis*)

Rust is known to occur in India, Sri Lanka and Philippines. In India, it has been recorded in Maharashtra and Tamil Nadu. It causes very little damage to the crop.

SYMPTOMS

Numerous, narrow, minute, brown pustules arranged in linear rows appear on the upper surface of the leaves.

PATHOGEN

The Uredia are brown and erumpent. The urediospores are brown, round, echinulate. The telia are black in colour. The teliospores are thick walled, smooth, globose, brown with persistant, long thick pedicels, fresh urediospores germinate readily but not the teliospores.

KODO MILLET DISEASES

Kodo millet is cultivated in many countries in the world, both for grain and fodder. In India, it is grown in Madhya Pradesh, Orissa, Tamil Nadu, Bihar, Karnataka and Andhra Pradesh, generally on soils, with poor fertility. Rust and smut are the important diseases of this crop in India.

Head smut (*Sorosporium paspali*)

In India, head smut has been recorded from Madhya Pradesh, Andhra Pradesh, Bihar, Tamil Nadu and Karnataka. According to Butler (1918), it causes heavy losses in some years.

SYMPTOMS

The entire panicle is transformed into a long sorus. When young, a cream coloured thin membrane covers the sorus. In some cases, it is enclosed in the flag leaf and may not emerge fully. The membrane bursts and exposes the black mass of spores.

PATHOGEN

The spores are held in loose ball-like masses of individual spores, which easily separate, are globose to angular, dark brown and with a thick smooth epispore. On germination a promycelium bearing terminal and lateral sporidia are formed.

INFECTION

The disease is mainly seed-borne. The spores adhere to the surface of the grains and infect the seedlings. Soil-borne infection is insignificant. After entering the seedling, the hyphae spread inter and intracellularly and fungus becomes systemic. It enters the meristamatic tissues and finally infects the ear.

CONTROL

Sattar (1930) reported that steeping the seeds in 1.5 per cent copper sulphate or dusting with copper carbonate at 6 g/kg of seed were equally effective. Even when the treated seeds are stored for two months, there is no reduction in germination of the seeds. Organomercurial dusts for seed treatment also control the disease.

Rust (*Puccinia substriata* Ellis and Barth)

Rust was first recorded on kodo millet from Kanaighat in Sylhet and from Kumaon hills as *Uredo paspali-scrobiculata*. Only the Uredial stages was seen at that time. Afterwards it was recorded from Coimbatore. Both the Uredial and the telial stages were recorded and it was renamed as *Puccinia substriata*.

SYMPTOMS

The Uredia are formed on the upper surface of the leaf blade and on the leaf sheath as oval, erumpent, brown sori. These are present throughout the year on the grass hosts from where they go to the cultivated millet. The telia are usually formed on the under surface of leaf sheath and leaf blade. They remain covered under the epidermis for a long time and are brown in colour.

INFECTION

The urediospores germinate readily in water drops within three hours, producing one or more stout germ tubes. The incubation period extends from 8 to 12 days. The rust readily infects the cultivated and the wild forms of hosts.

BARNYARD MILLET DISEASES

Barnyard millet is cultivated in India and in many parts of the Far-East. Of all the grain crops, it is one of the quickest growing and very popular in Japan and East Indies. Several smuts form the chief group of diseases infecting this millet.

Smuts

a) Head smut (Ustilago crus-galli)

The destructive head smut, is prevalent in India and the United States. The infected inflorescence is deformed or completely destroyed. In addition the smut produces gall like swellings on the stem, the nodes of the young shoots and in the axils of the older leaves. Sometimes, twisted, deformed clusters of leafy shoots with aborted ears may develop. The gall-like swellings are covered by a hairy tough membrane of host tissue and may be up to 12 mm in diameter (Mundkur, 1943).

Infection is seed borne and seed treatment with organomercurial will be useful in controlling the disease.

b) Kernel smut (Ustilago paradoxa)

Ustilago paradoxa causing the kernel smut, infects the ovaries only. In India, it has been observed in Bihar, Maharashtra and Tamil Nadu. All spikelets are not affected, some may escape infection. The infected ovary is transformed into a round, hairy, grey sac, not exceeding the size of a normal grain.

CONTROL

Since the disease is seedborne, it can be easily controlled by seed treatment with organomercurial fungicides.

c) Ustilago panici-frumentacei

This disease has also been reported from Uttar Pradesh, Bihar, Maharastra and Madhya Pradesh. It is ovaricolous but all the grains in an ear may not be affected. The affected ovaries enlarge two or three times their normal size and their surfaces become hairy. The spores escape through an opening at the tip.

CONTROL

Steeping the seeds for one hour in 3 or 4 per cent acetic acid mixed with 6 per cent milk of lime gave satisfactory control of the disease. Pall and Nema (1978) reported that seed treatment with Ceresan, Bavistin, Dithane M 45, MBC and Thiram controlled the disease.

REFERENCES

Anonymous. 1979. Annual report of the Virologist, AICRP on millets for the year 1978-79, UAS, Bangalore.

Anonymous. 1984. Annual report on Finger millet. All India Coordinated Millets Improvement Project for the year 1983-84.

Butler, E.J. 1918. *Fungi and Disease in Plants*. pp. VI + 547. Thaker Spink & Co., Calcutta.

Coleman, L.C. 1920. The cultivation of ragi in Mysore. Bull. Dep. Agric. Mysore, Gen. Ser. II.

Fischer, G.W. 1953. *A Manual of the North American Smut Fungi*. Ronald Press Co., New York.

Grewal, J.S. and M. Pal. 1965. Seed Microflora-1. Seed borne fungi of ragi, their distribution and control. *Ind. Phytopath.* **18**(1): 33-37.

Jagannathan, T., C. Padmanabhan and T.K. Kandaswamy. 1973. A mosaic disease of ragi in Tamil Nadu. *Madras Agric. J.*, **60:** 214-217.

Kulkarni, N.B. 1969. Comparative studies of four isolates of *Piricularia setariae*. Nishikado from *Setaria* spp. in India. *Mycopath. Mycol. app.* **38**(3): 299-303.

McRae, W. 1922. Report of the Imperial Mycologist. Sci. Rep. Agric. Res. Inst. Pusa. 1921-22.

McRae, W. 1932. Report of the Imperial Mycologist. Sci. Rep. Agric. Res. Inst. Pusa, 1930-31, pp. 73-86.

Mitra, M. 1931. Report of the Imperial Mycologist. Sci. Rep. Imp. Agric. Res. Inst. Pusa, 1929-30. pp. 59-71.

Mundkur, B.B. 1943. Studies in Indian Cereal smuts. VI. The smuts on Sawan, *Echinochloa frumentacea*. *Indian J. agric. Sci.* **13:** 631-3.

Nagaraju, S. Viswanath, K.A. Lucy Channamma and H.R. Reddy. 1981. Effect of streak virus on growth and yield of ragi. *Indian Phytopath.* **34**(3): 356-358.

Nagaraju and S. Viswanath. 1981. Studies on the relationship of ragi streak virus and its vector *Cicadulina chinai*. *Indian Phytopath*, **34**(4): 458-460.

Nagaraju, S. Viswanath, H.R. Reddy and K.A. Lucy Channamma. 1982. Ragi streak—a leaf hopper transmitted virus disease in Karnataka. *Mysore J. agric. Sci.* **16:** 301-305.

Paliniswamy, A., C.V. Govindaswamy and P. Vidhyasekaran. 1970. Studies on blast disease of Tenai. *Madras Agric. J.*, **57**(ii): 686-93.

Pall, B.S. and A.G. Nema. 1978. Chemical control of Sawan (*Echinochloa frumentacea*) Smut. *Food Farming and Agric.* Vol. X. N. 4. P. 123.

Ramakrishnan, T.S. 1963. *Diseases of millets*. Published by Indian Council of Agricultural Research, New Delhi 1. VI + pp. 152.

Rao, D.G., P.M. Verma and S.P. Capoor. 1965. Studies on mosaic disease of *Eleusine* in Deccan. *Indian Phytopath.*, **18:** 139-150.

Rath, G.C. and D. Mishra. 1975. Nature of losses due to neck blast infection in ragi. *Science and Culture*, **41**(7): 322-3.

Sattar, A. 1930. *Sorosporium paspali*. McAlp. on *Paspalum scrobiculatum* (Kodra). Bull. Imp. agric. Res. Inst. Pusa. 201. pp. 16.

Singh, B.P., B.S. Pall, A.G. Nema and M.V. Deshkar. 1976. Varietal reaction to blast of Kangni. *Curr. Res.* 1. 13.

Thirumalachar, M.J. and M.J. Narasimhan. 1949. Downy mildew of *Eleusine coracana* and *Iseilema laxum* in Mysore. *Indian Phytopath.* **2:** 47-51.

Venkatarayan, S.V. 1947. Diseases of ragi (*Eleusine coracana*). *Mysore Agric., J.* **24:** 50-7.

Wang, C.S. 1943. Studies on the cytology of *Ustilago crameri*. Ibid. **33:** 1122-33.

Yaraguntaiah, R.C. and K.V. Keshavamurthy. 1969. Transmission of the virus component of the ragi disease complex in Mysore. *Plant. Dis. Reptr.* **53:** 361-363.

23

INSECT PESTS OF SMALL MILLETS AND THEIR MANAGEMENT IN INDIA

T.K. Murthi and G. Harinarayana

INTRODUCTION

Small millets are small grained cereals, mostly grown under rainfed conditions in marginal soils. These form the staple food of the poorest of the poor and a larger section of tribal and hill peoples. The evolution of new and high yielding small millet varieties may encourage the build up of insects which were not predominant before. With the establishment of small millet centres of excellence since 1978, the entomological work gained momentum. Headway has since been made in identifying key insect pests on small millets and in evolving management practices. The major insect pest species that attack small millets and their management practices are presented. Additionally, potential insects and areas of future research are discussed.

PEST COMPLEX OF SMALL MILLETS

A list of insect pests associated with small millets is given in Table 1. The major pests are shootfly species on different small millets, stem borers, earhead worms and aphids on finger millet and army worms, leaf rollers and leaf beetles on foxtail millet. Other potential and occasional pests which infest the millets in limited areas are: gall midge in little and kodo millets, leaf miner in foxtail millet, termites in proso millet, stem borer in barnyard millet and leaf hoppers and root aphids in finger millet. Other minor pests, though present in the field do not cause economic losses and are of secondary importance (e.g. white grubs, bugs, and grasshoppers).

TABLE 1
Insect pests on small millets in India

Common name	Scientific name	Damaging stage	Plant part attached	Status
Kodo millet (*Paspalum scrobiculatum*, L.)				
Shootfly	*Atherigona simplex*, Thom.	Maggot	Growing point	High
Gall midge	*Orseolia* sp.	"	Spikelet	Moderate
Stem or Pink borer	*Sesamia inferens*, Wlk.	Caterpillar	Stem	"
Leaf roller	*Marasmia trapezalis*, Wlk.	"	Leaf	Low
Jassid	*Hecalus* sp.	Nymph & Adult	"	"
Gundhi bug	*Leptocorisa acuta*	" "	Ear	"
Army worm	*Mythimna separata*, Wlk.	Caterpillar	Leaf	"
Grasshopper	*Acrida exalta*, Wlk.	Nymph & Adult	"	"'
B. Foxtail millet —*Setaria italica* Beauv.				
Shootfly	*Atherigona atripalpis*, M.	Maggot	Growing point	Moderate
Flea beetle	*Chaetocnema basalis*, Baly.	Adult	Leaf	Mod-High
	Madurasia sp.	"	"	"
Army worm	*Mythimna separata*, Wlk.	Caterpillar	"	"
Leaf roller	*Marasmia trapezalis*, Wlk.	"	"	"
Stem borer	*Chilo partellus*, Swim·	"	Stem	Low
Surface grasshopper	*Chrotogonus* sp.	Nymph & Adult	Leaf	"
Ant	*Sima* sp. nr *longiceps*, Foral	Adult	Ear	"
Leaf miner	—	Maggot	leaf	Moderate
C. Little millet—*Panicum miliare*, Lam.				
Shootfly	*Atherigona miliaceae*, M.	Maggot	Growing point	High
Gall midge	*Orseolia* sp.	"	Spikelet	Moderate
Flea beetle	*Chaetocnema* sp.	Adult	Leaf	Low
Stink bug	*Nezara viridula*	Nymph & Adult	Ear	Low
Black pentatomid	*Dolycoris indicus*, Stal.	"	"	"
Jassid	*Kolla mimica*, Dist.	"	Leaf	"
Grass hopper	*Acrida exalta*	"	Leaf	"

D. Proso millet—*Panicum miliaceum* L.				
Shootfly	*Atherigona miliaceae*, M.	Maggot	Growing point	
Termites	*Odontotermes* sp.	Workers	Seed to seedling	Moderate
	Microtermes sp.			
Field cricket	*Brachytrypes* sp.	Nymph & Adult	Leaf & shoot	Low
E. Barnyard millet— *Echinochloa frumentacea* (Roxb.) Link				
Shootly	*Atherigona falcata*, Thom.	Maggot	Growing point	High
White grub	*Anomala dimidiata*, Burm.	Grub	Root	Low
	Holotrichia seticollis, Mos.	Grub	Root	Low
Pink borer	*Sesamia inferens*, Wlk.	Caterpillar	Stem	Moderate
Aphid	*Hysteroneura setariae*	Nymph & Adult	Leaf & Shoot	Low
Leaf caterpillar	*Euproctis* sp.	Caterpiller	Leaf	Low
Grasshopper	*Acrida exalta*, Wlk.	Nymph & Adult	Leaf	Low
F. Finger millet—*Eleusine coracana*, Gaertn.				
Pink or stem borer	*Sesamia inferens*, Wlk.	Caterpillar	Stem	High
White stem borer	*Saluria inficita*, Wlk.	"	"	Low
Sorghum stem borer	*Chilo partellus*, Swin	"	"	Moderate
Earhead caterpillar	*Cryptoblabes gnidiella*, M.	"	Ear	High
	Heliothis armigera, Hb	"	Ear	High
	Eublemma silicula, Swin	"	"	Moderate
	Cacoecia epicyrta, Meur.	"	"	High
Red hairy caterpillar	*Amsacta albistriga*	"	Leaf	Low
Aphid	*Hysteroneura setariae*, Them.	Nymph & Adult	Leaf, stem, ear	High
Root aphid	*Tetraneura nigriabdominalis*, Sasaki	"	Root	Moderate
Leaf hopper	*Cicadulina bipunctella bipunctella* Matsumra	"	Leaf	Mod-high
	Cicadulina chinai, Ghauri	"	"	"
Surface grasshopper	*Chrotogonus* sp.	"	"	Low

SEEDLING PESTS

Shootfly (*Atherigona* spp.)

NATURE OF DAMAGE

Shootfly is the major seedling pest of small millets. Infestation usually begins during the seedling stage (1-5 leaf stage) but may be associated with older plants. The maggot causes dead hearts and in some cases the seedlings are killed. Tillers are produced excessively after late attacks. Damaged tillers may produce earheads, but with no grains (white ears). Maximum incidence occurs during late July or early August. Extreme temperatures and continuous rainfall adversely affect fly activity.

ECONOMIC LOSSES

A number of shootfly species attack small millets (Jotwani *et al.*, 1969; Singh and Dias, 1972; Nageshchandra and Musthak Ali, 1983a). *Atherigona destructor* M. alone could bring an yield loss of 36 per cent in proso millet (Natarajan *et al.*, 1974) and 39 per cent in case of little millet (Selvaraj *et al.*, 1974). Per cent reduction in yield to the extent of 44.9 in barnyard, 90.9 in proso, 78.5 in little, 35.0 in kodo and 1.8 in foxtail millets was reported by Nageshchandra and Musthak Ali (1983b).

SOURCES OF RESISTANCE

Besides preliminary studies in which a few tolerant lines of little millet (Shirole *et al.*, 1982) and proso millet (Sharma *et al.*, 1980; Murthi *et al.*, 1982) are identified, a large number of germplasm and selected varieties were systematically field screened to different insect pests at multilocations across years and locations. Lines relatively resistant against shootfly are shown in Table 2.

Foxtail millet lines (Table 3) showing relative tolerance to flea beetles, army worms, and leaf rollers have also been identified by the same authors. Some of them also exhibited multiple resistance to different insect pests (Table 4).

CULTURAL METHODS OF CONTROL

Date of planting

Early plantings with the onset of monsoon recorded low infestation of shootfly in kodo, little and proso millets with concomitant increase in yield than late planting (Table 5). Shirole *et al.* (in press) have reported that in kharif damage could be minimised by early sowings in little millet.

TABLE 2
Promising small millets relatively resistant to shootfly in India, 1980-85

Kodo millet	
GPLM No.	Variety
6, 11, 20, 21, 29, 32, 39, 42, 45, 50, 60, 106, 110, 113, 117, 119, 120, 121, 131, 142, 155, 158, 160, 170, 172, 173, 178, 180, 185	RPS 40-1, RPS 40-2, RPS 62-3, RPS 61-1, RPS 69-2, RPS 72-2, RPS 75-1, RPS 102-2, RPS 107-1, RPS 114-1, RPS 120-1, IQS 147-1, CO 2, Keharpur
Foxtail millet	
GS No.	Variety
101, 107, 110, 112, 119, 124, 128, 129, 132, 142, 150, 151, 155, 156, 157, 160, 167, 170, 172, 174, 175	RAU 1, 2, 6 ISe 119, 185, 358, 700, 700, 702, 703, SIA 5, 36, 67, 242, 326, 395, SE 21-1, SIC 1, 2 CO 3.
Little millet	
GPMR No.	Variety
7, 17, 18, 20, 22, 26, 46, 53, 78, 84, 92, 98, 101, 104, 106, 107, 112, 114, 115, 116, 117, 124, 132, 134, 136, 141, 148, 149, 163, 169, 170, 171, 172, 175	PRC 2, 3, 7, 8, 9, 10, 11, 12 RPM 1-1, 8-1, 12-1, 41-1 RAU 1, 2, K 1, CO 2, Dindori 2-1
Proso millet	
GPMS No.	Variety
101, 102, 105, 108, 112, 114, 115, 117, 122, 123, 124, 125, 126, 135, 136, 138, 148, 152, 153, 155, 156, 157, 159, 164	RAUm 1, 2, 3 MS 1307, 1316, 1437, 1595, 4872. PM 29-1, BR 6, CO 1
Barnyard millet	
GECH No.	Variety
102, 106, 108, 111, 120, 123, 127, 142, 149, 151, 157, 180, 205, 210, 218, 224, 226, 227, 230, 235, 236, 240, 241, 246, 247, 248, 250, 260, 276, 288	VL 8, 13, 21, 24, 30, 31, 32 ECC 19, 18, 20, 21 RAU 7, KE 16, K 1, PUNE 2386 Bhageshwar Local-2

TABLE 3
Foxtail millet entries found resistant to insect pests in India, 1980-85

Flea beetles	Army worms	Leaf rollers
Germplasm GS No.		
2, 12, 33, 47, 62, 64, 73, 89, 101, 111, 116, 117, 118, 123, 125, 129, 157, 167, 168, 170, 179, 182, 201, 213, 219	12, 29, 39, 102, 103, 104, 116, 117, 123, 125, 138, 157, 167, 168, 169, 198, 201, 219	26, 39, 73, 101, 121, 123, 126, 128, 137, 144, 170
Variety		
SIA 1432, 1557, 1583, 1720, 2423, 2424, 2425, SE 21.1, TNAU 18, TNAU 82, Chitra	SIA 1557, 1583, 1720, 2423, 2424, 2425, 2425, SS 21-1, ITS 69, SIC 31	SIA 1432, 2423, 2424, 2425, SE 21-1, SIC 28

TABLE 4

Foxtail millet entries showing multiple resistance to many insect pests in India 1980-85

Entry	Insect pest
GS 123	Flea beetle, army worm and leaf roller
GS 12, 125, 157, 167, 201, 219	Flea beetle and army worm
GS 101, 170	Flea beetle and leaf roller
SIA 2425, 2424, 2425	Flea beetle, army worm and leaf roller
SE 21-1	Flea beetle, army worm and leaf roller
SIA 1557, 1583, 1720,	Flea beetle and army worm
SIA 1432	Flea beetle and leaf roller

TABLE 5A

Effect of planting date on shootfly and yield in kodo millet, 1980-81

Sowing date	Variety RPS 76			Variety 40-1		
	(a)	(b)	(c)	(a)	(b)	(c)
July 5	11.5	13.2	26.5	7.0	8.0	19.0
July 15	14.2	12.6	21.2	9.8	11.7	14.3
July 25	21.1	23.6	12.5	13.2	13.8	12.4
August 4	34.9	30.1	8.0	25.4	27.6	7.2

(a) percentage of dead hearts 14 days after germination, (b) 28 days after germination, and (c) grain yield in q/ha

TABLE 5B

Effect of planting date on shootfly incidence measured by dead hearts (DH), white ear, and grain yield

Little millet 1983-85			Proso millet 1982			
Sowing date	DH (%)	Yield (q/ha)	Sowing date	DH (%)	White ear (%)	Grain yield (q/ha)
July 10	18.9	4.1	March 19	10.1	6.1	16.7
July 20	26.9	2.7	March 27	10.0	11.5	15.3
July 31	36.3	1.5	April 2	10.8	11.8	12.4
August 10	39.7	1.6	April 9	11.8	12.3	11.5
August 20	42.2	0.8	April 15	16.8	15.0	11.1

Plant density

Closer spacing may either favour some pest species or may increase the effectiveness of the natural enemies in reducing the pest populations. Low plant densities contributed to decrease in the incidence of shootfly in kodo and little millets (Table 6). Davis and Seshu Reddy (1982) have reported that a higher plant density increases number of shootflies, eggs laid and plants attacked in sorghum. The density of plants therefore has a significant effect on oviposition by the fly.

TABLE 6
Effect of plant spacing on shootfly incidence

	Kodo millet 1981		Little millet, 1982-84	
Spacing (cm)	Dead hearts (%)	Grain yield (q/ha)	Dead hearts (%)	Grain yield (q/ha)
5.0	24.7	11.2	30.9	40.0
7.5	18.3	10.7	34.0	4.1
10.0	15.7	9.7	19.3	3.4

Fertilizer use

The use of fertilizers to enhance plant nutrition often influenced the longevity and fecundity of insects and mites and the damage they cause as reported by the U.S.A. National Academy of Sciences (1969). Sowing with no nitrogen recorded low incidence of shootfly in kodo and little millets in India (Table 7). Similarly, Singh and Shekavat (1964) found that the percentage of maize plants infested by *Chilo partellus* and *Sesamia inferens* was the least with no nitrogen but increased as the level of nitrogen increased.

TABLE 7
Effect of fertilizer on the occurrence of shootfly

Kodo millet 1984-85			Little millet 1984-85		
N-P (kg/ha)	Dead hearts (%)	Yield (q/ha)	N-P (Dose) (kg/ha)	Dead hearts (%)	Yield (q/ha)
0-0	26.0	7.9	0-20	9.5	2.4
10-0	30.3	8.6	10-20	12.0	4.7
20-0	36.0	9.3	20-20	13.8	4.7
20-20	32.2	10.6	4-20	20.4	5.4

Intercropping

Modification of the micro-environment in intercropping and differences in nutrient uptake by the intercrops may influence plant infestation and the development and movement of insect pests. Intercropping of soybean and radish (Table 8) in little millet substantially reduced the shootfly occurrence as compared to sole millet or millet mixed with french bean, ladies finger or pigeonpea. These findings are in conformity with the work of Singh and Singh (1974, 1977) who showed that the presence of mung bean (*Vigna radiata*) or urd bean (*Vigna mungo*) and pigeonpea reduced the succession and build up of insect pests in sorghum and pearl millet.

TABLE 8

Shootfly damage and grain yield of little millet in inter cropping, 1983

Little millet intercropped with:	Row ratio*	Dead hearts (%)	Grain yield (q/ha)
Pure Millet	—	38.1	1.9
Ladies finger	2:1	18.1	2.1
Ladies finger	3:1	13.4	2.5
Frenchbean	1:1	13.2	2.2
2R LM + 1R Frenchbean	2:1	16.1	2.4
1R LM + 1R Radish	1:1	12.7	1.4
2R LM + 1R Radish	2:1	13.6	1.7
1R LM + 1R Soybean	1:1	11.3	1.8
2R LM + 1R Soybean	2:1	11.6	2.0
1R LM + 1R Pigeonpea	1:1	25.3	1.7
2R LM + 1R Pigeonpea	2:1	22.7	2.0
3R LM + 1R Pigeonpea	3:1	16.2	2.1
SE		3.8	10
CD 5%		8.4	21

*Row ratio number of rows of millet followed by number of rows of the inter crop.

Weeding

This practice involves the removal or destruction of weeds to eliminate the pest or deny it food and shelter. Elimination of weeds reduced shootfly infestation in kodo millet (Table 9).

CHEMICAL CONTROL

Soil application of phorate 1 kg a. i/ha in furrow is effective in checking shootfly infestation in kodo and little millets and gave higher yields. Carbofuran (1 kg a. i/ha) and quinolphos 5 G (2 kg a.i./ha) as soil treatments were reported effective against shootfly in kodo millet (Raghuwanshi and Rawat, 1985). Carbofuran 3G @ 1.5 kg a.i./ha as soil application was most effective in reduc-

TABLE 9

Effect of number of weedings on shootfly and yield in kodo millet, 1981

	Dead hearts (%) Variety		Yield (q/ha) Variety	
Number of weedings	RPS 76	RPS 41-1	RPS 76	RPS 41-1
3	1.3	1.4	11.6	10.8
2	3.6	2.4	9.8	8.3
1	9.1	5.8	8.2	6.9
0	7.8	5.9	5.4	4.5
SE	0.7		0.6	

ing shootfly incidence in proso millet. Methyl demeton, phosphamidon and quinolphos effectively reduced shootfly infestation in little millet by 70, 65 and 63 per cent and recorded 42, 103 and 80 per cent more grain yield than the control. Endosulfan and BHC dusts also reduced shootfly incidence by 67 and 60 per cent and gave 300 per cent more grain yield than the control in little millet. In barnyard millet, quinolphos, methyl demeton and phosphamidon were effective in minimising shootfly infestation. Phosphamidon effectively reduced the incidence of dead hearts and white ears, besides recording 46 per cent more grain yield than the control in proso millet. Carbofuran 5 per cent as seed treatment besides checking shootfly in little millet was found compatible with biofertilizers (*Azotobacter* and *Azospirillum*). The shootfly infestation was reduced from 78 to 7.5 per cent and the grain yield was increased from 1.7 to 11.6 q/ha (Table 10).

TABLE 10

Effect of carbofuran seed treatment and biofertilizer on shootfly and grain yield in little millet, 1983-84

Treatment	Dead hearts (%)	Grain yield (q/ha)
5% Carbofuran + *Azospirillum* + *Azotobacter*	7.5	11.6
5% Carbofuran + *Azospirillum*	13.3	9.8
5% Carbofuran + *Azotobacter*	19.1	8.8
5% Carbofuran	21.6	8.6
Azospirillum + *Azotobacter*	33.2	3.8
Azospirillum	40.6	3.3
Azotobacter	46.1	3.0
Control	78.9	1.7
SE	1.4	0.5
CD 5%	4.2	1.4

BIOLOGICAL CONTROL

After a survey of natural enemies, Kundu and Kishore (1971) reared out an eulophid wasp, *Aprostocetus* sp. from *Atherigona falcata*.

Insecticides were reported to be the most effective tools for controlling sorghum shootfly by Gupta and Pareek (1976). However, a harmonious combination of early sowing, judicious use of available chemicals combined with resistant varieties and effective parasites would be the appropriate solution in shootfly management.

PLANT PESTS

Stem borers on finger millet

ECONOMIC LOSSES

There are three stem boring caterpillars—pink borer, white borer and sorghum stem borer—in central and southern India. The borer of economic importance is probably pink borer. Krishnamurthi and Usman (1952) made detailed studies of the life cycle and economic loss caused by this pest to the finger millet in Karnataka. Infestation to the extent of 1 to 6 per cent was observed by them.

NATURE OF DAMAGE

The borers cause damage by initial feeding on the leaves and leaf whorls of plants and then by tunnelling the stem they cause dead hearts.

SOURCES OF RESISTANCE

Lingappa (1979) screened 248 varieties of finger millet and found three of them (IE 932, 982 and 1037) to exhibit resistance to pink borer. He detected more vascular bundles in susceptible lines. In resistant screening programme, 8 genotypes viz., PES 9, 144, 224, KM 1, 14, HR 228, JNR 1008 and T 36-B were observed to be resistant against pink borer (Kishore and Jotwani, 1980). Germplasm screening had demonstrated differential susceptibility to pink borer attack. Twelve lines showed resistance to pink borer (Table 11). Late varieties had greater incidence of pink borer and grey weevils than early and mid-late varieties.

CULTURAL CONTROL

The cleaning up and destroying of crop residues as well as alternate hosts help in greatly reducing the incidence and economic loss from stem borers. Borer population could also be minimised by timely removal of infested plants.

CHEMICAL CONTROL

Borers being internal feeders are difficult to control. The eggs are laid on or under leaf blades. The young larvae are extremely vulnerable to insecticides

TABLE 11

Promising finger millet genotypes resistant to different pests, 1980-86

Pink borer	Earhead worms	Aphids
KM 1	Indaf 7	PES 176
RAU 1	Indaf 8	RAU 1
RAU 3	PR 202	HR 374
Indaf 7	PR 177	
Indaf 8	HR 374	
HR 374	HR 1523	
HR 1523	PES 110	
HR 154	PES 1877	
PES 110	TNAU 1877	
PES 400	TNAU 294	
WR 9	VL 110	
VL 110		

between hatching and stem tunnelling stages. Several chlorinated hydrocarbons, organo-phosphates and carbamates (e.g. BHC 10 per cent, endosulfan 4 per cent, malathion 4 per cent, carbaryl 5 per cent) have been shown to be highly effective in reducing borer infestation when properly applied during early stages of larval development especially in leaf whorls. In leaf whorl application only 12.5 kg/ha insecticidal dust is required as compared to 20 to 25 kg/ha recommended for foliar dusting. Granules of any one of the above mentioned chemicals could also be used for whorl application.

BIOLOGICAL CONTROL

A few parasites like *Apanteles flavipes*, *Bracon chinensis*, *Stenobracon* sp. have been reported (Krishnamurthi and Usman, 1952, 1954).

Earhead caterpillars on finger millet

NATURE OF DAMAGE

A complex consisting of a number of lepidopterous larvae infests the earheads at the maturity stage of the grains. Total damage varies considerably with the variety, the season and other factors. The more compact or tight-fisted cultures are generally more susceptible to the attack as they provide a congenial microclimate for the worms to multiply or to hide within the closed head.

ECONOMIC LOSSES

In Karnataka, Puttarudraiah and Channabasavanna (1950) identified *Cacoecia* sp. as causing severe damage to developing finger millet ears. David *et al.* (1962) recorded 5 different caterpillar species at Coimbatore, of which

Cryptoblabes sp., *Eublemma* sp. and *Heliothis* were found to cause extensive damage (26 to 38 per cent).

Management of earhead caterpillars

SOURCES OF RESISTANCE

Jotwani (1978) screened high yielding lines of finger millet and concluded that early maturing strains are less susceptible. A number of finger millet genotypes were field screened and 11 lines showing less susceptibility to ear worms were identified (Table 11). The earhead worms and leaf rollers were more on early than mid late and late varieties.

CHEMICAL CONTROL

Insecticidal trials carried out revealed that carbaryl, malathion, endosulfan, quinolphos and BHC dustings were quite effective in reducing the infestation of ear worms (Table 12).

TABLE 12
Chemical control of earhead worms in finger millet, 1980-84

Dust formulation (15 kg/ha)	Percentage reduction	
	2 days after	8 days after
Carbaryl 5%	70.9	92.2
Malathion 4%	69.6	73.2
Endosulfan 4%	81.7	98.8
Quinolphos 4%	78.3	87.6
BHC 10%	67.5	93.3
Control	6.9	15.7

BIOLOGICAL CONTROL

A few parasites were recorded viz., *Gonizious* sp., *Apanteles* sp. and *Phanerotoma* sp. (Anonymous, 1975) on *Cryptoblabes*. Gahukar and Jot wani (1980) reported that naturally occurring predators and parasites keep *Heliothis* population under check.

Aphids

NATURE OF DAMAGE

The rusty plum aphid, *Hysteroneura setariae*, is often found infesting the leaves, stem and shoots in large numbers. Yield loss could be as high as 50 per cent in nursery stage (Nageshchandra, 1981).

Management of apnids

SOURCES OF RESISTANCE

Field screening revealed three genotypes of finger millet to be relatively less susceptible (Table 11). Aphids were found in higher frequency on midlate than early and late varieties.

CHEMICAL CONTROL

Endosulfan, penthoate, carbaryl (dusts), phosphamidon, dimethoate and monocrotophos (sprays) checked aphid incidence by next day, besides enhancing grain yields by 6.8 to 36.1 per cent (Table 13). Carbaryl dust controlled aphids by 15th day and had given 30.3 per cent more yield.

TABLE 13
Chemical control of aphids in finger millet, 1979-84

Insecticide	Dose	Grading			Grain yield	
		Before	1 day after	15 days after	Q/ha	% control
Carbaryl 5% D	15 kg/ha	3.5	0.8	0.0	40.4	130.3
Endosulfan 4% D	15 kg/ha	2.8	0.0	0.0	38.4	117.4
Penthoate 4% D	15 kg/ha	3.8	0.0	0.0	37.8	121.9
Malathion 4% D	15 kg/ha	2.7	0.4	0.4	—	—
BHC 10% D	15 kg/ha	3.0	0.3	0.2	—	—
Quinolphos 4% D	15 kg/ha	2.6	0.2	0.1	—	—
Phosphamidon	0.03%	3.0	0.0	0.0	42.2	136.1
Dimethoate	0.03%	4.0	0.0	0.0	38.3	123.5
Monocrotophos	0.03%	2.8	0.0	0.0	33.1	106.8
Control	—	3.6	3.8	4.2	31.0	100.0

BIOLOGICAL CONTROL

Natural enemies which generally keep the aphid population under check include ladybird beetles (*Chilomenes sexmaculata*), the larvae of syrphid fly and the lacewing (*Chrysopa basalis*).

OCCASIONAL INSECTS WITH A POTENTIAL FOR BECOMING FUTURE PESTS

Gall midge (Orseolia sp.*)*

It is found on little and kodo millets. The spikelet infestation ranged from 3 to 50 per cent in little millet. The spikelets were found to be hypertrophied due to irritation caused by feeding of the maggots on the ovary. The pest is also of common occurrence in kodo growing areas of Madhya Pradesh.

Pink borer (Sesamia inferens)

Its occurrence on barnyard millet sometimes reached up to 30-35 per cent. Early planted crop suffered more damage. The pest is active from July to October.

Termites (Odontotermes sp. and *microtermes* sp.*)*

Proso millet suffers from the early stage when seeds are sown and even the portions above the ground are not spared. The attack is more pronounced in light sandy soils.

Leaf roller (Marasmia trapezalis)

The caterpillar, rolls the edges of the foxtail millet leaves and pastes them to form a tube and eats away the green matter from inside. Continuous wet weather seems to favour the multiplication of the pest.

Leaf hoppers (Cicadulina bipunctella bipunctella and *C. chinai)*

Leaf hoppers gained importance in finger millet in recent years because of their role in communicating finger millet mottle streak and streak viruses. The mottle streak virus is transmitted by both the leaf hoppers while the streak virus only by *C. chinai*.

Root aphids (Tetraneura nigriabdominalis)

This often appears on roots of finger millet during rabi/summer in southern India. The presence of root aphids at the base of the plant is usually indicated by the activity of ants. Up to 100 nymphs and 200 adults may be found on one plant. Dispersal is by alates and by ants (*Camponotus campressus* and *Solenopsis germinata*). The affected plants show water stress symptoms and seed setting is reduced.

A common feature of these occasional pests is that the incidence and population levels were fairly high in certain years. However, their economic significance is not fully understood. Therefore, more study is required on this aspect.

FUTURE STRATEGY

Shootfly species on small millets

The biology of different species should be studied and their population dynamics throughout the annual cycle must also be determined. The specificity of shootfly and the occurrence of alternate or collateral hosts need critical investigations. The available germplasm should be field screened in pest sick plots and then tested under artificial infestation to confirm field resistance.

Stem borers and earhead worms on finger millet

Work on different borer and ear worm species need to be intensified with regard to economic status of each species by determining the incidence and losses caused in different areas. Besides, research for natural sources of resistance must be continued.

Aphids on finger millet and gall midges and borers on small millets

Biology and ecology refinement and standardization of screening techniques and continued search for the sources of resistance deserve attention.

BIOLOGICAL CONTROL

Limited information is available on the natural enemies of pests infesting small millets. Critical studies are, therefore, required of key pests and their natural enemies. Surveys of natural enemies should be undertaken and their relative efficiency assessed.

CHEMICAL CONTROL

Under certain circumstances (epidermics), insecticides are necessary for obtaining desired level of control of pests. Obviously, insecticides will remain as a management component, but their use will have to be selective and judicious.

REFERENCES

Anonymous. 1975. Progress Report of the All India Coordinated Improvement Project.

David, B.V., T. Kumaraswami and M. Mahanasundaram. 1962. Caterpillar pests of ragi (*Eleusine coracana*) ears in South India. *Madras Agril. J.*, **49**(80): 265-267.

Davis, J.C. and K.V. Seshu Reddy. 1982. Shootfly species and their graminaceous hosts in Andhra Pradesh, India. Insect Science and Its Application. 2. In: *Sorghum in the Eighties*, **237:** 238.

Gahukar, R.T. and M.G. Jotwani. 1980. Present status of field pests of sorghum and Millets in India. *Tropical Pest Management*. **26**(2): 138-157.

Gupta, H.C. and B.L. Pareek. 1976. Shoo off the shootfly. *Intensive Agriculture*. **14:** 22-23.

Jotwani, M.G. 1978. Research Bulletin on sorghum and millets, Div. of Entomology, IARI, 102-103.

Jotwani, M.G., K.K. Verma and W.R. Young. 1969. Observations on shootflies (*Atherigona* sp.) damaging different minor millets. *Indian J. of Entomology*. **31:** 291-294.

Kishore, P. and M.G. Jotwani. 1980. Screening of some improved genotypes of the finger millet (*Eleusine coracana*) for resistance of *Sesamia inferens* and *Myllocerus maculosis*. *J. of Entomological Res.*, **4**(2): 221-223.

Krishnamurthi, B. and S. Usman. 1952. The ragi stem borer, *Sesamia inferens*, W. Bull. Dept. Mysore, Ent. Ser. No. 15: 60.

Krishnamurthi, B. and S. Usman. 1954. Some insect parasites of economic importance noted in Mysore State. *Indian J. Ent.* **16:** 327-344.

Kundu, G.G. and P. Kishore. 1971. New record of parasites of *Sesamia inferens* W. and *Atherigona nudista* R. infesting minor millets. *Indian J. Ent.* **33:** 466-467.

Lingappa, S. 1979. Development of artificial diet for mass rearing *Sesamia inferens* W. (Lapidoptera: Noctuidae) and screening for resistance to finger millet germplasm. *Mysore J. Agric. Sci.* **8**(3): 353.

Murthi, T.K., S.M. Shirole and G. Harinarayana. 1982. Screening of *Panicum* for shootfly incidence. *MILWAI Newsletter*, **1:** 7.

Murti, T.K., S.M. Shirole and G. Harinarayana. (In press). Effect of Carbofuran with *Azotobacter* on the incidence of shootfly in little millet. *J. Maha. Agric. Univ.*

Nageshchandra, B.K. 1981. Insects in ragi production. Paper presented at All India Coordinated Millets Workshop held at College of Agriculture, Tamil Nadu, Agri. Univ. Coimbatore, April 26-28.

Nageshchandra, B.K. and T.M. Musthak Ali. 1983a. Shootfly species on minor millets in Karnataka. *MILWAI Newsletter* **2:** 15.

Nageshchandra, B.K. and T.M. Musthak Ali. 1983b. Losses due to shootfly on minor millets. *MILWAI Newsletter* **2:** 16.

Natarajan, V.S., S. Selvaraj and A. Ragupathy. 1974. Assessment of loss in grain yield caused by shootfly, *Antherigona destructor* M. (Anthomyiidae: Diptera) in certain varieties of Panivaragu, *Panicum miliaceum. Science and Culture* **40:** 502-504.

Puttarudraiah, M. and G.P. Channabasavanna. 1950. Host range of *Cacoecia epicyria* M. *Indian J. Ent.* **12:** 267-268.

Raghuwanshi, R.K. and R.R. Rawat. 1985. Chemical Control of *Atherigona simplex* Thomas. On kodo millet with seed and furrow treatments. *Indian J. Agric. Sci.*, **55**(7): 468-470.

Selvaraj, S., V.S. Natarajan and A. Ragupathy. 1974. On the occurrence of shootfly and its damage in some varieties of little millet. *Indian J. Agric. Sci.*, **44:** 556-557.

Sharma, D.D., S.S. Yazdani and L.B. Choudhar. 1980. Note on the relative susceptibility of cheena (*Panicum miliaceum*) to shootfly (*Atherigona varia soccata* R.) incidence. *Sci. Cult.* **46**(11): 396-397.

Shirole, S.M., T.K. Murti and G. Harinarayana. 1982a. Shootfly incidence and its control in *Panicum miliare. MILWAI Newsletter,* **1:** 6.

Shirole, S.M., T.K. Murti and G. Harinarayana (In press). Effect of date of sowing on little millet shootfly. *J. Maha. Agril. Uni.*

Singh, V.S. and C.A.R. Dias. 1972. Occurrence of different species of *Antherigona* attacking some minor millets at Kanpur. *Entomologists' Newsletter,* **2:** 39-40.

Singh, V.B. and G.S. Shekhawat. 1964. Incidence of stem borers in maize under different fertility levels. *Indian J. Agron.*, **9:** 48-50.

Singh, K.M. and R.N. Singh. 1974. The population build up of Pyrilla perpusilla (Walker) on sorghum and pearl millet under dryland conditions at Delhi. Indian J. Ecol., **1:** 12-16.

Singh, K.M. and R.N. Singh. 1977. Succession of insect pests in green gram and black gram under dryland conditions at Delhi. *Indian J. Ecol.*, **3:** 365-370.

USA National Academy of Science 1969. *Insect pest management and control.* Publication No. 1695. Washington D.C. USA: National Academy of Sciences.

24

CROPPING SYSTEMS, PRODUCTION TECHNOLOGY, PESTS, DISEASES AND UTILIZATION OF SMALL MILLETS IN BANGLADESH

M.A. Majid, M.A. Hamid and Mannujan

INTRODUCTION

Kaon (foxtail millet) and Cheena (proso millet) are the two important small millets cultivated in Bangladesh. The acreage and production of these two crops were 36671 hectares and 31720 metric tonnes, respectively in the year 1983-84, with an average grain yield of 865 kg/ha. The low yield is due to poor soil, lack of high yielding varieties and inputs and non-adoption of improved cultural practices. Though the acreage under these millets has declined in recent years, they occupy an important position in the prevailing cropping patterns because of their short duration, wide adaptability and drought tolerance. Recognizing the role of small millets, Bangladesh Agricultural Research Institute has undertaken an integrated research programme on varietal improvement as well as improved cultural practices of these millet crops to develop a package of technology for increasing their production.

CROPPING SYSTEMS

Foxtail millet is grown both in *kharif* (monsoon season) and *rabi* (postmonsoon) either as a sole crop or mixed with other crops like chilli, aus (pre monsoon) rice or sesame. In Rangpur and Dinajpur districts foxtail millet is grown as a sole crop followed by transplanted rice. In Pabna, Kushtia and Faridpur districts foxtail millet is grown as a sole crop followed by jute. It is said that jute grows well in the soil where preceding crop was foxtail millet. Proso millet is grown sole or mixed with mustard. Millets can easily be fitted to any cropping pattern including multiple cropping system in view of their short duration and photoinsensitive nature. The common millet based cropping patterns followed in different districts of Bangladesh are:

Thakurgaon, Rangpur and Dinajpur districts: Fallow-Foxtail millet-Transplanted Aman, Potato-Foxtail millet-Transplanted Aman

Jamalpur, Tangail and Pabna districts: Foxtail millet-Jute-Wheat, Aus rice-Transplanted Aman-Proso millet

PRODUCTION TECHNOLOGY

Farmers practice

FOXTAIL MILLET

Normally, farmers plough their lands three times and use a seed rate of 9 kg/ha. The crop is sown in rabi in November/December and in kharif in February/March. Besides FYM, some farmers apply a little quantity of urea to supply nitrogen. Normally one to two weedings are done.

PROSO MILLET

The land is ploughed 3-4 times and sown using a seed rate of 8 to 10 kg/ha. Proso millet is sown in November/December as a *rabi* crop. Some farmers apply FYM if available and rarely a little quantity of inorganic fertilizer primarily in the form of urea. Normally one to two weedings are done.

IMPROVED PRODUCTION TECHNOLOGY

Field trials were conducted to find out the appropriate time of sowing, seed rate and fertilizer dose for millet crops. The findings are summarised below: Studies on different row spacings and fertilizer doses showed no significant yield differences among row spacings 15, 25, 35 and 45 cm. However, significant yield differences were observed for different doses of nitrogen. Highest grain yield of 2678 kg/ha was obtained in treatment with 90 kg N/ha. In the trial on the effect of different seed rates and fertilizer doses on the yield of foxtail millet, 10 kg seed/ha gave the highest yield of 1920 kg/ha. Highest grain yield was obtained by the application of 60:45:45 kg/ha of N, P_2O_5 and K_2O.

Seed rate and manurial studies in proso millet showed that 8 kg seed/ha with 40:30:30 kg/ha of N, P_2O_5 and K_2O as optimum and gave the highest grain yield of 2360 kg/ha.

No significant yield difference between February 15 and March 15 sowing was found in case of foxtail millet.

It has been found that application of fertilizers, optimum seed rate and seeding time considerably influence the yield of millets but their response varies with the region, soil type and season.

PESTS AND DISEASES

Survey and monitoring of insect pests damage on proso and foxtail millet revealed the following:

Stemfly (*Atherigona miliaceae*) is the major insect pest but the damage of this pest is greater on proso than on foxtail millet. The extent of damage may vary from 15-25 per cent depending on the year, location and genotype. The other minor insect pests found in the millet fields are —stripe borer, pink borer, flea beetle, aphid and pentatomid bug.

Similarly commonly found diseases on millets are: Footrot (*Sclerotium rolfsii*), leaf blast (*Pyricularia setariae*), leaf spot (*Helminthosporium* sp.), leaf and sheeth blight (*Drechslera* sp.) and grain spot (*Phoma* sp., *Fusarium* sp. and *Curvularia* sp.).

Foot rot caused by *Sclerotium rolfsii* is an important disease and mortality is high on millet. The cultivars BPm-52, Islampur and Telipara of proso millet and Parameshpur and Shibnagar of foxtail millet were found to be resistant to this disease in screening tests.

UTILIZATION

Millets are mainly grown by the poor farmers for home consumption as a supplement to meet up the rice deficit. Grains can be dehusked either fresh or after par boiling. It is cooked solo or with rice and consumed like rice with vegetables or pulses. It is also consumed in the form of chapati or bread prepared from flour. Whole fried grain or its flour is consumed after mixing with 'Gur' (molasses) and salt. Various types of delicious cakes are prepared from flour of millets. Palatable sweet dishes like '*Payesh*' and 'Firney' are made from dehusked foxtail millet mixed with milk and sugar.

The straw of millets are used as fodder, fuel and for thatching.

25

CROPPING SYSTEMS, PRODUCTION TECHNOLOGY, PESTS, DISEASES OF FINGER MILLET IN NEPAL

Deep Man Sakya

Finger millet (*Eleusine coracana* Gaertn.) is cultivated extensively in Nepal ranking fourth in terms of area and production after rice, maize and wheat. Most of the production occurs in the hills in rainfed areas up to about 2500 m above mean sea level but it is more commonly grown between 1,000 and 2,000 m. It is also cultivated to a lesser extent in the terai (lowland areas). In Nepal finger millet is most frequently grown in association with maize as a relay crop. In some areas it is cultivated as a sole crop. Yields have declined from 1100 to about 900 kg/ha in the past decade.

Agriculture in the hills is very intensive and most farmers traditionally follow intricate cropping patterns that involve sequence cropping, mixed cropping and relay cropping with a wide array of different crops to meet their subsistence needs. It is a challenge for the farmers and the agricultural researchers to devise cropping/farming systems that fit both the physical environment and of the needs of the people living in the Nepal hills.

The more commonly grown cropping patterns that include finger millet are indicated in Table 1.

1) Transplanted millet as a relay crop (maize/finger millet)

Transplanting as a relay crop in the maize field is the predominant pattern of millet cultivation. The date of transplanting varies depending on the elevation and other environmental conditions and on the cropping pattern in use. At lower elevations finger millet is transplanted in the standing maize crop when the maize is between the tasseling and the mature grain stages. Under these conditions maize is usually planted during the last week of April and harvested

TABLE 1

Predominant cropping patterns in the major ecological zones of Nepal that include finger millet

Ecological zone	Cropping pattern
Terai and inner terai (up to 600 m.a.s.l.)	Early millet-paddy
	Early maize-millet-wheat
	Maize-millet-brassica oilcrop
Mid hills (600-2000 m.a.s.l.)	Maize/millet-wheat or oilcrop
	Maize/millet-fallow
	Millet-wheat + oilcrop
High hills (2000-2500 m.a.s.l.)	Soybean + millet-wheat or barley
	Millet-barley-false cereal
	Millet-potato-wheat (in a 2 year pattern)
	Millet + amaranthus-buckwheat

Symbols and terms used
— followed by
+ mixed with
/ relayed with
false cereals are buckwheat, amaranthus and quinoa

Source: Sherchan *et al.*, 1986.

130-140 days later. At higher elevations the finger millet is frequently transplanted earlier, even before the initiation of tasseling of the maize plants. Such a procedure allows the millet to be harvested in time for the planting of a winter crop such as wheat, barley or oilseed brassica. When the maize crop is nearly mature, the lower leaves are frequently removed to allow maximum sunlight to reach the millet crop.

2) Transplanted millet as a monocrop

The growing of finger millet as a transplanted monocrop is a common practice in the terai and inner terai (1000 m.a.s.l. and below). Such a practice is more compatible with the cropping patterns in use in the terai. Monocropped finger millet yields are usually higher when grown as a mixed or relay crop. As soon as the finger millet is harvested a winter crop such as wheat, barley or oilseed is planted.

The yields of several finger millet varieties when grown as a monocrop and as a relay crop with maize are presented in Tables 2 and 3, respectively (Singh and Tamulonis, 1985).

3) Direct seeding of millet

Direct seeding is a much less popular practice of establishing finger millet crop and is practised either at low elevations or in the high hills of Nepal. In

TABLE 2

Yield of finger millet varieties when grown as a monocrop at two locations (1985/86 season)

Variety	Location		
	Khumaltar Yield (kg/ha)	Kavre Farm Yield (kg/ha)	Mean yield of both locations (kg/ha)
NE 1703-34	2422	1157	1789
NE 6401-26	1753	1341	1547
NE 1104-13	1365	1158	1261
NE 1001-1	1501	1329	1415
NE 1102-12	1142	1183	1162
NE 1304-1	901	1217	1059
Dalle-1	1762	1086	1424
Mean Yield	1549	1210	1380

TABLE 3

Yield of finger millet varieties when grown as a relay crop with maize at two locations (1985/86 season)

Variety	Locations		
	Khumaltar (kg/ha)	Kavre Farm (kg/ha)	Mean yield of both locations (kg/ha)
NE 6401-26	873	1528	1200
NE 3801-2	1559	1233	1396
NE 1304-1	664	1197	930
Okhale-1	1126	1309	1217
Mean Yield	1056	1317	1186

the terai, direct millet seeding is used in early millet-rice and millet-mustard cropping patterns.

FERTILIZATION

In the eyes of the Nepalese hill farmers, finger millet is considered as a crop that can be cultivated with low inputs and grown under stressful conditions. Fertilizer is generally not applied to the millet crop. If chemical fertilizer, farm yard manure or compost are used they will usually be applied to the maize crop. If finger millet is grown as a sole crop it is not uncommon for the farmer to apply farm yard manure at the rate of 5 to 10 tons per hectare.

HARVESTING AND POST-HARVEST OPERATIONS

Finger millet takes 120 to 200 days to mature from seeding to harvest depending on elevation, variety used and seasonal variations. Finger millet is usually harvested from late September to mid November. Harvesting is generally done in two stages. First the earheads are cut with a sickle and then the straw (stalk) is cut close to the ground. The earheads are put in piles, covered with cloth and left in the piles for 3 to 4 days. Heat develops in the piles and slight fermentation takes place. This helps the millet grain to separate easily from glumes.

The millet grain is removed from the cured earheads by hand threshing, by sticks or by driving bullocks over the earheads. Once the grain is well dried it is usually stored in locally made bins. Finger millet seed, if properly dried and stored, retains its viability for a long period of time, up to several years. It is not normally attacked by insects in storage.

DISEASES AND INSECTS

There have not been any systematic observations and assessments of finger millet diseases in Nepal because of the low priority thus far given to this crop. Based on limited surveys conducted by the Plant Pathology Division, blast (*Pyricularia* sp.) and leaf blight (*Helminthosporium nodulosum*) are the two important diseases in Nepal. Besides these two diseases, footrot and wilt, caused by *Sclerotium rolfsii,* were also recently recorded in the terai area and the incidence is sporadic.

Finger millet insect pests are also not well documented and control measures are virtually non-existent. The important insect pests are pink stem borer (*Sesamia inferens*) and grass hoppers (*Chrotogonus* sp.). Attack of army worms (*Mythimna separata*), resulting in significant yield losses, have also been occasionally observed.

REFERENCES

Singh, R.M. and John P. Tamulonis. 1985. Report on finger millet. Paper presented at the 13th Annual Summer Crop Workshop, March 10-13, Janakpur, Nepal.

Sherchan, K., T.B. Shrestha and K.O. Rachie. 1986. A Consultancy Report on Hill Crops Improvement Programme in Nepal. IDRC/ARPP, DOA, HMG, Nepal.

26

CROPPING SYSTEMS, PRODUCTION TECHNOLOGY, PESTS, DISEASES AND UTILIZATION OF SMALL MILLETS IN SRI LANKA

S. Ponnuthurai

CROPPING SYSTEMS AND PRODUCTION TECHNOLOGY

The cropping system in Sri Lanka is determined by the availability of water for crop production. Accordingly, there are rainfed and irrigated cropping systems and a third category, a combination of both. The source of water for irrigation is from (1) river diversion projects, (2) reservoirs, and (3) underground water from open dug or tube wells. Millets are cultivated in the highlands under rainfed conditions in the southern, southeastern parts of Sri Lanka and in the north central province; agro-ecologically these areas come under the dry zone, where the average annual rainfall received is from 1270-1900 mm. Likewise, millets are cultivated in parts of northwestern, central and Uva provinces, and these areas fall under the intermediate zone, where the average annual rainfall is 1900-3175 mm.

Millet cropping systems in the dry zone

In the dry zone a major rainy season occurs during the period September to January and a minor one in April to July. Finger millet and foxtail millet are grown during the major rainy season (rainfall received is 760-890 mm) under the shifting forest fallow (chena) system. The size of the holding per farming family is about one ha. The shrub jungle is burnt and the seed is broadcast sown. The soil is harrowed to cover the seed. The farmer does not use any fertilizers or practise any pest or disease control measures. Incidence of pests and diseases are rather low. Not only millets, a mixture of millets and cereals such as maize, sorghum, and vegetables are often grown under this system. Growth duration of millets is about four months. The earheads are collected as they mature in two or three harvests. Yield obtained ranges from 800-1000 kg/ha. In the minor (dry) rainy season common millet is cultivated following the same shifting forest fallow system. Rainfall received during the season is about 200 mm and not very dependable. The seed is broadcast sown and buried in soil by light harrowing. Minimum management practices are followed in this system of cultivation. Yield obtained varies from 400-700 kg/ha. Cultivation of sesame (*Sesamum indicum*) is also practised during this season. Cropping systems in these regions could be represented as follows:

Finger millet/ foxtail millet		Fallow		Common millet/ sesame		Fallow
Sept-Jan.	→	Feb.	→	Mar-July	→	August

Under the river diversion projects in the southern, southeastern parts of Sri Lanka and in the north central province; high cash value crops such as chillies, pulses (green gram), maize, soybean and vegetables are grown and millets do not generally find a place in these cropping patterns.

Millet cropping systems in northern areas

In northern Sri Lanka (Jaffna District) several cropping patterns are followed both under rainfed and irrigated conditions. Cropping intensity is relatively high in this region compared to other areas in the dry zone. Farmers grow cash crops—onions, chillies, vegetables, potato, tobacco, cassava, etc. In some of the cropping patterns cash crops are followed by one or more millets, viz., finger millet, common millet or foxtail millet. In this region, the farm holdings are just around one fourth of a hectare and several cropping patterns are followed to utilize the farm resources and the available technology in the best possible way to maximise income. There is also an animal component in most of their farming systems. A few (about 3-5) head of cattle are usually maintained on these lands to enhance the soil fertility by supplying animal wastes such as cow dung, etc. These herds are shifted at short intervals within the holding to benefit the whole area alike.

As the wet season commences a green manure crop such as sunnhemp (*Crotolaria juncea* L.) is planted. At the time of flowering, this is ploughed in and the land is planted with potato or chillies which occupies the land from about four to seven months. This is followed by vegetables, onions and often some millets. Incorporation of cattle manure and compost to the soil, in addition to the inorganic fertilizer is often practised in this intensive farming system. This practice improves moisture retention and the fertility of the soil. It is with this intention that a green manure crop is often included in the cropping pattern.

Finger millet seedlings are raised in nurseries. At the end of 25 days or so the seedlings are uprooted and transplanted at random with one plant per hill. Prior to transplanting the field is sown with the seeds of *Amaranthus* spp. Ten days after transplanting, the crop is dressed with 15 to 20 kg N/ha and this is followed with an irrigation. By this time, the Amaranthus is ready for harvest to be used as leafy vegetable. A basin system of irrigation is followed and the source of water is open dug well. The period of finger millet cultivation is from May to August.

Irrigation frequency is about once in ten days, and 7 or 8 irrigations are adequate to reap the crop. Growth duration of the crop is 3½ months. Ripening of the earheads is uneven and at least 2-3 harvests are made at weekly intervals. The earheads are well dried and hand threshed. The grain is cleaned by winnowing with the help of hand winnowers. Yield obtained is about 2000 kg/ha. At Rs. 8/- per kg of grain the income derived is Rs. 16,000/- and the net profit per ha (less cost Rs. 6,000/-) would amount to Rs. 10,000/- (1 US $ = Rs. 27/-). Millet though included in the cropping pattern is characterised by relatively lower net returns compared to that of cash crops such as potato, or chillies where the profit is Rs. 40,000/- to Rs. 50,000/- per ha. Millets straw is used as cattle feed. Finger millet is often planted as an intercrop when new banana plantations are established. This intercrop matures well ahead of the banana establishment and form a ground cover. (A. Senthinathan, personal communication).

Availability of much underground water in the region has made possible the growing of a number of crops the year round. Farm power, particularly for irrigation is costly. Land preparation and other operations are generally done manually with both family and hired labour. In the irrigable highlands, cropping pattern followed is as follows:

Green manure	→	Potato/chillies	→	Vegetables (Egg plant, okra, tomato, pole bean cabbage, etc.)/onions/tobacco /pulses/millets and cassava.
Sept-Nov		Nov-May		April-August

In the low lying areas rainfed rice is cultivated during the period September to January, As the rice crop is harvested, the residual moisture is often made use of to raise a crop of pulses such as green gram, cowpea, millets or sesame; sometimes a mixture of some of these crop species are grown, during the period February-April. Thereafter the land lies fallow. Possible cropping pattern for the low lying and less intensive cropping areas are as follows:

Rainfed Rice	→	Pulses, grain legumes, millets, sesame, etc	→	Fallow
Sept-Jan		Feb-April		May-Aug.

PESTS AND DISEASES

In the cultivation of millets, the damage or crop losses due to pests and diseases is generally low. One of the reasons for the low incidence of pests and diseases is the type of cultivation practised. In the shifting forest fallow system farmers pay less attention in growing this crop. Burning of the shrub jungle prior to the cultivation helps to get rid of the insect pests and the disease pathogens. Sporadic cultivation does not favour the multiplication and carry over of the pests and diseases. After the cultivation of millets in the shifting forest fallow system, the land lies fallow and the chances of survival of the insect pest or the disease pathogen is less. Adherance to basic agronomic practices often help to prevent or decrease the severity of the damage caused.

Pests

Millets grown in the wet season, namely finger millet and foxtail millet sometimes suffer damage caused by leaf eating caterpillars and grass hoppers. Army worms (*Spodoptera marutia*) feed on leaf and at times cut the seedlings at the base. Extent of defoliation caused by this pest depends on the severity of infestation. Both nymphs and adults of grass hoppers feed on leaves. They also damage the earheads. Dusting with gamma BHC 10 per cent D at the rate of 20 kg/ha or carbaryl 85 WSP at the rate of 2.0 kg/ha controls the pest effectively. Stem borer damage to the crop occurs occasionally and the losses are not appreciable. Presence of the pest is identified by appearance of "dead heart" at the vegetative growth stage and the chaffy earhead with near erect fingers at the reproductive growth stage. Incidence of this pest has been observed both in the wet and dry season. Application of a systemic insecticide provides adequate control. Presence of aphids (*Aphis* spp.) has been observed in finger millet. Sometimes the infestation is very high and the plant parts affected are earheads, culms and the base of the culm near the soil surface. This can be considered as one of the major pests of finger millet. Aphid damage is observed both on the wet and dry season crop. Higher relative humidity and overcast skies are favourable environmental conditions for their multiplication. Affected plant at times becomes stunted. Earheads damaged by aphids produce poor quality grains. Some species of ladybird beetle and other insects have been found feeding on aphids. This pest is controlled by the timely application of a systemic insecticide (P. Mylvaganam, personal communication).

Bird damage (especially parrots) is often observed on foxtail millet at the time of maturity. Damage to growing seedlings by wild rabbits is also observed when finger millet is cultivated near shrub-lands.

Diseases

Blast and helminthosporium fungal diseases are often seen on finger millet and foxtail millet, in the wet season when favourable environmental conditions prevail. Careful agronomic practices help to minimise their incidence. Seed treatment, plant spacing and regulating the amount of nitrogenous fertilizer applied to the crop are important measures to minimize the occurrence of these diseases. Smut disease is seen on the earheads of common millet and this can be controlled by treating the seed with copper fungicide (Seneviratne and Appadurai, 1966).

UTILIZATION AND FORAGE USE

Millets form an item of supplementary food and is a poor man's crop. The grain is ground to flour like rice or wheat and tasty preparations such as porridge, 'pittu', string hoppers and 'roti' are made. Prawns (shellfish) and sometimes vegetables and rice is added to the porridge. While making roti' and 'pittu', coconut scrapings are added to the millet flour. Millet flour is mixed with wheat flour and used for making cake. Diabetic patients sometimes eat finger millet preparations instead of rice. Common millet 'rice' served with curries is very tasty indeed, and so is the porridge prepared from common millet 'rice'. Common millet 'biryani" is a delicacy. Hoppers, string hoppers which are common Sri Lankan foods, and all kinds of local sweet meats can be prepared from common millet flour (White, 1943). Some of the millets are better balanced food than rice, by virtue of their higher protein, fat and mineral contents. Food value of different millets is presented in Table 1.

Millets are sometimes used as poultry feed. Millets straw is usually relished by cattle when it is green and tender. Foxtail millet grains are sometimes used to feed pets such as parrots and other small birds that are kept in captivity.

ACKNOWLEDGEMENTS

Grateful thanks are due to the Director of Agriculture and the Deputy Directors of Research, Peradeniya, Gannoruwa and Kilinochchi for the encouragement and affording this opportunity to participate in this workshop. Thanks are also due to the IDRC for providing the necessary financial assistance.

TABLE 1
Food value of millets (%) compared with rice

Name	Moisture	Protein	Carbohydrate	Fat	Fibre	Mineral	Calcium	Phosphorus	Calorific value (100 g)
Eleucine coracana Finger millet (Kurakkan)	13.0	8.0	72.0	1.3	3.0	2.7	0.3	0.3	332
Panicum miliaceum Common millet (Panichchamai/Panivaragu/Meneri)	11.1	13.71	72.26	1.76	0.10	1.07	0.01	0.2	341
Setaria italica Foxtail millet (Thinai/Thanahal)	11.9	9.7	72.4	3.5	1.0	1.5	0.04	0.3	353
Panicum miliare Little millet (Samai/Heen meneri)	11.1	13.4	72.3	1.8	0.1	1.1	0.02	0.3	360
Paspalum scrobiculatum Kodo millet (Varagu/Amu)	11.6	10.6	59.2	4.2	10.2	4.4	0.04	0.3	346
Eragrostis tef Teff	11.2	9.1	74.3	2.2					
Echinochloa frumentacea Japanese barnyard millet	11.9	6.2	65.5	2.2	9.8	4.4			

Digitaria exilis									
Hungry rice	6.0	8.7	81.0	1.1	1.1	2.1			
Oryza sativa									
Rice (polished)	13.2	7.5	76.7	1.0	0.3	1.6	0.01	0.17	348

Source: Purseglove, J.W. (1972 revised 1985); Food Technologist's report: Department of Agriculture, Sri Lanka.

REFERENCES

Seneviratne, S.T. and R.R. Appadurai. 1966. *Field Crops of Ceylon*, Lake House, Colombo.

White, J.S.I. 1943. *Panicum miliaceum* (meneri). A promising short aged food crop. *Tropical Agriculturist*, Vol. XXIX, No. 4, pp. 207-211.

Purseglove, J.W. (1972, revised, 1985). *Tropical Crops, Monocotyledons.* Longman.

Food Technologists Report, Department of Agriculture, Sri Lanka (1986).

27

CROPPING SYSTEMS, PRODUCTION TECHNOLOGY, PESTS AND DISEASES OF FOXTAIL MILLET IN CHINA

Chen Jiaju

INTRODUCTION

Foxtail millet has been grown in China for thousands of years. The technical aspects of production are well known to Chinese farmers and in the past 30 years, new technology has been introduced into agricultural production. Replacement of traditional cultivation methods can only succeed when proven economic effects are significant. The following are some of the main improved management practices in the production of foxtail millet in China.

CROPPING SYSTEMS

In China, foxtail millet is distributed in temperate and cool regions, where one crop in a year, 3 crops in two years and two crops in a year are possible. In order to increase the multiple crop index (cropping intensity), summer millet was popularized during the past 20 years. Of the total millet-growing area the share of spring millet is 85 per cent and summer millet is 15 per cent. Spring millet is usually grown after corn, sorghum, soybean or spring wheat; while summer millet is grown after winter wheat, barley, pea or rape. Repeated cropping of foxtail millet will cause decrease in yield. Therefore one crop of foxtail millet at an interval of three years has proven to be beneficial. Intercropping is not very popular but when adopted, millet occupies at least 2/3 of the area in the field to ensure sufficient light for the millet growth.

CULTIVATION PRACTICES

Land preparation

In order to facilitate the emergence of seedlings from the tiny seed, the plough layer should be finely harrowed. In semi-arid regions, soil moisture is the main restricting factor in the spring. The major techniques for keeping the soil moisture are deep ploughing to enhance seepage of autumn rain, soil compression in the winter and surface soil harrowing in the early spring. This important practice is the result of experience of farmers of Shanxi Province. Non-ploughing of the fields for summer seeding is necessary to avail an early seeding date.

Seeding

The choice of a proper seeding date is based on soil moisture and the growth durations of the varieties planted. In the past, say 30 years ago, the seeding dates began in April, but now they are generally delayed up to May. Tools for sowing vary with different cultural backgrounds. Drill sowing machines are widely used. Some new sowing techniques have been developed in Northeast China where traditionally wide row spacing is practised. One of the techniques is to sow seeds on the ridges in a wide strip or in 2-3 narrow strips. Hole (hill) seeding was developed for selected seeds to reduce the seed rate. Generally, 11 to 15 kg seeds per hectare is required. Seed placing depth is between 3 and 5 cm and, surface soil compression is needed after seeding.

Thinning

Thinning of seedlings is necessary as relatively large quantity of seeds are sown. Thinning is usually done about 20 days after emergence when seedlings have 4 to 5 leaves. Hand thinning is very slow and only 100-200 m^2 can be done per day. Seedlings are thinned leaving mounds of 3-5 plants spaced at 15 to 20 cm, or single plants in rows. Simplification of thinning method is one of the most important problems that researchers are working with.

Planting density

In the past, it was commonly accepted that low density would produce large ears. But when higher yield is required, higher density is beneficial, so adequate increase in planting density has been encouraged. The population maintained per unit area in different regions are summarized in Table 1.

Management during growing stage

The critical stage for applying fertilizer and water is from floret differentiation to microsporogenesis. Ammonium nitrate (NH_4NO_3) is applied at the rate of 150 kg/ha for normal production and for high yield field, more than 225 kg/ha can be applied at two stages, first at the start of internode elongation

TABLE 1
Planting density followed in different regions of China

Region	Mode of planting and fertility status	Plant population per hectare
Northeast	Ridge space 60 cm, 2 to 3 strips, low to moderate fertility	600-680 thousand 750-900 thousand 1.4-1.5 million
North and Northwest China	Row space 30-50 cm, low fertility	300 thousand
Inner Mongolia	Moderate fertility Summer sown, early, very early variety	370-500 thousand 750-900 thousand 1-1.2 million

and again during the booting stage. Manuring and watering are combined with inter-tillage. The detection of associations among growth, development and leaf index have helped in developing efficient management methods.

High yield and integrated cultural technology

There are many successful examples; the Jinzhuany brigade which is located in a mountainous area in Shanxi Province has 26.7 hectares of spring millet. The yield has been stabilized at 6 tonnes/ha since the 1970's. In some limited areas, 8.8 tonnes/ha has been obtained. In 1985, this experience was extended to 200 thousand hectares in Shanxi Province and the average increase of millet was 300 kg/ha. In the eastern part of inner Mongolia, the average yield of spring millet was 6.3 tonnes/ha for a total of 86.6 hectares in irrigated fields. In Shandong Province, standardized technology for foxtail millet cultivation was extended to summer cropping and the average yield reached 3.8 tonnes/ha for a total area of 267 thousand hectares. A book named "The Cultivation of Foxtail Millet in China" contains all the achievements and efforts of thirty years and written by a group of researchers from all over China. This book will soon be published.

DISEASE AND PEST CONTROL

Foxtail millet is susceptible to many kinds of diseases and pests. In epidemic years, the yield losses are severe. The main diseases and pests reported in China are given below:

Diseases

Blast (*Piricularia setariae* Nishik)
Leaf rust (*Uromyces setariae-italiae* (Peit) Yoshina)
Brown stripe (*Pseudomonas setariae* (Okabe) Savalesum)

Leaf spot (*Helminthosporium setariae* Sawada)
Downy mildew or green ear (*Sclerospora graminicola* Sacc. Schrect)
Kernel smut (*Ustilago crameri* Koern)
Red leaf virus disease
Nematode (*Aphelenchoides besseyi* Christic)

Pests

Mongolian mole cricket (*Gryllotalpa unispina* Saussure)
African mole cricket (*Gryllotalpa africana* Palisot et Boauvois)
False wireworm (*Opatrum sabulos* Linnaeus)
European corn borer (*Ostrinia furnacalis* Guenee)
Millet stem borer, yellow sugarcane borer (*Chilotrate infuscatellus* Snellen)
Millet flea beetle (*Chaetocnema ingenus* Baly)
Millet stem maggot (*Atherigona biseta* Karl)
Army worm (*Mythimna separata* Walker)
Indian mallow bug (*Liorhyssus hyalinus* Fabricius)
Foxtail millet webworm (*Mampava bipunctella* Ragonot)

Integrated control of diseases and pests is the main traditional method. However, chemical control is also used extensively. Biological control has been just started. Some of the methods followed in integrated pest and disease control are as follows:

1) Rotation

Crop rotation not only checks the soil borne diseases by preventing their multiplication, accumulation and spread but also helps in weed control particularly giant foxtail (*S. viridis* var. *major*).

2) Seeding date adjustment

Delaying the seeding date up to May is effective in controlling green ear, millet stem borer, etc. Summer millet should be seeded on an earlier date to avoid millet stem maggot.

3) Growing resistant varieties to diseases and insect pests.

4) Seed treatment and field control

Seeds are treated after drying under the sun, clearing all soot by fungicides. Metalaxyl (Ridomil) is a highly effective and low toxic fungicide for controlling green ear at an efficiency of more than 90 per cent. Phoxim, Bayer 5621 and Phorate 3911 can be used to fumigate the seed and Lindane mixed with seeds controls soil borne pests.

Phoxim and carbofuran are used to protect millet from stem borer, trichlorofen is used against army worm and dimethioate Acc-12880 for aphids.

The red eye bee is now commercially cultured and released into fields to control European corn borer and millet stem borer.

V

CROPPING SYSTEMS, PRODUCTION TECHNOLOGY, PESTS AND DISEASES OF SMALL MILLETS IN AFRICA

28

CROPPING SYSTEMS, PRODUCTION TECHNOLOGY, PESTS AND DISEASES OF FINGER MILLET IN UGANDA

J.P.E. Esele

IMPORTANCE AND DISTRIBUTION

Finger millet is an important cereal in Africa and Asia. In Uganda, finger millet is the second most important cereal after maize with an estimated annual area of over 330,000 hectares and annual production of over 450,000 metric tonnes (Table 1). It is grown throughout the country but the major production areas are in the interior plateau, between Mt. Elgon and Lake Kyoga, Northern and Western Uganda. Production reaches its maximum in Tesco, Bukedi and in the highlands of Kigezi. The crop is produced purely at subsistence level and has not attained commercial importance.

TABLE 1

Area and production of the major food crops in Uganda (1981-1984)*

Food crop	1981 A	1981 P	1982 A	1982 P	1983 A	1983 P	1984 A	1984 P
	Area (A) and production (P) in '000 ha and '000 mt							
Maize	260	342	285	393	295	413	347	281
Finger millet	300	480	330	528	341	545	332	223
Sorghum	170	320	200	258	207	407	206	164
Cassava	310	300	331	313	372	324	401	188

**Source:* Ministry of Agriculture and Forestry, Entebbe, Uganda.

Finger millet is not important in the dry Karamoja area where sorghum is the principal food crop. Finger millet is the principal food grain of the Nilotic and Nilo-Hamitic tribes. It also forms an important item of the diet for the Bantu people in the western region and along the Nile in the north (Thomas, 1970). It is consumed as unleavened bread (ugali), porridge and used in making home brewed beer. Beer is regularly consumed by family members but also served to guests and in all ceremonial functions.

Although there is a close relationship between finger millet distribution and its association with different tribal groups, the extent of its production is decided by the availability of other foods. In areas of high rainfall, especially around Lake Victoria, crops like banana, sweet potato and maize predominate. The main zone of finger millet production is not suitable for other crops where conditions are more harsh either in terms of rainfall reliability or soil fertility. However, finger millet is gradually getting replaced by sorghum. This change is common in the north and northeast, but the proportion of each cereal will vary in the transitional periphery from a pure finger millet stand to a pure sorghum stand. In Kigezi, finger millet is gradually getting replaced by sorghum mainly because of higher yields of sorghum than finger millet on the fertile volcanic soils.

AGRONOMIC PRACTICES

Land preparation and place in rotation

Finger millet does best on relatively fertile soils but not on clay loams which have poor drainage. It grows best on well drained sandy loams particularly in areas where rains are well distributed during the growing season without prolonged droughts. It is grown in areas of 500-1000 mm rainfall and up to an elevation of 2400 m. It requires a fine seedbed.

Most commonly, finger millet follows cotton in crop rotation. In areas where cotton is not grown, millet follows sweet potato or groundnut or cowpea or sesame. In the cotton-millet rotation, the system is as follows:

May-December	Cotton
January-May	Millet
June-January	Cotton
February-June	Millet
July-September	Cowpea
March-August	Groundnut

Cassava normally follows groundnut. In a few cases especially on swampy land, finger millet is sown on newly opened land. In the western highlands, finger millet is grown during the second rains following sorghum, maize, peas or beans or a mixture of all these crops or sweet potato. In northern Uganda, the first crop in the rotation is sometimes sesame followed by cotton and then finger millet. In Bugisu and western Uganda where population pressure is very high, continuous cropping is common. Finger millet may be grown in the first rains of every year and a second crop of beans or groundnut follows during the second rains.

Sowing, spacing and time of planting

All the finger millet in the country is broadcast. The cultivator usually uses a mixture of three to six distinct varieties but such a mixture is usually uniform for height and maturity period. This is done as a sort of insurance cover as no prediction can be made on the yield potential of any one variety in a given season at a given location. In cotton plots, finger millet is dry sown. The millet is broadcast on the standing cotton stalks, normally during the dry season in December/January. The seed is then scuffed in with a hoe. The seed remains dormant until the first showers are received, and then it germinates. However, this crop is subjected to a prolonged dry season after germination. Sometimes, the seedlings may die, necessitating a second sowing after the rains resume. The cotton stalks are uprooted soon after finger millet seedlings are established. If millet seedlings survive this moisture stress successfully, the yields are normally high. Also, the crop is harvested early enough to allow a following crop of cotton or sorghum or cowpeas to be planted.

More recently, the common practice is to skim plough the land after the rains are received and subsequently sow finger millet. To provide a firm seedbed and to cover the seed, a herd of cattle is driven over the seeded field. Alternatively, a tree branch with leaves may be used to cover the seed with soil. The seed rate is about 2.2 kg/ha.

The spacing of the seed depends on the ability of the farmer who is sowing. Normally dense populations give poor yields, while, a widely spaced crop tillers more and bears larger heads. Farmers are advised to space their millet at hoe width.

The timing of planting is very important. The millet grown from early December-February gives the highest yields.

Intercropping

Finger millet is always intercropped with two or more other crops. However, finger millet is always the predominant crop in the mixture. In the high rainfall areas, millet is intercropped with maize where maize is always eaten when cobs are green. In the low rainfall areas finger millet is intercropped with sorghum. This is a precaution against drought; if millet fails, sorghum will at least give some produce. In the northern part of the country, finger millet is always intersown with pigeon peas, sesame, cucumber, cowpeas or sorghum or a mixture of all these.

Weeding

This is the most laborious of all the operations. The millet is usually weeded twice depending on weed population. The first weeding is done 3-4 weeks after germination and the second weeding just before booting. However, most farmers weed only once because of scarcity of labour and other cultivation commitments. The weeding is normally done with small hoes, adzes or knives. Sometimes, weeds are removed by hand. Weeding is normally done by women. A weeding operation when done communally becomes faster where a few families join together and weed their fields in turn. Quite often, a family would prepare a meal or 'Ajono' (local drink) and call the neighbours to weed their field and they have the meal or Ajono in return.

During weeding, some thinning and transplanting may be done to reduce the plant density or gap fill as the case may be.

Harvesting and storage

The crop takes about four months to mature. It is harvested soon after ripening, as most local cultivars shatter, and are subject to bird damage.

The ears are cut with about two centimetres of stalk. A sharp hand or finger knife is used for harvesting, The harvested ears are kept in a pile for a few days to ripen the grain further and to give the desirable taste. They are then sun-dried.

Storage is done in granaries made out of reeds and mud walls.

Marketing

Finger millet is not produced for marketing and farmer will sell only the minimum amount that is in excess of his needs or when he is in urgent need of money. It can also be bartered for meat or other foods.

PRODUCTION CONSTRAINTS

Pests

Except birds and aphids, finger millet does not have any serious pest problems and apparently no research has been conducted on the control of pests on finger millet. Birds cause most damage just before harvest, especially on the white-seeded cultivars. The most common species are the *Quelea* and weaver birds. Mirid bugs, *Taylorilygus* sp. may damage the compact ears and bring in grain discolouration.

Among leaf eaters, grasshoppers, notably *Chrotogonus* spp. and *Zonocerus elegans* and black beetles can be a problem. The maize aphids, *Rhopalosiphum maidis* is quite important especially during the dry spells. Spora pests, the army worm, *Spodoptera exempta* is the most important pest of finger millet. If the attack by this pest is early immediately after establishment, there may be a total crop loss.

The stalkborers notably *Busseola* sp., *Sesamia* sp. and *Chilo* sp. affect the crop. The sorghum shootfly, *Atherigona* sp. can be serious, especially on later plantings and during dry spells.

Diseases

Until recently, finger millet suffered from few diseases (Leakey, 1970). However, today with intensive production, diseases are attaining economic importance (Esele, 1982). The most important disease is blast caused by *Pyricularia grisea*. The disease occurs throughout the country and in all other African and Asian countries where finger millet is grown. All stages of the plant are susceptible to its attack but the ear and the neck infection are the most important in Uganda.

In the farmers fields, the disease causes not less than 10 per cent yield loss (Emchebe, 1975). On some collections at Serere, up to 80 per cent loss in yield was noticed (Esele, 1984). A programme at Serere is presently screening for resistance to this disease. Resistance exists in many materials developed at Serere. These are mainly the materials derived from the disease resistance breeding programme.

Another disease which has recently gained economic importance is *Cylindrosporium* leaf spot. The disease occurs around 60 days after planting and progresses towards maturity. Serious leaf spotting and lesion coalescing is observed which may impair grain development through the destruction of chlorophyll necessary for the synthesis of plant foods.

Tar spot (*Phyllachora eleusines*) consisting of small jet-black and slightly raised spots on the leaves and neck of the plant occurs towards maturity. Where tar spot and *Cylindrosporium* leaf spot occur together, serious defoliation is observed.

Virus-like infections also occur especially on later plantings. These are characterised by leaf streaks, stunting, yellowing of leaves and mosaic symptoms. Other diseases of minor importance include *Helminthosporium* leaf spot, bacterial blight and *Sclerotium* wilt.

A trial conducted at Makerere University on the fungicidal control of finger millet diseases in Uganda revealed that Benlate was the best fungicide for the control of blast and other leaf diseases (Adipala and Mukiibi, 1985).

Quality seed availability

Many farmers keep their own seed which is invariably a mixture of local cultivars. Farmers have experienced outcrossing in the field both between varieties and with 'Ekitu' (*Eleusine africana*) (Tribe, 1965). In addition, local varieties are variable yielders, doing better in some years. Besides, mixtures normally exhibit uneven ripening. Farmers therefore exercise care and select uniform ears and preserve this as seed for the next season. Farmers select the best heads in the field, cut them, dry and store them separately in long strawed bundles. This practice has helped the farmer to carry forward varieties which he feels superior.

Today, there is an increasing trend towards adoption of improved finger millet varieties developed at Serere. In the north of the country, Engeny, Serere I and Gulu E are being grown by farmers. In eastern parts P 224 is getting widely adapted especially in areas around the Research Station. This is because of higher yields of the improved varieties over the local cultivars. However, the major constraint in the spread of these varieties is the availability of seed. The Uganda Seed Project under the Ministry of Agriculture, has the mandate to multiply the seed and make it available to farmers through extension and cooperative services. Unfortunately, the project has had some economic problems and is presently unable to function satisfactorily. The only improved seed currently available to farmers is the one that is multiplied on a limited scale at Serere. In 1983, Serere made available to farmers over 5,000 kg of improved finger millet seed.

Adoption of new technologies

Finger millet depletes soil fast. It is therefore necessary to follow strict rotational regimes or application of fertilizers and other recommended practices. Owing to the weakening of the extension service, it has been difficult for farmers to pick up the newly developed technologies. As a result, broadcasting of seed, late and/or staggered planting, interplanting, scant or late weeding and absence of fertilizer application, continue to be done. The dissemination of information on new crop varieties has been poor. In surveys carried out in 1975, 1981 and 1985, not many of the farmers interviewed had heard of improved varieties and even less actually grew them. All the farmers had heard of row planting but were not practising it except for cash crops (Oryokot, 1985). However, since 1985, Serere researchers have started a Farming System Research Programme aimed at extending the developed materials and technologies to the farmers. Useful results are expected.

The Labour Bottleneck

Finger millet is a crop whose production requires plenty of labour especially during weeding and harvesting. Owing to the nature of its production at subsistance level, the only labour force available to the farmer is his household. In addition to this, the farmer also produces a number of other crops at the same time such as groundnut, sesame, cassava, cotton, etc. The net result is that the millet crop does not receive the second weeding or timely weeding.

Aggressive Weeds

The most troublesome weed in finger millet is its relative *Eleusine africana*. The aggressiveness of *E. africana* is compounded by its close resemblance to finger millet in the vegetative phase, which allows it to grow vigorously in cultivated plots till a very late stage by which time it is too late to weed it out (Thomas, 1970). The weed also matures early—at least three weeks earlier than the crop and shatters immediately, disseminating its seed. It is especially important to check this weed in less fertile soils and on continuously cropped land.

LIMITED USE

Finger millet is only produced for food and beer. There are no proposals for its diversified utilization. If its use can be diversified in poultry and livestock feed formulations and as forage, then its production would definitely increase. Diversified uses in making recipes, biscuits and pastries would further increase the level of importance of this crop.

REFERENCES

Adipala-Ekwamu and J. Mukiibi. 1985. Fungicidal control of fungal diseases of finger millet. In: *Proceedings of the Fourth Regional Workshop on Sorghum and Millets Improvement in Eastern Africa,* 22-26th July 1985. Soroti, Uganda.

Emchibe, A.M. 1975. *Some aspects of crop diseases in Uganda.* 43 pp. Department of Crop Science, Makerere University, Uganda.

Esele, J.P.E. 1982. A survey of finger millet diseases in Uganda, UAFRO, Serere (Unpublished).

Esele, J.P.E. 1984. Finger millet pathology. Progress report on screening for resistance to blast disease in finger millet. UAFRO, Serere (Unpublished).

Leakey, C.L.A. 1970. Diseases of finger millet. In: J.D. Jameson (ed.). *Agriculture in Uganda,* 151-152. Oxford University Press.

Oryokot, J.O.E. 1985. Adoption of sorghum and finger millet technology in Uganda. In: *Proceedings of the Fourth Regional Workshop on Sorghum and Millets Improvement in Eastern Africa,* 22-26 July 1985, Soroti, Uganda.

Thomas, D.G. 1970. Finger millet (*Eleusine coracana* (L.) Gaertn). In J.D. Jameson (ed.) Agriculture in Uganda 123-152. Oxford University Press, London.

Tribe, A.J. 1965. Finger Millet, *Eleusine coracana.* In: *Technical Communication* No. 20, Dept. of Agriculture, Entebbe, Uganda.

29

CROPPING SYSTEMS, PRODUCTION TECHNOLOGY, PESTS, DISEASES AND UTILIZATION OF SMALL MILLETS IN ZIMBABWE WITH SPECIAL REFERENCE TO FINGER MILLET

F.R. Muza

CROPPING SYSTEMS AND PRODUCTION TECHNOLOGY

After the improved cultivars are developed, the emphasis will be on refinement of agronomic practices as traditional practices followed contribute to low yields.

Finger millet requires a longer growing season than most currently recommended sorghums or pearl millet. In Zimbabwe's highveld a growing period of up to 150 days is usual (Johnson, 1968). In drier conditions, however, it can ripen in about 120 days (Masefield, 1949). For maximum yield finger millet requires unrestricted moisture supply for about 10 weeks after its germination in January or February. That is why in Zimbabwe it is grown most widely in the wetter areas in Natural Regions II and III. It can withstand moist weather but dry sunny weather is required during ripening (Johnson, 1968).

As regions II and III happen to be maize growing areas, finger millet tends to be pushed to poorer soils of low fertility, where reasonable yields can be obtained as compared to maize. The best soil for the crop are fertile soils ranging from medium sand to fairly heavy clays. The crop also seems to be tolerant to both acidity and alkalinity.

In Zimbabwe, traditionally finger millet is intercropped with other crops. However, it is often grown in pure stands. It is intercropped with maize, sorghum, groundnuts and beans or cowpeas. One factor which must be considered while rotating finger millet is its susceptibility to witchweed (*Striga asiatica*). It is usually grown after a well fertilized crop since it uses residual fertility efficiently.

When grown as a sole crop, finger millet sometimes is put on virgin land and usually no fertilizer is applied. Better land preparation resulting in fine seed-bed will ensure better crop stand. The common traditional practice is to broadcast the seed and a brushwood is drawn over the field to cover the seed.

Finger millet responds to both fertilizers and manures (Johnson, 1968). Nitrogen increases tillering and number of ears per plant and tends to accelerate the growth of early and intermediate tillers more than that of late ones. Table 1 shows the fertilizer recommendations made for finger millet in Zimbabwe. For better plant stand, seed treatment is recommended before planting with fungicide Deildrin. Pre-soaking in water up to 24 hours, is sometimes done and then the seed is dried in shade. This results in quicker germination. Early planting is recommended as finger millet requires a long growing period.

TABLE 1
Recommended doses of fertilizers to finger millet

	Suggested applications in pounds per acre based upon probable available nutrient status in the soil			
Nutrients	High	Moderate	Low	
N	0-30	30-50	50-80	Nat. Region II
	0	0-30	0-30	Nat. Region III
P_2O_5	20-30	30-60	60-90	
K_2O	0	30-50	50-90	

Dry seeding is possible and convenient. The crop can also be transplanted. In Zimbabwe, transplanting is done only to fill up the gaps. It is recommended that the crop be row-planted in a continuous stream in 20-40 cm rows, using a wheat drill or a maize planter with suitable plates or by dribbling out the seed from a bottle in pre-marked rows. It is best planted when sown from 0.6 to 1.5 cm deep, using 5-8 kg seed per ha and covered with a roller or brush-harrow.

Weed Control

When seed is broadcast, weeding has to be done by hand and this is time consuming. In early stages, it is difficult to distinguish rapoko grass (*Eleusine indica*) and finger millet. So, it is necessary to wait for three to five weeks till the weed becomes darker than the crop (Howder, 1965). It is also difficult to apply herbicides because of this rapoko grass. However, 2, 4-D can be used, and this can be effective on weeds like striga.

Harvesting

Harvesting is done manually by cutting the ears and drying them before threshing. Finger millet stores well, resisting insect damage up to four years in the granary when the ears are stored unthreshed. If grain is required for planting, or for malting it should not be kept for more than 18 months as viability reduces to 50 per cent after 2 years of storage. Finger millet will not shatter even if harvesting is delayed due to rains. However, lodging can be a problem. It can also be harvested using combines.

Agritex (Agricultural Technical and Extension Services) is educating the farmers on the improved management practices. Demonstrations in the communal areas have helped to make the farmers accept these practices. The present market price is $300/tonne and this is bound to boost the production.

Pests and diseases

In Zimbabwe, pests are not so serious on finger millet, both during the growing period and in storage. The red-billed quelea bird can be a problem towards harvest at late stage. Army-worm (*Spodoptera exempta*) will attack finger millet in preference to virtually all other crops (Johnson, 1968).

Maize ladybird (*Epilachna simitis*) can be controlled by giving 50 per cent Malathion as a cover spray, rapoko bug (*Ischnodemus congoensis*) and grasshoppers can be controlled by spraying 85 per cent carbaryl at 0.5 per cent concentration. Rats can be a problem in storage.

Although no diseases of importance are recorded on finger millet in Zimbabwe, the streak virus and leaf blight may become significant in the future.

Utilization

In Zimbabwe, finger millet is mainly used for brewing beer (hwahwa) as well as for preparation of thin (maheu) and thick (sadza) porridges. The farmers retain most of the produce and as the crop is a controlled product, the excess produce is sold to the Grain Marketing Board.

Finger millet has lower nutritional value than maize or other tropical small grains (Table 2). It is however, much richer in calcium and has the same effects as pearl millet and barley in counteracting the softening effects of maize in pig rations. Although not recommended for growing pigs, the ground grain of finger millet can constitute up to one half of the grain ration of breeding pigs and baconers after 54 kg live weight (Calder, 1966).

Food preparations

i) *Flour preparation:* The mature dry grain is pounded to remove the glumes and whole grain is ground into flour in a stone grinder or more often these days, in a grinding mill. Roasting before milling improves the flavour of the flour. No dehulling is necessary.

ii) *Thin porridge (maheu):* The flour is mixed with water to form a thin porridge, and a small quantity of fermented flour is added as a starter and kept in a warm place for a day. This is added to boiling water with constant

TABLE 2

Comparative analyses of several Zimbabwean grains (Percentage by weight)

	Bulrush millet	Finger millet	Sorghum	Maize
Dry matter	89	87	90	90
Carbohydrate	70	72	72	74
Crude protein	10	7	10	8.5
Fat	4.4	1.3	2.7	2.1
Fibre	1.8	3.4	2.1	2.1
Ash	2.6	4.0	1.7	1.5
Digestible protein	8.0	5.5	7.2	6.8
Total digestible nutrients	80	72	78	80

stirring to obtain a smooth thin porridge. This is allowed to cool and drunk as a refresher. Other thin porridges (bota) also have medicinal uses.

iii) *Thick porridge (sadza):* Prepared by adding flour to boiling water with constant stirring until a thick consistency is obtained. This is served with meat, vegetables or milk. Pregnant women eat this in preference to maize sadza because of the high calcium content.

iv) *Traditional beer:* The malted grain is fermented for up to seven days. It is preferred to pearl millet beer as it is not accompanied by head-aches and there is little hangover.

Alternative uses

Some food processing companies in Zimbabwe are carrying out analytical tests on finger millet grain to explore possibilities of using it to make products like baby cereals, flakes and biscuits.

Forage use

Although finger millet is never grown primarily for forage in Zimbabwe, after harvest the cattle are let into the fields before the rest of straw is ploughed in.

REFERENCES

Calder, A. 1966. The value of Rapoko (millet) for pig-feeding. *Rhod. Agric. J.*, **57:** 116.

Howder, R.H.C. 1965. *Rapoko Extension Officer's Handbook*. Govt. Printer, Harare.

Johnson, D.T. 1968. Finger millet (*Eleusine coracana*)—Rhodesia's neglected crop. *Rhod. Agric. Bull.*, **65:** 87-91.

Masefield, G.B. 1949. *A handbook of tropical agriculture*. p. 30. Oxford University Press.

30

CROPPING SYSTEMS, PRODUCTION TECHNOLOGY AND UTILIZATION OF SMALL MILLETS WITH SPECIAL REFERENCE TO FINGER MILLET IN KENYA

C. Mburu

CROPPING SYSTEMS

Millets in most parts of Kenya are grown traditionally and rarely as major food crops. Among millets, sorghum is an important crop. However, in some areas around Lake Victoria (in striga-stricken areas) and in the eastern parts of country which is hot and dry, farmers grow millets other than sorghum as major cereals. For example, in Busia district bordering Uganda, many farmers grow finger millet as their sole cereal. In Kitui and Machakos districts of the eastern province some farmers plant pearl millet as the sole cereal. Otherwise maize has dominated most areas which earlier used to be under sorghum and millets.

The present practice of growing small millets is to plant them in small patches in pure stands. For example, farmers in western Kenya plant small areas, less than half an acre with pure stands of finger millet just outside their homesteads. In the eastern province and in parts of central province it is common to see small patches of pearl millet or small millets scattered in farmers fields amidst other crops.

Often millets are intercropped with other crops. Finger millet which is common in western Kenya is often grown mixed with maize, sorghum or both. Sometimes, it is intercropped with cassava, or relayed with cotton; the first crop being millets. Pearl millet, which is common in eastern Kenya is intercropped with sorghum and/or maize or with cassava in coastal regions. Generally the intercrops are of no defined ratios.

The practice is to plant finger millet either with the onset of the long rains in March-April, or dry planted in January-February. Finger millet is rarely planted during the short rains which occur in December. In the western side of the country, with the onset of the short rains, land may be planted with sorghum, cotton or even maize. The other millets being more drought tolerant can be grown during the long rains or short rains or both. As the rainfall received in the short rains is low in the eastern side, sorghum, pearl millet and the other millets can follow the long rain crop.

In most parts of the Rift Valley, the rains are monomodal in nature and thus the millets especially finger millet mixed with maize can only be grown during that season. Due to the cold weather prevailing in these areas these crops take a long time to mature.

PRODUCTION TECHNOLOGIES

Most of the millet growing farmers are small farmers who use manually or oxen drawn ploughs for land preparation. Fields meant for millets are rarely adequately prepared to the required tilth.

The local cultivars that have been selected over the years by farmers are grown. They are medium to tall in stature, late in maturity and poor yielders compared to the improved cultivars which give seed yield up to 2500 kg/ha. However, with good management some local cultivars namely Ekalakala, Ikhulule and Gule-E have given as much seed yield as improved varieties in research stations.

Sowing is done by broadcast after digging or ploughing and re-dug to cover the seeds or left as it is after sowing. Often farmers use old seeds which fail to germinate resulting in poor plant population.

No fertilizer or chemical is used in millet cultivation. After germination, no thinning is done. The crop is often very much weed infested as only one weeding is done. Control of insect pests and diseases are very rarely done. Birds are a menace to millet crops and the only remedy is to scare them manually which is tedious and laborious. The decrease in area under millet has resulted in enhanced bird damage on these crops.

Harvesting is done by removing ears using a curved knife. The ears are then dried and threshed. Threshing is done by beating with sticks. This is true for all millets including sorghum. The produce is stored in baskets or tins or bags or in granaries.

Occasionally some farmers use ash as seed dressing in grain storage. However, small millets do not have problems of storage pests.

The Government of Kenya has of late been putting a lot of emphasis on sorghum and millets development under a National Programme for Research and Improvement of Sorghum and Millets (PRISM). Research is being carried out to improve the varieties and to develop improved agronomic practices.

Many varieties of finger millet have been developed and evaluated. Some of the promising varieties are—P 283, P 224, P 221 and Serere 1 and a few others introduced from Uganda. Similarly in foxtail millet and proso millet the varieties ISE 285 (from India) and N 40101 (from USSR) respectively have shown promise.

Agronomic work is in progress to develop suitable packages for both eastern and western Kenya taking into account the major problems of these areas, the constraints of the farmers and the over-riding need to minimise the risk rather than maximizing yields. Mostly the work is on determining optimum plant population, fertilizer doses, sowing time and timely weeding.

UTILIZATION

Millets in Kenya are grown for both home consumption and for sale in its local market. Grain is ground to flour and used in making porridge and cake like preparations (*Ugali*); a common food for the people of Kenya.

Finger millet is used in brewing a local drink 'Busaa' which is very common in western parts of Kenya especially during festivals. It is fermented after germination, it is then added to a mixture of water and flour of either maize or sorghum.

The flour of finger millet, depending on the availability, may be used alone or with other flours like maize or sorghum or cassava. Taste of millets including finger millet usually is not very appealing compared to other more favoured flours such as maize or wheat. Kenya's Industrial Research and Development Institute is looking into ways of processing for preparing better food stuffs from millets. It is anticipated that millet can be used to prepare better food products than that is prepared from maize or wheat in view of its higher mineral content, especially in the preparation of infant and baby foods.

The stalks of finger millet are rarely used for feeding animals. Otherwise, they are left in the field or used as firewood.

CONCLUSIONS

It is evident that the small millets have not lost their popularity with the Kenyan peasants. There is a lot of germplasm to be collected and conserved. The low yields presently realised is mainly due to genetically poor varieties and poor agronomic practices accompanying their cultivation. Areas marginal for maize, the hot dry areas in the east, the waterlogged, striga sticken areas in the west around Lake Victoria, the coast, and the cold highlands of the Rift Valley are cropped with millets. With better processing and utilization avenues; evolution of improved cultivars and methods of cultivation the small millets production is likely to increase.

31

CROPPING SYSTEMS, PRODUCTION TECHNOLOGY, PESTS, DISEASES, UTILIZATION AND FORAGE USE OF MILLETS WITH SPECIAL EMPHASIS ON TEFF IN ETHIOPIA

Seyfu Ketema

CROPPING SYSTEMS

Teff is the most important cereal crop in Ethiopia. Like many cereal crops, it is cultivated predominantly under a monocropping system. However, in a few areas it is cultivated under multiple cropping system with *Brassicas*, safflower or sunflower as intercrops. Also it is relay cropped with maize. This is done by removing lower leaves from maize after the cob is formed and then teff is sown in between maize plants. Not much research has been done on multiple cropping systems using teff. This needs to be studied more thoroughly as multiple cropping has potential to enable the farmer to maximize the utilization of resources in his farm.

PRODUCTION TECHNOLOGY

At the present moment teff is produced mainly by farmers and not in state farms. Farmers practise traditional methods of land preparation using an oxen drawn plough.

Seed bed preparation

Teff seeds are very small and should not be sown deeply. They have to be left on the surface or covered very thinly. This means teff needs a very smooth seed bed. If harvesting is to be done by combine, the practice of packing the seed bed before sowing has been found useful since most of the present day teff cultivars lodge at maturity.

Seed rate

Fifteen to 55 kilograms of seed can be sown per hectare. If sowing is to be done by hand broadcasting, it is difficult to evenly distribute 15 kilograms of seed per hectare. Therefore, 25-30 kg seeds per hectare are recommended. If a manually or motor-driven broadcaster or drill is available, lower seed rates of around 15 kg/ha are preferable.

Fertilizer rate

Systematic studies on the fertilizer requirements of teff under different conditions have not yet been conducted and completed. For the time being the following recommendations are made in Ethiopia.

On heavy clay soil (vertisol) — 60 N : 26 P_2O_5 kg/ha

On sandy clay loam soil — 40 N : 26 P_2O_5 kg/ha

Sowing date

Teff can be sown in a season when the rainfall is reliable and well distributed. For most cultivars 300-500 mm of rainfall per growing season is adequate. Early maturing varieties (60-80 days) can do with less than 300 mm of seasonal rainfall. It germinates and establishes quickly on lighter sandy clay loam soils than on heavy clay soils.

Weeding

Hand weeding once at early tillering stage (25-30 days after emergence) is ideal and adequate if the weed population is low. Under heavy weed population hand weeding has to be done for the second time at the stem elongation stage. Hand weeding after heading is not recommended since heavy damage may be caused to the plant.

Application of presowing herbicide Gesatan 500-FW (Ametryne 25 per cent + Prometryn 25 per cent) at the rate of 2 to 4 kilograms per hectare two weeks before sowing teff is recommended. It gives good control over both annual grasses and broad leaved weeds.

Post-emergence herbicides — 2, 4-D 720, MCPA 625, 2-4-DDP and Brito × 52 (Bromoxynil + Ioxynil + mecopropester) at the rates of 1.5, 3, 5 and 2.5 litres per hectare, respectively give satisfactory control of broad leaved weeds.

Harvesting

The crop is harvested when the green colour of the plant disappears and the vegetative parts turn yellowish or straw colour. This depends on the maturity period of the varieties which varies from 60-120 days. Harvesting before the plant is too dry helps to prevent field loss due to seed shattering.

Post-harvest management

Teff seeds are not attacked by weevils or fungi. Therefore, they do not need protective chemicals in storage.

PESTS AND DISEASES

In the main teff growing areas of Ethiopia pests and diseases are not serious problems. And teff is considered to suffer less from diseases in its major production centres than most other cereal crops in Ethiopia (Stewart and Dagnatchew, 1967). However, teff rust (*Uromyces eragrostidis* Tracy) and head smudge (*Helminthosporium miyakei* Nisikado) are reported as the most important diseases on teff (Stewart and Dagnatchew, 1967; Tareke, 1981). These diseases are considered to cause significant loss in yield in humid south western parts of Ethiopia. No control measures for these diseases has so far been developed. One pest that can at times be severe is central shootfly (*Hylemya arambourgi*). Seed dressing before sowing with 40 per cent Aldrin WP at the rate of 50 grams for 10 kilograms of seed is recommended.

UTILIZATION

In Ethiopia, teff is mainly grown for its grain. The flour is most widely used for making a pancake called 'enjera'. It is also used for making porridge and native alcoholic drinks called 'tella' and 'katikalla'. Its straw is used for reinforcing mud plastered walls of tukuls (Tadese, 1969; Rouk and Melak Hail, undated). Its straw is used as feed for cattle.

The nutritive value of teff grain compares well with some of the major cereals such as wheat, barley, rice, maize and sorghum. According to Rouk and Melak Hail (undated), the Ethiopian nutrition survey (1959) reported that four unspecified varieties when analysed biochemically were found to contain an average of 300 calories, 11.6 grams protein, 0.65 grams fat and 70.56 grams carbohydrate per 100 grams and that teff supplies on an average two-thirds of the total protein in the Ethiopian diet. Also, Melak Hail (1966) reported that some samples of teff when analysed contained higher amounts of calcium, copper, zinc, aluminium, sodium and barium than that is present in samples of winter and spring wheats, barley and grain sorghum. Thus, though not consumed world-wide, the nutritive status of teff is comparable to that of the major cereals.

Teff is used for making "enjera" and alcoholic drinks too. Often, teff is mixed with other cereals before making "enjera" but enjera from pure teff is greatly preferred. The procedure of preparing 'teff enjera' is described on p. 312.

Teff engera preparation—Ethiopian Nutrition Institute (1980)

Ingredients	*Weight*
Teff flour	3 kg
Ersho* (yeast)	480 g
Water for dough	6000 L
Water for baking	4000 L
Cooked product	
Number of enjeras	18
Weight of one enjera	450 g

(*Note:* *Ersho = A starter which is a dough saved from previous fermentation and it is used for starting fermentation in new dough)

Preparation of the dough

1) Sift the flour into a container large enough to hold the entire recipe.

2) Stir in the ersho. Use a large wooden spoon or the hands.

3) Add three litres of water and mix well. Add three more litres of water gradually stirring all the time. Use either a wooden spoon or hand. Mix well.

4) Cover the dough and allow it to stand for three days to ferment.

Baking of the enjera

1) Pour away the water that has settled on top of the dough

2) Add ½ litre of dough to 1 litre of water and boil.

3) Mix 3 litres of cold water into the dough. Stir in the hot mixture and allow to stand for 30 minutes.

4) In the mean time heat the metad* (enjera oven)

5) To bake enjera, pour the batter on to the hot greased metad using a circular motion from outside towards the centre to make a circular enjera. When holes begin to form on top of the enjera, cover with the akenbalo** (enjera oven lid) and bake for 2-3 minutes. Use about ½ litre of batter for each enjera.

6) Grease the metad with oil between each baking. Repeat the above process until all the dough is used.

The comparison of nutrient content of 100 g enjera made of teff and wheat is given in Table 1. The other important points to be followed in enjera preparation are:

1) Although no longer customary, oil can be added to the dough at the same time as the boiled water. Oil improves the appearance of the underside of the enjera. It becomes more shiny.

(*Note:* * Metad = A flat round clay griddle of enjera oven that is used for baking enjera.
**Akenbalo = The lid of the enjera oven (metad). It is used to cover the oven when the enjera is baking. It is usually made of clay and should be air tight).

TABLE 1
Comparison of nutrient composition between teff and wheat enjera

	Composition in 100 g														
	Energy (Calories)	Moisture (%)	Protein (g)	Fat (g)	Carbo-hydrate (g)	Fibre (g)	Ash (g)	Calcium (g)	Phos-phorus (mg)	Iron (mg)	B-Car. Eguiv. (mg)	Thiamin (mg)	Ribo-flavin (mg)	Niacin (mg)	Ascorbic acid (mg)
Teff	162	59.8	4.2	0.6	33.9	1.7	1.5	64	129	30.5	0	0.21	0.07	0.8	1
Wheat	172	57.4	5.4	0.9	35.6	0.9	0.7	28	135	3.3	0	0.14	0.09	—	1

Source: Ethiopian Nutrition Institute, Addis Ababa.

2) Teff flour is most commonly used but barley, wheat, corn, finger millet and sorghum flours may also be used.

3) The length of fermentation depends on the climate, where the climate is hot, enjera is baked on the same day as the dough is mixed.

4) The dough should never be fermented for more than 3 days or it will become too sour and might cause gastritis. Besides, the longer the fermentation time higher the destruction of nutrients.

FORAGE USE

Teff is not mainly grown for forage nor is the production of forage of any kind widely practised in Ethiopia. Hence, there is a shortage of feed during the dry season. When teff is grown for its grain, its straw is a very important source of feed especially during the dry season. Cattle prefer teff straw more than the straw of any other cereal.

According to Burt-Davy (1913), "the chief value of teff as a hay crop lies in its palatability, high nutritive value, narrow albumaid ratio (for a grass hay), high yield, rapid growth, drought resistance and ability to smother weeds".

According to Tadese (1969), teff is found to produce more than twice as much forage as weeping lovegrass (*Eragrostis curvula*) producing an average of 14.5 tonnes of green material per hectare within 3 months. This shows that teff has a great potential to be improved through breeding and put to use for forage production in addition to its use as a cereal crop.

REFERENCES

Burt-Davy, J. 1913. Teff (*Eragrostis abyssinica*). *Agronomy J. Union of South Africa,* **5:** 27-37.

Ethiopian Nutrition Institute, 1980. Ethiopian Traditional Recipes, Addis Ababa, Ethiopia, 176 pp.

Melak Hail, M. 1966. Chemical composition of Teff (*Eragrostis tef*) compared with that of wheat, barley and grain sorghum. *Economic Botany,* **20:** 268-273.

Rouk, H.F. and M. Melak Hail. (undated). An introduction to teff (*Eragrostis abyssinica* Schard.) a nutritious cereal grain of Ethiopia: Debre-Zeit Agricultural Research Centre Bulletin, No. 26, Alemaya University of Agriculture, Dire Dawa, Ethiopia, 21 pp.

Stewart, R.B. and Y. Dagnatchew. 1967. Index of plant disease in Ethiopia. *Experiment Station Bulletin* No. 30: Alemaya University of Agriculture, Ethiopia, 95 pp.

Tadese, E. 1969. Teff (*Eragrostis tef*) The cultivation, usage and some of the known diseases and insect pests: Part-I Debre-Zeit Agricultural *Experiment Station Bulletin* No. 60. Alemaya University of Agriculture, Dire-Dawa, Ethiopia, 56 pp.

Tareke, B. 1981. Inheritance of lemma colour, seed colour and panicle form among four cultivars of *Eragrostis tef* (Zucc.). Trotter. Ph.D. Thesis, University of Nebraska; Lincoln, Nebraska, USA, 84 pp.

32

FINGER MILLET CROPPING SYSTEMS AND MANAGEMENT PRACTICES IN TANZANIA

R.O.F. Mwambene

INTRODUCTION

Millets and sorghum are the traditional grain crops in East Africa. Although they have been displaced by maize in many places, they are still the staple food crops in some areas. The term millet refers to several species of grain plants and, the usage changes from place to place. In East Africa, the major millet crops of importance are:

Name of the crop		
English	Latin	Swahili
Finger millet	*Eleusine coracana*	Wimbi/Ulezi
Bulrush millet	*Pennisetum typhoides*	Mawele/uwele
Sorghum	*Sorghum bicolor*	Mtama

Finger millet is the only small millet on large scale production in Tanzania. The main areas are Rukwa and Mbeya regions in the southern highlands of Tanzania and Mara region on the eastern side of Lake Victoria. It is also grown to a lesser extent in Kilimanjaro, Kondoa, Iringa and Ruvuma districts.

The exact area under finger millet in Tanzania is not precisely known as all government records include finger millet along with sorghum and other millets. A case study of finger millet production in the Ufipa Plateau of Rukwa Region, the major finger millet growing area in Tanzania revealed that nearly half of the total area is always cultivated with finger millet (Ulvund and Mkindi, 1976).

CULTIVATION PRACTICES

The natural condition of land used for cropping differs from place to place. Table 1 shows the frequency of different types of vegetation present before the cultivation of finger millet. The cropping duration of the agricultural land which varies from village to village is given in Table 2. Most land is allowed to regenerate natural vegetation 3 to 5 years after clearing.

TABLE 1
Frequency of different types of vegetation before finger millet cultivation in Ufipa Plateau

Village	Bush	Forest	Grassland	Fallow	Total no. of fields
	Number of fields				
Kapewa	6	3	1	0	10
Mwazye	7	1	1	1	10
Chipapa	1	4	5	0	10
Singiwe	1	1	11	0	13
Sintali	1	0	10	0	11
Chala	5	5	2	0	12
Total	21	14	30	1	66
%	31.8	21.2	45.5	1.5	100

TABLE 2
Number of years fields kept under cropping

Village	Number of years									Total no. of fields
	1	2	3	4	5	6	7	8	9 or more	
Kapewa	5	4	1	0	0	0	0	0	0	10
Mwazye	0	3	5	0	1	0	0	0	1	10
Chipapa	2	2	5	1	0	0	0	0	0	10
Singiwe	1	2	5	3	1	1	0	0	0	13
Sintali	2	7	1	0	1	0	0	0	0	11
Chala	1	2	3	1	1	1	0	1	2	12
Total	11	20	20	5	4	2	0	1	3	66
%	16.5	30.5	30.5	7.5	6.0	3.0	0	1.5	4.5	

Clearing and cultivation of fields

In forest and bushy lands the trees are cut and burnt during the dry season. Long stumps and big trees are left upright. In grasslands, no clearing is necessary. After clearing the field, the soil is prepared using a plough or hoe. In grasslands, moulding is widely practised at the end of the rainy season. In both situations, soil preparation and tilling start soon after the soil becomes soft after the first rains are received.

Planting time

Planting time depends upon the rains and the time taken to prepare the seed bed. Normally the rains start by the end of November and finger millet is planted from December to February with highest frequency of plantings made in January.

Planting method

All farmers broadcast finger millet. Ridges and rows are not made and different methods followed for covering the seed are as follows:

By hoeing	.. 19.7%
Trampling by cattle	.. 66.7%
Sweeping with twigs	.. 1.5%
Twings pulled by oxen	.. 12.1%

Seed rate

Most farmers use a very high seed rate, ranging from 10 to 30 kg/ha for finger millet.

Weeding

Weed control is one of the most difficult problems in finger millet production. All farmers in Tanzania weed finger millet only once. Weeding is done either by hand or by using a big hoe.

Intercropping with finger millet

Intercropping is very common in finger millet production. The frequency of farmers mixing millet with other crops are presented in Table 3. In most cases finger millet is mixed with maize, but quite often only a few plants of maize are seen almost making finger millet a pure crop.

POST PRODUCTION PRACTICES

Harvesting and threshing

Ears are usually harvested using a small knife. In spite of the highly developed skill in using the knife, harvesting is too laborious and is one of the discouraging factors in finger millet production in Tanzania.

TABLE 3

Frequency of intercropping with finger millet in Ufipa Plateau

Village	Intercropping with finger millet	Pure stand of finger millet	Total no. of observations
Kapewa	11	0	11
Mwazye	9	1	10
Chipapa	6	4	10
Singiwe	4	8	12
Sintali	5	6	11
Chala	12	0	12
Total	47	19	66
%	71.2	28.8	100

In the old Rungwe district (except Bulambya Division) finger millet is cut by sickles and piled on flat ground to ferment and later on washed in water by rubbing the loosened heads. This method is good, but a lot of grain is lost in the process.

Yield

While determining the yield of finger millet one has to take into account the intercrop also. However, grain yield of 0.5 t/ha to 2.0 t/ha is quite common in the southern highlands of Tanzania in different cultivators holdings.

Storing

Compared to other cereals, finger millet is easy to store. Most of the farmers store the threshed grain and a few store it in unthreshed sheaves. Threshed grains are kept in 'Vihenge' (like a large basket) with the walls made firm by plastering with soil.

Food uses

In the Ufipa plateau, finger millet has remained the most important food crop. In other regions its use is confined to making local beer (Pombe). As food finger millet is used for making *ugali* and *uji* (hard and soft porridge).

REFERENCE

Ulvund, K.A. and G.I. Mkindi. 1976. An agro-economic study in Ufipa Plateau. Yole Agric. Centre Research Report No. 19.

VI

FOOD AND FODDER USES

33

UTILIZATION OF SMALL MILLETS IN ANDHRA PRADESH (INDIA)

P. Pushpamma

INTRODUCTION

Small millets, which are produced mainly by subsistence farmers as rainfed crops, continue to play an important role in the diets of people living in interior rural and tribal areas in the semi-arid tropics. In these areas, small millets are usually grown where agroclimatic conditions are too severe for other crops. Since most of the consumers of these crops are producers in rural areas, small millets rarely occupy a place in urban food markets. As a consequence, information regarding their food value is scarce.

In 1983, a study was conducted by the Home Science Department of Agricultural University, Andhra Pradesh, India. The objective of this study was to gain a better understanding of the post harvest problems and utilization of sorghum and millets. Information was collected from 2160 households, covering six districts in rural Andhra Pradesh. The main focus of the study was sorghum and pearl millet, but three small millets were also included *viz.* foxtail millet (*Setaria italica*), Proso millet (*Panicum miliaceum*), and kodo millet (*Paspalum scrobiculatum*).

PRODUCTION AND CONSUMPTION

In the region of study, 27 per cent of households were producing foxtail millet, 13 per cent were producing proso millet and 5 per cent kodo. Most of them were subsistence farmers and the grain was used mainly for home consumption. Only 1-2 per cent of them sold the grain. Farmers who did sell grain, sold it either in their own village or in a nearby village. Among the three millets, foxtail millet was purchased by 11 per cent of families surveyed for consumption. The other small millets were not purchased by the households in the survey.

STORAGE

Millets are usually stored in gunny bags or in a storage structure made with bamboo and plastered with mud and cow dung. Insect damage during storage is minimal compared with heavy insect damage in sorghum and pearl millet (Table 1). This might be one of the reasons why households store small millets in larger quantities (Table 2). The hard pericarp and small seed size could be contributing factors to the greater resistance to insect damage found in small millets.

TABLE 1
Insect damage in sorghum and millets in households surveyed

Foods	Percentage with damage
Sorghum	93
Pearl millet	90
Finger millet	5
Proso millet	2
Foxtail millet	3
Kodo millet	6

TABLE 2
Association between farm size and quantity of sorghum and millets stored (Average per farmer in quintals)

Foods	Small farmer	Medium farmer	Large farmer	Over all average
Sorghum	3.6	6.5	11.4	6.4
Pearl millet	4.9	5.4	6.6	5.5
Finger millet	6.7	8.0	12.8	8.7
Proso millet	6.7	9.3	20.7	9.5
Foxtail millet	5.4	6.6	15.8	8.7
Kodo millet	11.6	5.5	4.7	6.0

PROCESSING

Unlike sorghum and pearl millet, which are consumed both as whole or dehulled grain products, small millets must be dehulled before cooking. The dehulling of proso millet is often done in the rice mill. The dehulled grain can be stored for more than a month. Kodo is processed mostly by the traditional method of dehulling, using a stone mortar and wooden pestle (Table 3). Foxtail millet is often dehulled by the traditional abrasive method using a big boulder-like stone which is rolled from one side to the other by two women. The traditional method of processing the small millets surveyed takes 20 to 22 minutes per kg of grain. This method is very tiring as this work demands high energy. The extraction rate of kodo millet is very low being only 54 per cent compared to other small millets, which were found to have extraction rates ranging from 70 to 80 per cent (Table 4).

TABLE 3
Methods of dehulling proso, foxtail and kodo millets used by households surveyed (Percentages in brackets)

Foods	Traditional methods used for dehulling		
	By milling	Dry abrasive method	Wet abrasive method
Proso millet	103 (45)	118 (51)	10 (4)
Foxtail millet	95 (19)	413 (81)	0 (0)
Kodo millet	5 (10)	41 (79)	6 (11)

TABLE 4
Time taken for dehulling sorghum and millets and extraction rate

Foods	Time taken for dehulling		Extraction rate (percentage)
	Min.	Sec.	
Sorghum	5	07	79
Pearl millet	6	20	74
Foxtail millet	21	20	71
Kodo millet	22	10	54

UTILIZATION

All these millets are usually cooked as rice after dehulling. In addition foxtail millet is consumed as stiff porridge called sargati, or as an leavened bread known as roti, after the dehulled grain has been milled into flour. Proso millet flour is also used as a substitute for rice flour in various snack foods.

NUTRITIVE VALUE

The protein contents of the small millets are reported to range between 10 and 13 per cent, which is comparable to any cereal or millet. In terms of protein quality, as indicated by lysine content, (Table 5) proso millet is better than the other two millets. The level of lysine in proso millet is comparable to that in rice or wheat. However the quality of protein of foxtail millet is lower as its lysine content is less than 3 per cent, comparable to lysine levels in sorghum protein.

TABLE 5
Protein, fat and lysine content of small millets

Foods	Protein g/100 g	Fat g/100 g	Lysine g/100 g Protein
Proso millet	11-12	1.8-2.7	4.35-4.45
Foxtail millet	10-12	4-5	2.29-2.7
Kodo millet	11-13	3.5-4.9	3.2 -3.7

No information is available on the biological value to assess the human food value of these millets compared to other cereals and millets. However, from the chemical composition, they appear to be not inferior to other more popular food grains.

34

PROCESSING OF SMALL MILLETS FOR FOOD AND INDUSTRIAL USES

N.G. Malleshi

INTRODUCTION

Small millets such as finger millet or ragi (*Eleusine coracana*), proso millet or panivaragu (*Panicum miliaceum*), foxtail millet or navane (*Setaria italica*), kodo millet or varagu (*Paspalum scrobiculatum*), barnyard millet or banti (*Echinochloa colona*), little millet or same (*Panicum miliare*) form staple foods for a large segment of the population in India. Rajakeera (*Amaranthus paniculatus*) and Job's tears (*Coix lacryma-jobi*) grown in India are also classified as small millets. Besides India, small millets are grown in Russia, China, Japan, USA and a few other tropical African and East Asian countries. Teff (*Eragrostis tef*) and Fonio (*Digitaria exilis*) are the small millets cultivated in Ethiopia and West African countries respectively. Almost all the grain produced is used for food in India and in other developing countries, whereas in the USA and other developed countries grain of small millets are mostly used as feed to calves and birds. The potential of production of these millets in India is high as, they are superior to other cereals in their performance under moisture stress and low soil fertility, and have good capacity to respond to improved inputs. They are relatively short duration crops also. However, more efforts are needed to improve and optimise methods for utilizing small millets in diversified ways.

By virtue of their composition, small millets are quite comparable to rice or wheat in their nutritive value (Table 1). Some of them are even better in protein, oil and mineral content than rice. The protein content of finger, kodo, little and barnyard millets varies from 6 to 10 per cent whereas those of proso and foxtail millets ranges from 9 to 14 per cent. Proso and foxtail millets contain about 4 per cent fat. Finger millet contains the highest level of calcium among cereals whereas teff contains a high level of iron. As regards to the

TABLE 1
Proximate composition of small millets (per 100 g)

Name	Protein (g)	Fat (g)	Minerals (g)	Fibre (g)	Carbohydrates (g)	Calcium (g)	Phosphorus (g)	Thiamin (g)	Edible matter (g)
Finger millet	7.3	1.3	2.7	3.6	72.0	344	283	420	100
Proso millet	12.5	3.1	1.9	7.2	70.4	14	206	400	59
Foxtail millet	12.3	4.3	3.3	8.0	60.9	31	290	590	79
Little millet	7.7	4.7	1.5	7.6	67.0	17	220	300	66
Kodo millet	8.3	1.4	2.6	9.0	65.9	27	188	330	58
Barnyard millet	6.2	2.2	4.4	9.8	65.5	11	280	300	65
Teff	9.5	2.5	2.2	2.5	82.0	140	415	500	—
Digitaria exilis	8.3	3.5	3.8	8.5	73.6	30	170	300	80
Job's tears (milled)	17.5	6.0	1.8	0.5	63.4	23	480	310	100
Rajkeera seeds	15.5	—	3.6	2.0	66.8	221	646	—	80
Rice (milled)	6.8	0.5	0.6	0.2	78.2	45	160	—	100
Wheat	11.8	1.5	1.5	1.2	71.2	41	306	—	100

Source: Nutritive Value of Indian Foods, Ed. Gopalan *et al.*, National Institute of Nutrition, Hyderabad.
RP*

essential amino acids of the millet proteins, they are however, deficient in lysine, tryptophan and threonine (Table 2). Finger millet protein is unique among cereals to possess very high levels of sulphur amino acids. Generally the leucine to isoleucine ratio of millet protein is not favourable and, in the case of foxtail millet, the higher level of arginine is reported to affect the lysine availability. Diets based on millets as the sole source of protein have been reported to produce poor growth, however, when they are supplemented suitably with lysine rich materials such as legumes, oilseed cakes, animal proteins or synthetic amino acids, their growth promoting value was enhanced significantly. The subject of nutritional quality and processing of millets has been reviewed exhaustively by Rachi and Peters (1977); Hulse *et al.* (1980) and Hoseney *et al.* (1982).

Small millets are considered as coarse grains and are used as articles of food in situations where other food grains generally cannot be raised, or purchased at economic prices. Therefore small millets have largely remained as food of the poor and less privileged section of the population. The outer tough seed coat and the characteristic flavour associated with these millets are, the main reasons why they are less popular among rice and wheat eaters. Except for finger millet, all the small millet seeds have a slight resemblance with paddy (rough rice) in their morphological features and have an outer husk, bran and starchy endosperm whereas the finger millet seed coat is tightly bound with soft endosperm. As compared to major cereals like maize, sorghum and pearl millet, the information on processing of small millets for food and industrial uses is very limited (Desikachar, 1975, 1976, 1977). The relevant information available and possible industrial uses of millets are discussed in this paper.

MILLING

Milling is the primary processing of food grains. In the case of rice, the term milling is confined to dehusking and debranning whereas wheat milling includes debranning and sizing the endosperm into semolina or flour. Milling of small millets is done by adoption of both wheat and rice milling techniques.

The finger millet seed coat, contains coloured pigments tightly bound with the soft and friable endosperm. Efforts made to debran finger millet using abrasive type milling machinery have not been successful. Generally, finger millet seeds are powdered, and the whole meal (sometimes a small portion of coarse bran is sieved off) is utilised for food preparations. However, moistening the seed with 3-5 per cent moisture, tempering for about 30 min (moistening toughens the bran and reduces its friability without affecting the endosperm property), grinding followed by sieving separates most of the bran and yields fairly white flour (Kurien and Desikachar, 1962). The roller flour mill, universally used for wheat milling, could be used to obtain fully refined finger millet flour. However, the yield of the flour is hardly 60 per cent (Kurien and Desikachar, 1966). Malleshi and Desikachar (1981b) obtained refined flour

TABLE 2

Essential amino acid contents of small millets (g/gN)

Name of millet	Protein content (g/100 g)	Argi-nine	Histi-dine	Lysine	Trypto-phan	Phenyl-alanine	Tyro-sine	Methio-nine	Cystine	Threo-nine	Leucine	Isoleusine	Valine
Finger millet	7.3	0.30	0.13	0.22	0.10	0.31	0.22	0.21	0.14	0.24	0.69	0.40	0.48
Foxtail millet	12.3	0.22	0.13	0.14	0.06	0.42	—	0.18	0.10	0.19	1.04	0.48	0.43
Little millet	7.7	0.25	0.12	0.11	0.06	0.33	—	0.18	0.09	0.19	0.76	0.37	0.35
Kodo millet	8.3	0.27	0.12	0.15	0.05	0.43	—	0.18	0.11	0.20	0.65	0.36	0.41
Proso millet	12.5	0.29	0.11	0.19	0.05	0.31	—	0.16	—	0.15	0.76	0.41	0.41
Barnyard millet	8.3	—	—	0.18	0.03	0.20	—	0.12	—	0.14	0.04	0.55	0.40
Job's tears	17.5	0.27	0.13	0.13	0.03	0.30	0.27	0.16	0.11	0.20	1.02	0.30	0.36
Fonio	8.5	0.24	0.13	0.16	0.09	0.32	0.24	0.35	0.18	0.25	0.61	0.25	0.36
Teff	9.5	0.22	0.13	0.19	0.08	0.44	—	0.33	—	0.21	0.48	0.25	0.33
Amaranthus seeds	15.0	0.91	0.18	0.51	0.05	0.30	—	0.15	—	0.27	0.51	0.43	0.38
Wheat	11.8	0.29	0.13	0.17	0.07	0.28	0.18	0.09	0.14	0.18	0.41	0.22	0.28
Rice	7.0	0.48	0.13	0.23	0.08	0.28	0.29	0.15	0.09	0.23	0.50	0.30	0.38

Source: Nutritive Value of Indian Foods, Ed. Gopalan *et al.*, National Institute of Nutrition, Hyderabad.

from malted finger millet using moist conditioning and grinding technique. A mini millet mill has been developed recently by the Central Food Technological Research Institute in Mysore (CFTRI) which consists of a common plate grinder with water mixer and sifter attachments. It is a versatile mill where debranning and sizing the endosperm take place in one operation (Shankara *et al.*, 1985). The mill can be used to obtain semolina and flour from wheat, maize, sorghum and millets (Fig. 1).

In the case of other small millets, milling process includes dehusking and debranning (decortication) and grinding if desired. The traditional method of dehusking and debranning by hand operated pestle or *denki* still persists. The Rice milling machinery, such as disc sheller, rice huller and centrifugal sheller also dehusk the small millets effectively (CFTRI Ann. Report, 1976). Dehusking can be carried out even in the plate mill by suitably adjusting the clearance between the plates. The dehusked or decorticated grains ('brown rice') are polished in a rice polisher to get 'millet rice' (Fig. 2). Polished grains may be pulverized in a plate mill or hammer mill to obtain semolina, or flour, as required. Alternately the grains could be used for cooking like rice, or can be processed for flaking.

Co-milling of small millets with wheat and other cereals to produce composite flours has also been reported by Crabtree and Dendy (1977), and Lorenz *et al.* (1980).

It may be worthwhile to mention here that the millet bran contains nearly 15-20 per cent oil. In case milling of these grains is carried in an organised sector, the bran could be used as an extender of rice bran for extraction of oil.

CONVENTIONAL FOOD PRODUCTS

Roti (unleavened pan cake), mudde (dumpling) and porridge are the main food products prepared from millets. Millet protein lacks gluten, hence it is unsuitable as the sole material for preparation of bakery products. For preparing roti, millet flour is mixed with hot water to partially gelatinize the starch. This imparts the necessary binding of particles and helps to roll the dough into thin sheets. The flattened dough is baked on a hot plate. Roti resembles wheat chapathi or maize tortilla. Mudde from millet flour is prepared by steaming the dough and making it into balls. Mudde is similar to 'TO' of Africa. Roti and mudde are eaten with dhal (legume soup) and vegetables. Millet flours suspended in cold water containing a little butter milk is left overnight for mild fermentation. Next morning the slurry is cooked to prepare porridge. Millets can substitute rice completely in the preparation of idli and dosa (steamed and baked preparations) (CFTRI Ann. Report, 1976). Millets and blackgram (*Vigna mungo*) mixed in the ratio of 3:1 are wet ground, and the mixed batter is fermented overnight. The batter is steamed to make idli or baked on hot pan to prepare dosa or wet pancakes.

1. Main hopper
2. Water tank
3. Water mixer
4. Tempering hopper
5. Plate mill
6. Sifter
7. Aspirator

Fig. 1. Mini grain mill

1. Millet grains
2. Dehusked grains
3. Husk (Hulls)
4. Dehusked and debranned millet
5. Bran

Fig. 2. Milled samples from proso millet

Enjera is a popular food item prepared in Ethiopia (Gebrekidan and Hiwot, 1981). *Enjera* is prepared by wet grinding teff, fermenting the batter and baking on hot pan similar to *dosa*. *Enjera* is nutritionally inferior to *dosa* as the former is prepared out of teff only. It is advisable that cowpea or other protein rich material should be mixed with teff to prepare *enjera*. This aspect deserves the attention of the concerned in Ethiopia. Millets can substitute wheat up to 20 per cent in bakery flours. Incorporation of higher levels of millet flours affects the texture of the products, without affecting the nutritive value (Awadalla and Slump, 1974).

NON-CONVENTIONAL FOOD PRODUCTS

Debranned small millets, when dropped in boiling water, cook soft within 5-10 min. This beneficial property of millets needs to be exploited for developing quick cooking cereals.

Flakes

Pearled grains are soaked in water, steamed or cooked under pressure to effect complete gelatinisation of the starch, dried to about 18 per cent moisture and pressed to requisite thickness between heavy duty rollers and dried to prepare flakes. Flakes hydrate quickly when added to warm water or milk and are used to prepare sweet or savoury dishes. Flakes, when deep fried, expand and form crispy products. The relatively smaller size and quick hydration of millets, make them most suitable for the production of flakes.

Extruded products

Noodle-like products could be prepared from millet flours. Noodles prepared from blends of millet and legume flours form nutritionally balanced food which could be used as supplementary or weaning foods. The pearled grains soaked in water for 1-2 days, wet ground and the mash cooked, extruded and dried, make excellent crispy product when deep fried. The quality of these products is equal to that prepared from rice. These products could be economically produced as a cottage industry, as the equipment needed are very simple, and the capital investment required is also low (Kumate, 1983).

Parboiling of millets

Parboiling of rice is a well known traditional process of processing of rice. Desikachar (1976) reported that steam treatment of finger millet hardens the endosperm, enables the production of grits, and reduces the sliminess of *mudde*. Shreshta (1972) reported that parboiling of kodo millet improved its milling quality. It is well known that parboiling of rice improves milling quality and reduces the loss of thiamine during milling. Parboiled rice is also used to prepare expanded rice, which is a precooked ready to eat product. The same may hold good with millets too. Research work in this direction may be highly fruitful.

Popping

Popping or puffing is a simple processing technique of cereals to prepare ready to eat products. Popped grain is a crunchy, porous, precooked product. Popping invariably improves taste and flavour. Among the cereals finger millet develops highly agreeable flavour on popping. The volume of popped millets ranges from 8-10 ml/g (Table 3). Popped grains find extensive usage as snacks. Popped flour blended with puffed chickpea or toasted green gram forms a nutritious food (Desikachar, 1984). Popped finger millet flour is often consumed after mixing with jaggary (brown sugar) and milk is traditionally called hurihittu. Popped finger millet flour is now produced and is marketed at cottage industry level in some places. Popped millets are also used as adjuncts in brewing. Malleshi and Desikachar (1981a) reported that to obtain fully expanded millets, the grain moisture content should be around 19 per cent and popping temperature of about 250°C. They also studied the varietal differences in popping (Malleshi and Desikachar, 1985). It is difficult to debran popped grains. Hence, popped millet meals have slightly higher fibre content.

TABLE 3

Comparative expansion of different cereals and millets during popping

Sample	Expanded Volume (ml/g)
Maize	35
Sorghum	22
Paddy	18
Proso millet	12
Kodo millet	11
Pearl millet	10
Finger millet	8
Foxtail millet	7
Little millet	7
Barnyard millet	7

Ref: Malleshi (1984).

Malting

Malting of barley in temperate countries and that of sorghum in African countries is practised on industrial level for brewing. Malting of finger millet has been a traditional process in certain parts of India (Chandrasekhara and Swaminathan, 1953a). Malted finger millet is mostly used for feeding young children and also for use in milk based beverages. Among the tropical cereals finger millet possess superior malting characteristics (Malleshi and Desikachar, 1986a). Finger millet malt possesses a highly agreeable flavour, with adequate starch hydrolysing enzymes. It is rich in calcium and sulphur amino acids and forms an ideal base for weaning food formulations (Malleshi and Desikachar, 1986a).

Recently, a weaning food of low dietary bulk and high calorie density was developed by the Central Food Technological Research Institute Mysore, using malted finger millet and green gram (*Vigna radiata*) (Fig. 3) (Malleshi and Desikachar, 1982). The food was nutritionally comparable to proprietary weaning foods (Table 4), and was readily accepted and tolerated by children (Venkat Rao *et al.*, 1985). The pilot scale production of the weaning food was standardized and the process know-how has been provided to a few enterpreneurs. Use of malted finger millet for preparation of malt extract and malt syrups (Chandrasekhara and Swaminathan, 1953b) and in brewing (Venkatnarayana *et al.*, 1979) has also been reported.

Malts from other small millets are also acceptable, however the development of amylases is low in them as compared to finger millet (Table 5). Refined finger millet malt flour blended with milk powder, sugar and a flavouring agent, forms a thickener for milk-based beverages. In addition, malt flour blended with barley malt hydrolysed to dextrin, flavoured with cocoa and vacuum

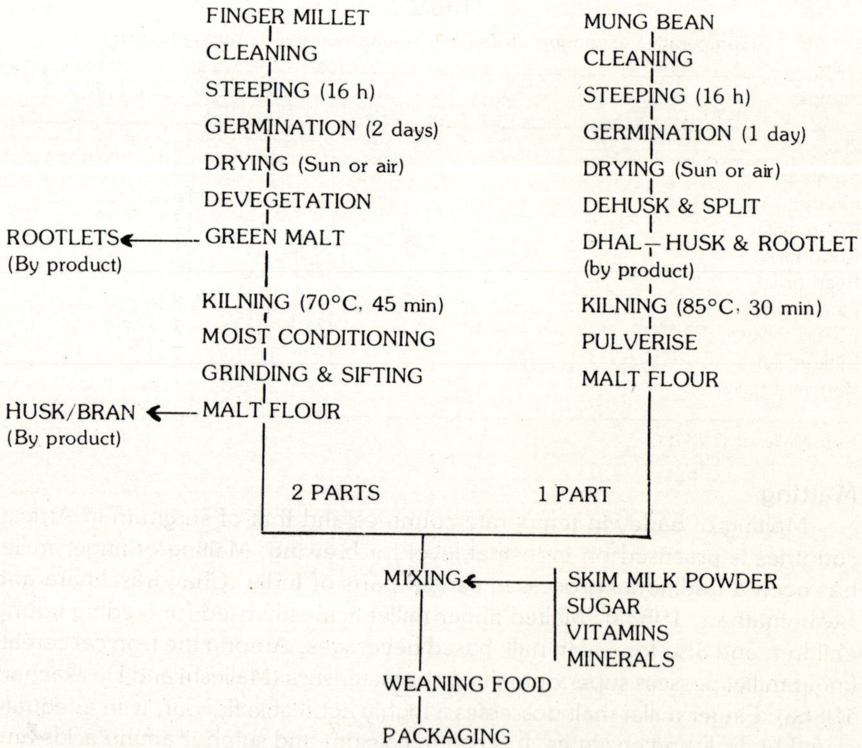

Fig. 3. Flow sheet for preparation of malted weaning food

TABLE 4
Composition of malted weaning food (MWF) as compared with proprietary weaning foods

(per 100 g)

	MWF	Farex*	Nestum*	Cerelac*	Balamul*
Moisture (g)	6.0	3.5	5.0	2.2	4.0
Protein (g)	11.5	12.0	7.5	11.0	22.0
Fat (g)	1.5	3.0	—	7.8	3.5
Crude fibre (g)	1.8	0.9	—	—	—
Total ash (g)	2.3	3.5	3.0	2.0	—
Calcium (mg)	240	750	690	275	800
Phosphorus (mg)	210	400	570	225	690
Calorie	396	350	366	422	380
PER	2.4	2.3	—	—	—
Cooked paste viscosity (15% slurry) cpu	250	9000	12000	3500	900
Market price (Rs./kg)	Rs. 15/- (estimated)	33.75	28.50	32.25	17.10

*Extract from the display on unit pack tins (1983 batches).

self-dried, forms a fluffy product, for use in milk-based beverage formulations (CFTRI Ann. Report, 1982). Thus, finger millet malt is an extremely valuable raw material in malt industry. If it is fully exploited, the import of barley malt can be reduced substantially.

STARCH PRODUCTION

Small millets are rarely used to produce starch for industrial uses. However, starches were isolated from millets and their physico-chemical properties were studied (Modi and Kulkarni, 1976; Lorenz and Hinze, 1976; Wankhede *et al.*, 1976; Paramhans and Tharanathan, 1980; Muralikrishna *et al.*, 1986; Malleshi *et al.*, 1986). Millet starches generally exhibited higher gelatinisation temperature, higher water binding capacity and slow in enzymatic hydrolysis than wheat or rice starches.

Millets contain a relatively higher proportion of unavailable carbohydrates (Kamat and Belavady, 1980) and the release of sugars from millet based diets is slow (Gopalan, 1981). These factors could be best utilised in developing special foods for diabetics.

The finger millet seed coat which contains pigments is a byproduct of the finger millet milling industry. The bran could be economically exploited for extraction of food grade colours, as there is a growing demand for natural food colours.

TABLE 5

Comparative malting characteristics of some tropical cereals and millets

Malting period	2 days		4 days		Sensory acceptability
	Amylase activity (maltose units)	Malting loss (g%)	Amylase activity (maltose units)	Malting loss (g%)	
Rice	50	3.6	105	7.0	Tastes bitter
Wheat	120	6.8	230	11.5	Acceptable
Triticale	145	4.5	260	9.2	Acceptable
Maize	50	2.9	150	6.8	Strong maize flavour acceptable
Sorghum*	67	4.0	170	8.6	Acceptable
Finger millet	165	5.5	200	16.2	Highly acceptable
Pearl millet*	170	4.0	154	9.5	Acceptable when fresh; develops bitter taste on storage within a week
Proso millet	100	3.5	145	5.2	Bland but acceptable
Foxtail millet	85	5.0	132	9.2	Bland but acceptable
Barnyard millet	48	3.9	70	7.4	Bland but acceptable
Kodo millet	10	3.4	55	7.4	Not acceptable

*Incidence of mould observed during germination.

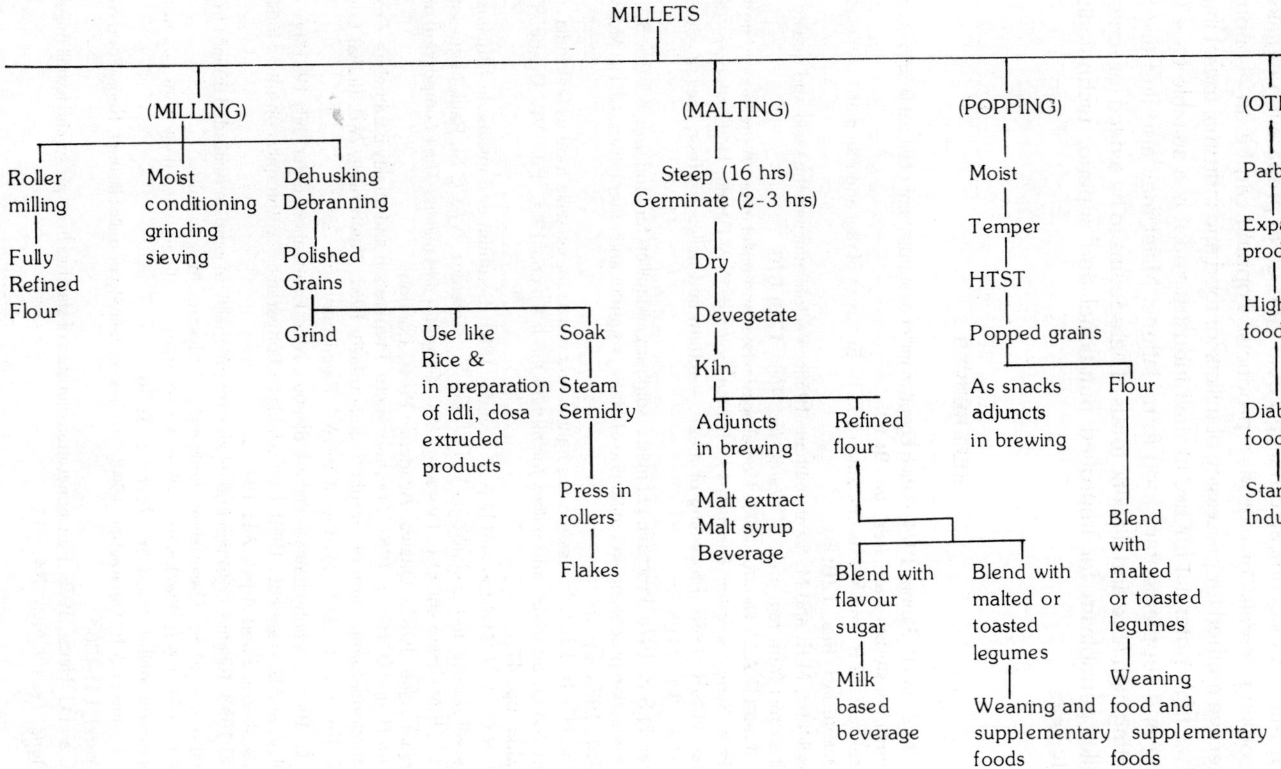

Fig. 4. Processing of small millets for food and industry

In conclusion, it may be stated that small millets can be processed to diversify their uses, to improve their nutritive value and consumer acceptability (Fig. 4). Milled or decorticated millets could be used in preparation of flakes, quick cooking cereals, or extruded products. Popping can be an economic and effective method for processing of millets for food and industrial uses. Finger millet has high potential for use in malt industry and it is a suitable base for weaning and supplementary food formulations. Millet may also find use for formulating high fibre and diabetic foods. There seems to be a need for screening millet germplasm for improved nutritional and superior technological characteristics.

REFERENCES

Awadalla, M.Z. and P. Slump. 1974. Native Egyptian millet as supplement of wheat flour in bread. I-Nutritional studies *Nutr. Rep. Int.* **9:** 59.

Chandrasekhara, M.R. and M. Swaminathan 1953a. Enzymes of ragi and ragi malt I. Amylases *J. Scient. ind. Res.* **12B:** 51.

Chandrasekhara, M.R. and M. Swaminathan. 1953b. Factors affecting the yield and quality of malt extract from ragi malt. *J. Scient. Ind. Res.* **12B:** 610.

Crabtree, J. and D.A.V. Dendy. 1977. Comilling of wheat and millet grain. A preliminary study. In: *Proc. Symp. Sorghum and millets for human food.* Ed. DAV Dendy. TPI London pp. 125.

Desikachar, H.S.R. 1975. Processing of maize, sorghum and millets for food uses. *J. Scient. ind. Res.* **34:** 231.

Desikachar, H.S.R. 1976. Processing of maize, sorghum and millets for food uses in India. Symp. on production processing and utilization of Maize, sorghum and other millets. CFTRI Mysore (India), 1976, pp. 38.

Desikachar, H.S.R. 1977. Processing of sorghum and millets for versatile food uses in India. *In: Proc. Symp. Sorghum and millets for human food,* Vienna 1966, Ed. DAV Dendy. TPI London. pp. 41.

Hoseney, R.C., E.V. Marston and D.A.V. Dendy. 1981. Sorghum and millets. In: *Advances in cereal science and Technology,* Vol IV, Ed. Y. Pomeranz, A ACC St. Paul, Minnesota.

Hulse, J.H., E.M. Laing and O.E. Pearson. 1980. *Sorghum and millets: Their composition and nutritive value.* IDRC. Ottawa, Academic Press, London.

Gebrekidan, B. and B. Hiwot. 1981. Sorghum injera. Preparation and quality parameters. *Proc. International Symposium on Sorghum grain quality* 1981, Patencheru, A.P. (India) Eds., L.W. Rooney and D.S. Murthy, ICRISAT, Patencheru. pp. 55.

Gopalan, C. 1981. Carbohydrates in diabatic diet bull. Nutr. Foundation, India (July 1981) pp. 3.

Kamat, M.V. and B. Belavady. 1980. Unavailable carbohydrates of commonly consumed Indian foods. *J. Sci. Food Agric.* **31:** 194.

Kumate, J. 1983. Relative crispness and oil absorption quality of *sandige* (extruded dough) from cereal grains. M.Sc. Dissertation, University of Mysore, Mysore (India).

Kurien, P.P. and H.S.R. Desikachar. 1966. Preparation of refined white flour from ragi using a laboratory mill *J. Food Sci. Technol.* **3:** 56.

Kurien, P.P. and H.S.R. Desikachar, 1962. Studies on refining of millet flours I. Ragi. *Food Sci. (Mysore)* **11:** 136.

Lorenz, K. and G. Hinze. 1976. Functional characteristics of starches from proso and foxtail millets. *J. Agric. Food Chem.* **24:** 911.

Lorenz, K., W. Dilsaver and L. Bates. 1980. Proso millets, milling characteristics, proximate compositions, nutritive value of flours. *Cereal Chem,* **57:** 16.

Malleshi, N.G. and H.S.R. Desikachar. 1981a. Studies on the suitability of roller flour mill, hammer mill and plate grinder for obtaining refined flour from malted ragi. *J. Food Sci. Technol.* **18:** 37.

Malleshi, N.G. and H.S.R. Desikachar. 1981b. Varietal differences in puffing quality of ragi. *J. Food Sci. Technol.* **18,** 30.

Malleshi, N.G. and H.S.R. Desikachar. 1982. Formulation of a weaning food with low hot paste viscosity based on malted ragi and green gram. *J. Food Sci. Technol.* **19:** 193.

Malleshi, N.G. 1984. Studies on malting of millet grains. Ph.D. Thesis, University of Mysore (India).

Malleshi, N.G. and H.S.R. Desikachar. 1985. Milling, popping and malting characteristics of some minor millets. *J. Food Sci. Technol.* **22:** 400.

Malleshi, N.G. and H.S.R. Desikachar. 1986a. Studies on comparative malting characteristics of some tropical cereals and millets. *J. Inst. Brew.* **92:** 174.

Malleshi, N.G. and H.S.R. Desikachar. 1986b. Influence of malting conditions on the quality of finger millet malt. *J. Inst. Brew.* **92:** 81.

Malleshi, N.G., H.S.R. Desikachar and R.N. Tharanathan. 1986. Physicochemical properties of native and malted finger millet, Pearl millet and foxtail millet starches. Die Staerke (In Press).

Modi, J.D. and P.R. Kulkarni. 1976. Studies on the starches of ragi and redgram. *J. Food Sci. Technol.* **13:** 9.

Muralikrishna, G., N.G. Malleshi, H.S.R. Desikachar and R.M. Tharanathan. 1986. Effect of popping on the properties of some millet starches. *Die Staerke* **38:** 48.

Paramhans, S.V. and R.N. Tharanathan. 1980. Scanning electron microscopy of enzyme digested varagu. *Die Staerke* **34:** 73.

Rachie, K.O. and L.V. Peters. 1977. *The Eleusines.* ICRISAT, Hyderabad, India.

Shankara, R., N.G. Malleshi, H. Krishnamurthy, M.N. Narayana and H.S.R. Desikachar. 1985. Development of mini grain mill for dehusking and grinding of cereals. *J. Food Sci. Technol.* **22:** 91.

Shreshta, K.B. 1972. Dehusking of varagu (*Paspalum scrobioulatum*) and its utilization for edible purposes. M.Sc. Dissertation, University of Mysore, Mysore, India.

Venkatnarayana, S., V. Sreenivasmurthy and B.A. Sätyanarayana. 1979. Use of ragi in brewing. *J. Food Sci. Technol.* **16:** 204.

Venkat Rao, S., S. Kurien, D.N. Swamy, V.A. Daniel, I.A.S. Murthy, N.G. Malleshi and H.S.R. Desikachar. 1985. Clinical trials on a weaning food of low bulk based on ragi and green gram. Paper presented at 'International Workshop on Weaning Foods', Iringa, Tanzania.

Wankhede, D.B., A. Shehnaj and M.R. Raghavendra Rao. 1976. Preparation and physicochemical properties of starches and their fractions from finger millet and foxtail millet. *Die Staerke* **31:** 153.

35

SCOPE FOR USING SMALL MILLETS AS FORAGE IN INDIA

S.R. Sampath

INTRODUCTION

Time and again it has been emphasized that the animal production industry in the country can only be developed to yield results when sufficient feed input is available. In the absence of sufficient feed, there is considerable risk in improving animal production.

The development and spread of dwarf varieties of food crops has played a significant role in increasing the grain production of the country in recent years. But, this has reduced the bulk and fibrous portions which otherwise would have been available for meeting the needs of large animals.

In the absence of sufficient land for pasture or forage production, the only alternative, is to augment the feed resources by developing small millets as crops for livestock feeding. There is also a need for obtaining this feed in increased quantities and with higher nutrient content. An attempt has been made in this paper to examine the utility of small millets forage and crop residues as feed for small and large animals.

Finger millet forage

Locally known as ragi (*Eleusine coracana* Gaertn.), finger millet is considered to be a cultigen of the wild species *E. indica,* a species which is native to India and Africa and now seen throughout the warm regions of the world. It is an important crop in India and east Africa, and is the staple food of a large section of population. The grain is also used for malting and brewing.

Finger millet is grown extensively in India in dry regions as a rainfed crop. Finger millet grain contains 7-8 per cent protein, about 73 per cent carbohydrates, 0.33 per cent calcium and 0.27 per cent phosphorus. This is considered to be poor man's staple food and is consumed by people more in the rural parts, particularly by the working community than by the elite. Finger millet is also used in feeding infant calves and growing animals. The sick and convalescing animals are given a gruel made out of finger millet flour.

In some parts, finger millet is grown exclusively for forage and gives 13-15 tonnes of green forage per acre in three cuttings. The chemical composition of finger millet silage and straw is given in Table 1.

Irrigated crops of finger millet yield 6 to 10 tonnes of dry straw per hectare while rainfed crops yield 2-3 tonnes per hectare. The straw obtained from the rainfed crop is relished more by cattle compared to straw obtained from the irrigated crop. Straw of irrigated crop is tough and fibrous and hence is less palatable. Under irrigation with proper management, 12-14 tonnes of green fodder per hectare can be obtained in about 50 to 60 days.

Evaluation of 50 genotypes from different states in India has revealed the superiority of the varieties from Uttar Pradesh, Madhya Pradesh and Karnataka for fodder yield, internodal length, leaf length and plant height. The varieties from Tamil Nadu, Bihar, Orissa and Andhra Pradesh showed high tillering.

Studies in Japan have shown that the fodder yield, crude protein content, cell wall constituents, non-structured carbohydrates and dry matter digestibility of finger millet is higher compared to that of Rhodes grass and Italian rye grass. Several studies in India have shown that application of inorganic nitrogen, along with mycorrhizal fungus and phosphorus solubilising bacterium increased the mineral contents in the finger millet straw.

Finger millet straw is used in many parts of the country for feeding all categories of animals, such as working animals, milch animals and dry animals. This serves as an important source of dry fodder which is a must in the daily ration at least in small quantities. It is said that for all kinds of cattle, ragi straw is superior to that of rice.

Little millet forage

Little millet (*Panicum miliare* Lank.) locally known as *same* is grown on poor soils and the grain is a famine reserve. The straw is thin stemmed and liked by cattle. The average yield of straw is only 8 to 12 q/ha. It is also poor in quality although cattle eat it readily on account of its thin and leafy nature. It is grown in India on 666,000 hectares as a dryland crop on both black and red soils in the *kharif* season in areas where the annual rainfall is less than 750 mm. The production of little millet in the country is about 79,000 tonnes, with Karnataka state producing about one-eighth of this. The straw is a byproduct after removal of the grain. It is estimated that about 16,700 tonnes of little millet straw is available in Karnataka state annually for feeding the cattle. The nutrient utilization of straw was evaluated at NDRI, Bangalore. Six adult dry non-producing and non-pregnant crossbred cows of comparable body

TABLE 1

Chemical composition and digestibility coefficient of finger millet, kodo millet, foxtail millet and proso millet (per cent on dry basis)

Stage/type of forage	Composition				Digestibility				Reference
	Crude protein	Ether extract	Crude fibre	Nitrogen free extract	Crude protein	Ether extract	Crude fibre	Nitrogen free extract	
Finger millet (*Eleusine coracana*)									
Fresh late vegetative	7.6	1.1	33.6	42.6	—	—	—	—	
Fresh dough stage	7.1	1.7	28.8	49.9	—	—	—	—	
Silage	3.6	1.5	38.8	46.2	8	44	69	52	Mudgal and Sampath (1970)
Straw	3.4	1.3	37.2	50.0	12	47	70	61	Mudgal and Sampath (1970)
Kodo millet (*Paspalum scrobiculatum*)									
Fresh early vegetative	11.4	1.4	28.8	44.1	—	—	—	—	Sen and Ray (1971)
Fresh dough stage	5.7	1.5	31.6	49.1	—	—	—	—	Sen and Ray (1971)
Straw	3.5	1.5	34.3	48.4	—	—	—	—	Sen and Ray (1971)
Foxtail millet (*Setaria italica*)									
Green	9.7	2.7	31.4	47.8	61	59	67	67	Morrison (1948)
Hay	8.2	2.7	25.3	44.7	60	64	62	57	
Setaria intermedia (Hay)	8.6	1.0	37.1	39.9	—	—	—	—	Amrith Kumar and Sampath (1973)
Setaria italica (straw)	8.2	1.2	32.9	42.1	—	—	—	—	Mudgal and Sampath (1970)
Proso millet (*Panicum miliaceum*)									
Green	8.1	2.4	29.9	52.2	—	—	—	—	Morrison (1948)
Hay	12.5	2.5	33.9	44.5	56	41	60	61	Morrison (1948)
Straw	4.8	1.2	35.5	49.6	—	—	—	—	Morrison (1948)

weights were fed with straw of little millet for a period of 30 days and during the last 7 days, a metabolism trial was carried out. The straw contained 4.13 per cent crude protein, 37.86 per cent crude fibre, 0.41 per cent calcium and 0.06 per cent phosphorus on dry weight basis. The dry matter intake by the animals ranged from 1.08 kg to 2.04 kg with an average of 1.52 kg/100 kg body weight during the preliminary feeding period. During metabolism period, the intake was 1.61 kg. The balance of nitrogen was marginally negative. The nutritive value in terms of digestible crude protein (DCP), starch equivalent (SE) and total digestible nutrients (TDN) of the material worked out to 0.55, 27.92 and 47.07 kg respectively for 100 kg dry matter. The straw is a good source of roughage for cattle (Shivaramaiah and Sampath, 1981).

Kodo millet forage

Kodo millet or scrobic millet (*Paspalum scrobiculatum*) is known locally as "Varagu" or "Haraka". This millet is of considerable importance in the Deccan region. It is an annual, drought resistant crop well suited to drier conditions and gravelly soils. The straw yield is low, one to two tonnes per hectare with 2.3 per cent protein. The grain is used as standby food in famine years.

The wild form is smaller and the grain sometimes reported to be poisonous. The nutrient content in different stages of crop growth is shown in Table 1.

Foxtail millet forage

Italian millet or foxtail millet (*Setaria italica*) locally called as 'Navane' is reported to have been cultivated in China as early as 2700 B.C. and introduced later to Europe. In India, it is raised on poorer soils in drier regions. The crop is hardy and less prone to diseases or pests.

The crop can be cut at 45 days after sowing for converting into hay for feeding cattle and buffaloes. About 70 to 90 quintals of hay per hectare may be obtained from an irrigated forage crop cut at the flowering stage. It is not reported to be suitable for introducing in pasture. The chemical composition of foxtail millet straw is shown in Table 1.

Straw of foxtail millet is thin stemmed and liked by cattle in south India. But, it is considered injurious to horses if they are fed exclusively on this.

Proso millet forage

This millet, locally known as 'Baragu', is a catch crop grown for grain in many parts of tropics and subtropics. It is a millet of the old world and since pre-historic times has been an important grain crop for human food.

The grain serves as poor man's standby and famine reserve food. The grain yield ranges from 2.8 to 5.6 q/ha under rainfed conditions and the straw yield is about a tonne per hectare. Under irrigation the yield is doubled.

Proso millet gives satisfactory results when used as a substitute feed for other grains. It should be ground before use except in poultry feeding. The ground proso millet is reported to be worth from 75 to 90 per cent as much as corn for fattening cattle, lambs and fully 90 per cent as much for laying hens. The chemical composition of proso millet fodder is given in Table 1.

In retrospect

The country's economy depends on agricultural production and animal production is a complementary enterprise which when dove-tailed to the former in an integrated way, can bring in considerable relief to the rural poor. Animal production programme requires quality inputs in terms of feed in order to be viable. In this context, the millet crops which are now confined to small areas could be developed to supply the forage needs for sustaining animal production industry, economically.

The nutritive value of the forage depends upon its chemical composition, digestibility and the nature of digested products. In addition, the amount of forage consumed by the animal is important as it affects the total nutrient intake and the animal's response to such factors as acceptability, rate of passage and presence of undesirable substances that may be contained in forage is to be reckoned with.

From the work carried out on various millet crops and varieties developed, there is considerable scope for utilizing the millet crop as forage by harvesting it at the appropriate time and feeding the same to the animals. It is also possible to take 2 to 3 cuttings of the millet crop so as to obtain an increased tonnage and thus render it economical. For instance, a forage yield in finger millet crop was estimated to be 13.4 tonnes per acre in 3 cuttings.

By using improved technologies and high yielding varieties small millet forage of acceptable quality and nutrients can be produced. The response to application of small quantities of fertilizer and harvesting at appropriate stage for maximum nutrient utilization needs to be studied. The institutions are to carry out intensive research in this direction.

REFERENCES

Amrith Kumar, M.N. and S.R. Sampath. 1973. Seminar on recent advances in Milk Production Technology. Feb. 27-28, 1973, U.P. Inst. of Agric. Sci., Kanpur.

Morrison, F.B. 1948. *Feeds and Feeding*. 20th Edn. 1948. Ithaca, New York, Morrison Public Co.

Mudgal, V.D. and S.R. Sampath. 1970. Review of research work carried out in Animal Nutrition, Southern Regional Station, National Dairy Research Institute, Bangalore.

Sen, K.C. and S.N. Ray. 1971. *Nutritive Value of Indian Cattle Feeds and the Feeding of Animals*, Sixth Edn. Indian Council of Agricultural Research, New Delhi.

Shivaramaiah, M.T. and S.R. Sampath. 1981. *Mysore J. Agric. Sci.*, **15:** 549.

36

UTILIZATION OF SMALL MILLETS IN CHINA

Chen Jiaju

Both foxtail and proso millets are native to China and have been cultivated for thousands of years. Millet products are popular and used in many ways. The main purpose of their cultivation is for food, but the straw is also used as feed.

The foxtail millet grain is husked for human consumption and most cultivars are non-glutinous. Millet gruel has a pleasant flavour and is not considered as a coarse diet. They have been used by royal families of old China, such as Qinzhouhany, Jinmi, Lonshanmi and Taohuami. Millet flour combined with legumes or by itself is used for making cakes. Glutinous millet is mainly used for making cakes and it can replace glutinous rice. Proso millet can also be used in place of glutinous rice.

Husked millet is used in the food industry also. Beer made from millet has a special flavour. It is also an ingredient for making vinegar. The grain is used in Chinese medicine and in the preparation of medicinal food. Its effect may partially be explained by its chemical composition.

Most foxtail millets are of high nutrient quality. Their protein content ranges from 10 to 15 per cent and they are rich in unsaturated fatty acids. Among the 2684 varieties analysed, 13 varieties had protein content higher than 15 per cent. For example the variety Hebai Haigu has protein content of 17 per cent, Hebai Maogu has 16 per cent protein in which lysine content is 0.27 and 0.26 per cent respectively. The lysine contents of 0.3 per cent are also found in some varieties. The oil and fat content of most accessions are 3-4 per cent while some have more than 5 per cent. The results of anlaysis for 6 kinds of fatty acids in 260 varieties showed that 171 varieties have more than 75 per cent linoleic acid content. The linoleic acid content in non-glutinous varieties is little lower than that of glutinous varieties (Table 1).

TABLE 1
Nutritional quality analysis of foxtail millet

Constituent	Number of accessions analysed (land races)	Result
Protein	2684	Higher than 15% (13 lines)
Lysine	2684	Higher than 0.3% (some)
Lipids	2684	Higher than 5% (some)
Linoleic acid	260	Higher than 75% (171 lines)
Vitamin A	100	Average 74 I.U.* highest 200-394 I.U.
Vitamin B_1	100	Average 0.63 mg/100 g, highest 1.03 mg/100 g
Vitamin E	100	Glutinous 13.2 μg/100 g, non-glutinous 10.4 μg/100 g

*I.U. = International Unit

Vitamin contents of 100 varieties are tabularized in the book "Manual of Cultivars of Chinese Foxtail Millet". Table 1 shows that the average content of vitamin A is 74 Iu. While the average content of vitamin B-1 is 0.63 mg per 100 grams, the highest is 1.03 mg per 100 grams. The average content of vitamin E is 10.4 μg/100 gram in non-glutinous varieties and 13.2 μg/100 gram in glutinous varieties. Selenium contained in foxtail millet was found valuable for improvement of human heart blood vessel diseases.

The grain is very suitable for storage. The hard glumes and high content of silicon in grains protect them from moisture and insect damage.

In addition, the grain is a good concentrated feed. According to farmers' experience, hens fed with millet grain lay more eggs. It is also a favourite feed for the cage birds.

Millet is mainly grown for self-consumption by local farmers and very little of the produce enters trade channels. Nevertheless, in the market, the price of millet is equal to rice.

During 1930's, the export of millet grains was considerable but it is negligible today. It is because the millet grain is recognized only as feed by the importing countries, whereas in our opinion, it is a good food grain.

Apart from the grain, the straw is also very important for the grower. Millet straw is soft and stores well. It is an indispensable forage for livestock in northern China. In the growing region of foxtail millet, livestock like horses and oxen are used as draft animals. Generally one horse consumes a hectare of millet straw annually. By value the straw makes up 1/4 to 1/2 of the millet income. In special cases, when in shortage, the price of straw by weight equals that of the grain.

TABLE 2
Composition of straws of certain crops

Crop	Water (%)	Protein (%)	Fat (%)	Fibre (%)	N-free extract (%)	Ash (%)	Calcium (%)	Phosphorus (%)
Foxtail millet	10.8	3.16	1.35	3.15	44.30	9.02	0.32	0.14
Wheat	8.6	2.48	1.64	35.16	43.07	8.50	0.21	0.36
Rice	12.6	3.07	1.65	22.90	44.20	15.60	0.18	0.09
Beans	11.5	9.36	1.72	3.19	38.50	5.75	0.15	0.10

According to the information of Institute of Animal Science, Chinese Academy of Agricultural Sciences, the nutrition of millet straw is superior to that of other major crops, such as wheat and rice, but inferior to the stalk of beans (Table 2).

We notice in literature, that the foxtail millet hay fed to horses may cause injurious effects. This injury is caused by a glucoside called setarin. But such an injury has not been detected in our country.

37

PROCESSING AND UTILIZATION OF FINGER MILLET IN UGANDA

J.P.E. Esele

Finger millet has a high nutritive value, being especially high in carbohydrates. Although it is less rich in protein than other cereals, its biological value is high. According to Purseglove (1975), the prolamine (or storage protein) of finger millet, called eleusinin, has high biological value, with a good content of cystine, tyrosine, typtophane and methionine, which are important in the prevention of kwashiorkor disease, but eleusinin is low in lysine. The grain is a rich source of calcium, containing 0.33 per cent compared with 0.01-0.06 per cent in many other cereals. It is also rich in phosphorus and iron.

Bredon (1962) found wide variations in the chemical composition of the two major types of finger millet in Uganda—white and brown (Table 1). The carbohydrate component is high and the protein content varies from 6 to 11.3 per cent.

TABLE 1
Composition (%) of finger millet grown in Uganda

Type of finger millet	Protein	Fat	Carbohydrate	Fibre	Ash
White grained	7.2-11.3	4.1-7.7	61.0-72.7	0.7-7.8	2.0-5.0
Brown-black grained	6.6-9.2	1.1-1.5	69.6-76.1	2.4-5.0	2.4-4.7

Finger millet in Uganda is mainly utilized as food, in making local beer and, to a limited extent, fed to livestock both forage and grain and its by-products.

Utilization as food

As food, finger millet is consumed in the form of stiff porridge (Ugali) and thin porridge. White coloured grains are more preferred for food than the dark coloured ones. After harvesting the millet heads are kept in a heap for 3-5 days to further ripen the grain and to give it a desirable taste. The millet is dried under the sun and stored. It is taken out in required quantities as and when needed.

The grain is separated by beating with sticks or, if required in small quantities, dehusking is done with a wooden mortar and pestle; then winnowed. Almost invariably, the millet is mixed with dried cassava or sweet potato chips or with sorghum and ground. In some places in the north and in Karamoja finger millet is eaten sole. The ratio of cassava/sweet potato to finger millet is about 5 to 1. Grinding is done by hammer mill, but most commonly in rural areas, the millet is ground between two stones. The flour is mixed with boiling water until it becomes stiff and this is the ugali. The ugali is eaten along with a vegetable such as cowpeas, groundnuts, sesame, meat, chicken, etc. This is the food habit in eastern and western Uganda. The porridge is also made with boiling water usually with sole millet and it is not mixed with cassava/sweet potatoes. In the east and north, millet porridge is a minor meal, being mainly served to lactating or pregnant women and to children. In the south and west, porridge is an important meal served to guests and visitors.

Utilization for beer

Millet for beer is not mixed with cassava or sweet potatoes. It is normally utilized sole. However, further north and in Karamoja, it is mixed with sorghum but millet forms the major component in the mixtures. The dark coloured grains are preferred for beer since they tend to make stronger beer. The millet is ground into a rough (not as fine as for food) flour which is mixed with cold water and left underground or in old tins or saucepans for 7-10 days. It is then roasted on a large pan with plenty of fire. The roasted millet malt can be eaten straight away especially by children. It is either mixed with sugar or with fresh milk and eaten. It is a delicious meal. The roasted malt is dried under the sun. The beer is made by adding yeast to the malt. Yeast is made by germinating millet grains. The grains take 3-5 days to sufficiently germinate depending on the variety. After germination, the seed (yeast) is dried and ground.

The dried roasted malt is mixed with the yeast at the ratio of 5 to 1 in cold water. The next day, the mixture is sweet and non-alcoholic. At this stage it can be eaten especially by children or drunk in hot water. This sweet mixture is often served to non-alcoholics. If more yeast is added, the mixture becomes alcoholic the following day. The beer is drunk with hot water using straw. This is an important drink in all the rural areas of the country and is also becoming important in urban areas. It is the major drink for family at home also. It is also served to visitors and on all ceremonial functions including marriages and funerals. Depending on the variety, the beer can be kept up to five days before becoming sour. It is discarded if it goes sour. The residue is normally thrown out but also fed to chickens and pigs. In famine years, this residue is dried and mixed with cassava or sweet potatoes and made into food. Finger millet beer is sold in both rural and urban areas and is a source of income.

Utilization as forage

Finger millet in Uganda is not generally produced as a livestock feed. In the rural areas, the utilization is almost 100 per cent as a human food. Chickens may also be fed with grain. After harvesting, cattle, goats and sheep are grazed on the straw in the field. Very long strawed types are cut in western Uganda for thatching of roof of granaries.

REFERENCES

Bredon, R.W. 1962. Chemical composition of some foods and feeding stuffs in Uganda, Dept. of Veterinary Services, Entebbe, Uganda.

Purseglove. 1975. Tropical crops. Monocotyledon, I and II. English Language Book Society and Longman, London.

VII

DISCUSSIONS AND RECOMMENDATIONS

REPORT OF GENETIC RESOURCES DISCUSSION GROUP

Chairman: G. Harinarayana
Discussant: G. Harinarayana

1. The need for updating the lists of available small millets germplasm accessions has been recognised.
2. The participating scientists felt the need for free exchange of small millets genetic resources.
3. Each of the participating countries is requested to identify the gaps in collecting small millets in their agroclimatic regions and/or provinces and arrange for collecting missions. It is desirable to associate breeders in collecting expeditions.
4. Besides collecting from farmers fields and markets, it is desirable to collect land races of cultivated small millets species in wild, as well as wild relatives of cultivated species.
5. Recognising the significance of environmental diversity, it is proposed to evaluate the germplasm in multilocational, National/International nurseries.
6. To obtain uniformity in evaluation, the participating countries are requested to use descriptors for various small millets as published by IBPGR, Rome.
7. Biochemical and processing characterization of small millets germplasm appears worthwhile. Identifying sources of useful genes from the already available and conserved germplasm should receive top priority since utilization of germplasm is very limited in small millet crops.
8. As the descriptor lists are likely to be voluminous and may have limited circulation, the utilization of the germplasm could be accelerated by circulating pocket size editions of descriptor lists with limited characters like plant height, maturity, pest and disease resistance, grain yield, etc.

REPORT OF BREEDING AND VARIETAL IMPROVEMENT DISCUSSION GROUP

Chairman: H. Doggett
Discussant: A. Seetharam

1. The small floret size in all small millets has limited the artificial hybridization and recombination breeding. The contact method and hot water emasculation method used to some extent have their own limitations. So, there is need to overcome this problem by possibly studying induced male sterility using gametocides and there is need to standardize these methods.

Genetic male sterile systems and mechanisms like protogyny may also be investigated and confirmed. We may also look for outcrossing systems wherever available.

2. All small millets are inbreeders. There is not much work done on the application of various breeding procedures and assessing their relative efficacy.
3. Mutation breeding could be one of the methods thought of in small millets since artificial hybridization is difficult. The possible application of this breeding procedure in different areas of varietal improvement needs to be fully exploited.
4. Single plant selections from germplasm accessions could be one of the simplest and effective means of obtaining superior genotypes.
5. The application of biotechnology particularly anther culture and ovule culture techniques and exploitation of somaclonal variability in callus cultures could be thought of as a method for evolving varieties rapidly.
6. Quality breeding in small millets is also important although it is a very difficult area to make any headway in short period of time. Seed protein, mineral content and malting quality are some of the areas for consideration. Screening of available germplasm for consumer, nutritional and processing quality characters may be the first step in this direction.
7. There is a need to understand more about the genetic control of various yield and yield contributing characters.
8. Identification of varieties with wide adaptation, high yield stability, and differing photoperiod sensitivity is important particularly in finger millet, foxtail millet and proso millet as these crops are grown in varying rainfall areas, temperature regimes and day lengths. Besides, evolution of varieties for different cropping systems, both relay and mixed cropping should also be attempted.
9. As all small millets are essentially rainfed crops confining themselves to semi-arid tropics, breeding of drought tolerant varieties is important. The initial step in this direction will be screening of all available germplasm and identifying the useful lines.
10. All small millets are low input crops and grown by poor and marginal farmers. Under such situations use of pesticides for the control of pests, diseases and weeds is neither feasible nor practicable. Inbuilt resistance is the best way of tackling this situation and this should receive high priority. Some of the important diseases and pests that could be considered for resistance breeding are blast in finger millet, shootfly in proso millet and little millet, smut in kodo millet, foxtail millet and proso millet.
11. Following identification of new practices like seeds, fertilizers, weedicides, fungicides, pesticides, etc. testing on farmers' fields for obtaining their acceptance, laying of large scale demonstrations for exhibiting the production as well as economic potential and scientists involvement in lab-to-land programmes for mutual information transfer require attention.

12. The concern for supplying certified seed of high quality small millets including seed treatment with pesticides and fungicides has been expressed. The mechanisms of seed production and distribution were discussed. Seed village concept and distribution of seeds through exchanging 'new' for 'old' varieties to farmers can be explored.

REPORT OF PRODUCTION TECHNOLOGY AND CROPPING SYSTEMS DISCUSSION GROUP

Chairman: K. Krishnamurthy
Discussant: H. Doggett

1) Intercropping

There are benefits from intercropping for small farmers, the use of legumes may help yields. Intercropping may be a practical way of growing a variety of crops needed for the household.

2) Line sowing

This could be an efficient improvement especially where weeds are serious. In Africa, this may require mechanization using animal draught. There is a whole technology developed in India which needs to be transferred to Africa, including the village maintenance services provided by blacksmith and carpenters.

3) Sowing time

In Africa, time of planting is often of critical importance for yield; working up a fine seed bed could take too much time, especially without animal draught power.

4) Weed control

In Nepal, transplanting is always used and it is useful in filling gaps in the stand. Transplanting deserves a lot more emphasis, as there is evidence of yield benefits. Most important, transplanting provides a way of seeding in time into a nursery and of a weed control by thorough cleaning of the remaining (5/6th) of the area being cultivated. Wild finger millet in Africa presents a daunting weed problem for other types of control. In Africa, *Striga* can become a serious problem, especially in pure stands of finger millet.

5) Organic manure

The use of farmyard manure and compost was mentioned by several delegates, and this important traditional practice deserves study to determine most effective methods of preparation and application. Results in India suggest benefits from incorporating legume residues *in situ*.

6) Transfer of technology

Recommended practices, should be developed in close consultation with the farmers, giving special attention to the limiting factors in their existing systems. Where recommended practices have been developed, surveys should be made to determine which factors are being adapted and which are not being adapted. The various constraints preventing the adaption of the latter should be carefully evaluated and alleviated as far as possible.

REPORT OF PESTS AND DISEASES DISCUSSION GROUP

Chairman: K. Krishnamurthy
Discussant: J.P.E. Esele

A. DISEASES

1. Blast is the most serious disease of finger millet. It appears at different stages of plant growth. Different biotypes of blast fungus also could be present as in other crops, and this area of racial differentiation of blast is worth investigating. Development of varieties with effective genetic resistance is a high priority.
2. Viral diseases on finger and foxtail millets have been reported from many countries. Insects are generally associated with viral transmissions. There is need to understand the biology of vectors and their relationship with viruses in the transmission of viral diseases.
3. Leaf spot diseases are important in several African countries on finger millet. Resistance breeding is the best and most effective in preventing yield losses.

B. PESTS

1. Birds and aphids have been identified as serious pests of finger millet in Africa. Stem borers occur on finger millet in Asia. Finger millet earhead pests deserve attention in Africa and Asia.
2. Shootfly has been identified as the most predominant pest on small millets in Asia and the USSR. Shootfly escape mechanisms by manipulating sowing time and other cultural practices, if identified for various regions of the production, would be very helpful.
3. The biology and dynamics of most important pests of small millets, the occurrence of alternate and/or collateral hosts and screening of the available germplasm in pest sick plots deserve investigations.

4. Besides inbuilt resistance and cultural control of pest populations, critical studies on the occurrence of natural enemies, their breeding and release in farmers' fields are required. The occurrence of pests and losses in mixed, inter and relay cropping systems is to be investigated.

C. PEST AND DISEASE CONTROL

1. Considering the economic value and status of small millets, it can be said that integrated disease/pest control measures are to be adapted with emphasis on in built resistance, cultural control, biological control and chemical control in that order.

REPORT OF FOOD USES DISCUSSION GROUP

Chairman: P. Pushpamma
Discussant: N.G. Malleshi

1. Screening of germplasm for malting and popping characteristics and breeding varieties for improved malting and popping characteristics.
2. Development of simple milling machinery and making it available in millet growing areas.
3. Diversification of uses of small millets and development of health or speciality foods from millets: diabetic foods, high fibre foods, weaning foods, flakes, quick cooking cereals, etc.
4. While breeding varieties, attention should be paid to retain the desirable qualities of millets such as good storage quality and high mineral content.
5. Analysis of varieties grown under different agroclimatic conditions for nutrient content and polyphenols.
6. Studies on fine structure of millet grains.
7. Setting up of a permanent millet quality laboratory for analysing or testing the varieties and taking research work on millet product formulations.
8. Improving status of millets by substituting rice and other major cereals with millets in prestigious foods.
9. Publication of booklets on millet products and giving wide publicity for these.
10. Consumer preference checks for different millet products and improving their quality and consumer acceptability.

REPORT OF FORAGE USES DISCUSSION GROUP

Chairman: P. Pushpamma
Discussant: B.R. Hegde

1. Millet straw especially from small millets is highly relished by cattle. The cost of straw in several millets compensates the cost of inputs in cultivation. Moreover the peasant farmers depend on this by-product to a large extent in maintaining their milch and draught animals. In many countries the farmers are willing to sacrifice grain but not fodder. This point has to be kept in mind in future breeding programmes.
2. Efforts are essential to identify suitable small millet varieties for fodder purpose. Millets as green forage crops assume more importnce especially in scarce rainfall years. Carefully selected varieties should be able to produce large amount of green fodder in a very short time.
3. Types suitable for multi-cuttings have to be identified which can be grown even under irrigated conditions.
4. Suitable millet varieties have to be identified for establishing in waste lands and eroded lands. Methods of establishing such crops also need to be worked out.
5. The dry straw of many millets may not be nutritious. Methods to fortify such low value fodder by use of urea, molasses, etc. have to be worked out and popularised.
6. The other possibility of improvement of the quality of the straw would be to include a legume along with the millet crop itself.

GENERAL RECOMMENDATIONS

Chairman: H. Doggett
Discussant: K.W. Riley

1. The papers presented at the International Workshop along with the discussions and the recommendations be published for wide circulation. The International Development Research Centre is requested to fund the publication of the proceedings of the First International Workshop on Small Millets.
2. Among the small millets, finger, foxtail and proso millets appear to have wider clientale and their importance and development aroused considerable interest. The reasonable consistency in performance, the ecological range and the production potential of kodo and barnyard millets have been noted.

The role of teff in the Ethiopian economy with its possible extensions elsewhere has been recognised. Little millet with limited production potential has its areas of specific adaptation. The growing demand for food and a variety of food products calls for interest and investment in their development.

3. The need for International Small Millets Research Institute was felt. The CGIAR system may consider this proposal.
4. A steering committee of 5 members has been elected and charged with the responsibilities of identifying a network for exchanging germplasm, transferring information and look into other service facilities. The members of the steering committee are:
5. The need for obtaining definite figures of area, production and productivity from different countries has been pointed out. This may help the planners and the policy makers for deciding priorities. This may also attract the attention of the international community of nations.
6. The need for organizing international small millets adaptation trials was felt.
7. The need for exchanging published literature on small millets through a nodal agency has been recognised.
8. The need for encouraging scientific visits and providing expertise where necessary has been felt.
9. The need for providing training facilities to scientific and technical personnel has been expressed.

Dr. A. Seetharam	Chairman
Dr. Seyfu Ketema	Co-chairman
Dr. K.W. Riley	Secretary
Mr. Chen Jiaju	
Mr. Figuhr Muza	

GENERAL INDEX

INDEX OF SCIENTIFIC NAMES